T0211348

LONDON MATHEMATICAL SOCIETY LECTURE NOTE SERIES

Managing Editor: Professor E. Süli, Mathematical Institute,
Woodstock Road, University of Oxford, Oxford OX2 6GG, United Kingdom

The titles below are available from booksellers, or from Cambridge University Press at
www.cambridge.org/mathematics

London Mathematical Society Lecture Note Series: 456

Surveys in Combinatorics 2019

Edited by

ALLAN LO
University of Birmingham

RICHARD MYCROFT
University of Birmingham

GUILLEM PERARNAU
Universitat Politècnica de Catalunya, Barcelona

ANDREW TREGLOWN
University of Birmingham

CAMBRIDGE
UNIVERSITY PRESS

CAMBRIDGE
UNIVERSITY PRESS

University Printing House, Cambridge CB2 8BS, United Kingdom

One Liberty Plaza, 20th Floor, New York, NY 10006, USA

477 Williamstown Road, Port Melbourne, VIC 3207, Australia

314-321, 3rd Floor, Plot 3, Splendor Forum, Jasola District Centre, New Delhi - 110025, India

79 Anson Road, #06-04/06, Singapore 079906

Cambridge University Press is part of the University of Cambridge.

It furthers the University's mission by disseminating knowledge in the pursuit of education, learning and research at the highest international levels of excellence.

www.cambridge.org
Information on this title: www.cambridge.org/9781108740722
DOI: 10.1017/9781108649094

© Cambridge University Press 2019

First published 2019

A catalogue record for this publication is available from the British Library

ISBN 978-1-108-74072-2 Paperback

Contents

Preface

The Twenty-Seventh British Combinatorial Conference is to be held at the University of Birmingham from 29th July to 2nd August 2019. The British Combinatorial Committee had invited eight distinguished combinatorialists to give survey lectures in areas of their expertise, and this volume contains the survey articles on which these lectures were based.

In compiling this volume we are indebted to the authors for preparing their articles so accurately and professionally, and to the referees for their rapid responses and keen eye for detail. We would also like to thank Tom Harris and Clare Dennison at Cambridge University Press. Finally, without the previous efforts of editors of earlier Surveys and the guidance of the British Combinatorial Committee, the preparation of this volume would have been somewhat daunting.

This conference is organised in partnership with the Clay Mathematics Institute, the Heilbronn Institute, the Institute of Combinatorics and its Applications, the London Mathematical Society and the University of Birmingham; we thank each of these organisations for their generous involvement and support.

Allan Lo
Richard Mycroft
Andrew Treglown
University of Birmingham

Guillem Perarnau
Universitat Politècnica de Catalunya

February 2019

Clique-width for hereditary graph classes

Konrad K. Dabrowski, Matthew Johnson and Daniël Paulusma

Abstract

Clique-width is a well-studied graph parameter owing to its use in understanding algorithmic tractability: if the clique-width of a graph class $\mathcal{G}$ is bounded by a constant, a wide range of problems that are NP-complete in general can be shown to be polynomial-time solvable on $\mathcal{G}$. For this reason, the boundedness or unboundedness of clique-width has been investigated and determined for many graph classes. We survey these results for hereditary graph classes, which are the graph classes closed under taking induced subgraphs. We then discuss the algorithmic consequences of these results, in particular for the COLOURING and GRAPH ISOMORPHISM problems. We also explain a possible strong connection between results on boundedness of clique-width and on well-quasi-orderability by the induced subgraph relation for hereditary graph classes.

1 Introduction

Many decision problems are known to be NP-complete [84], and it is generally believed that such problems cannot be solved in time polynomial in the input size. For many of these hard problems, placing restrictions on the input (that is, insisting that the input has certain stated properties) can lead to significant changes in the computational complexity of the problem. This leads one to ask fundamental questions: under which input restrictions can an NP-complete problem be solved in polynomial time, and under which input restrictions does the problem remain NP-complete? For problems defined on graphs, we can restrict the input to some special class of graphs that have some commonality. The ultimate goal is to obtain complexity dichotomies for large families of graph problems, which tell us exactly for which graph classes a certain problem is efficiently solvable and for which it stays computationally hard. Such dichotomies may not always exist if P$\neq$ NP [129], but rather than solving problems one by one, and graph class by graph class, we want to discover general properties of graph classes from which we can determine the tractability or hardness of families of problems.

1.1 Width Parameters

One way to define a graph class is to use a notion of "width" and consider the set of graphs for which the width is bounded by a constant. Though it will not be our focus, let us briefly illustrate this idea with the most well-known width parameter, *treewidth*. A *tree decomposition* of a graph $G = (V, E)$ is a tree T whose nodes are subsets of V and has the properties that, for each v in V, the tree nodes that contain v induce a non-empty connected subgraph, and, for each edge vw in E, there is at least one tree node that contains v and w. See Figure 1 for an illustration of a graph and one of its tree decompositions. The sets of vertices that form the nodes of the tree are called *bags* and the width of the decomposition is one less than the size of the largest bag. The treewidth of G is the minimum width of its tree decompositions. One can therefore define a class of graphs of bounded treewidth; that is, for some constant c, the collection of graphs that each have treewidth at most c. The example in Figure 1 has treewidth 2. Moreover, it is easy to see that trees form exactly the class of graphs with treewidth 1. Hence, the treewidth of a graph can be seen as a measure that indicates how close a graph is to being a tree. Many graph problems can be solved in

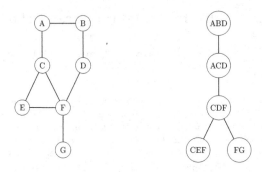

Figure 1: A graph, and a tree decomposition of the graph.

polynomial time on trees. For such problems it is natural to investigate whether restricting the problem to inputs that have bounded treewidth still yields algorithmic tractability. An approach that often yields polynomial-time algorithms is to apply dynamic programming over the decomposition tree. A disadvantage of this approach is that only sufficiently sparse graphs have bounded treewidth.

We further discuss reasons for focussing on width parameters in Section 1.2, but let us first note that there are many alternative width parameters, each of which has led to progress in understanding the complexity of problems on graphs.

Clique-width, the central width parameter in our survey, is another well-known example, which has received significant attention since it was introduced by Courcelle, Engelfriet and Rozenberg [56] at the start of the 1990s. Clique-width can be seen as a generalisation of treewidth that *can* deal with dense graphs, such as complete graphs and complete bipartite graphs, provided these instances are sufficiently regular. We will give explain this in Section 3, where we also give a formal definition, but, in outline, the idea is, given a graph G, to determine how it can be built up vertex-by-vertex using four specific graph operations that involve assigning labels to the vertices. The operations ensure that vertices labelled alike will keep the same label and thus, in some sense, behave identically. The clique-width of G is the minimum number of different labels needed to construct G in this way. Hence, if the clique-width of a graph G is small, we can decompose G into large sets of similarly behaving vertices, and these decompositions can be exploited to find polynomial-time algorithms (as we shall see later in this paper).

We remark that many other width parameters have been defined including boolean-width, branch-width, MIM-width, MM-width, module-width, NLC-width, path-width and rank-width, to name just a few. These parameters differ in strength, as we explain below; we refer to [95,111,116,164] for surveys on width parameters.

Given two width parameters p and q, we say that p *dominates* q if there is a function f such that $p(G) \leq f(q(G))$ for all graphs G. If p dominates q but not the reverse, then p is *more general* than q, as p is bounded for larger graph classes: whenever q is bounded for some graph class, then this is also the case for p, but there exists an infinite family of graphs for which the reverse does not hold. If p dominates q and q dominates p, then p and q are *equivalent*. For instance, MIM-width is more general than boolean-width, clique-width, module-width, NLC-width and rank-width, all of which are equivalent [42,114,151,154,164]. The latter parameters are more general than the equivalent group of parameters branch-width, MM-width and treewidth, which are, in turn, more general than path-width [59, 155, 164]. To give a concrete example, recall that the treewidth of the class of complete graphs is unbounded, in contrast to the clique-width. More precisely, a complete graph on $n \geq 2$ vertices has treewidth $n - 1$ but clique-width 2. As another example, the reason that rank-width and clique-width are equivalent is because the inequalities $\mathrm{rw}(G) \leq \mathrm{cw}(G) \leq 2^{\mathrm{rw}(G)+1} - 1$ hold for every graph G [151]. These two inequalities are essentially tight [150], and, as such, the latter example also shows that two equivalent parameters may not necessarily be linearly, or even polynomially, related.

1.2 Motivation for Width Parameters

The main computational reason for the large interest in width parameters is that many well-known NP-complete graph problems become polynomial-time solvable if some width parameter is bounded. There are a number of meta-theorems which prescribe general, sufficient conditions for a problem to be tractable on a graph class of bounded width. For treewidth and equivalent parameters, such as branch-width and MM-width, one can use the celebrated theorem of Courcelle [51]. This theorem, slightly extended from its original form, states that for every graph class of bounded treewidth, every problem definable in MSO_2 can be solved in time linear in the number of vertices of the graph.[1] In order to use this theorem, one can use the linear-time algorithm of Bodlaender [17] to verify whether a graph has treewidth at most c for any fixed constant c (that is, c is not part of the input). However, many natural graph classes, such as all those that contain graphs with arbitrarily large cliques, have unbounded treewidth.

We have noted that clique-width is more general than treewidth. This means that if we have shown that a problem can be solved in polynomial time on graphs of bounded clique-width, then it can also be solved in polynomial time on graphs of bounded treewidth. Similarly, if a problem is NP-complete for graphs of bounded treewidth, then the same holds for graphs of bounded clique-width. For graph classes of bounded clique-width, one can use several other meta-theorems. The first such result is due to Courcelle, Makowsky and Rotics [58]. They proved that graph problems that can be defined in MSO_1 are linear-time solvable on graph classes of bounded clique-width.[2] An example of such a problem is the well-known DOMINATING SET problem. This problem is to decide, for a graph $G = (V, E)$ and integer k, if G contains a set $S \subseteq V$ of size at most k such that every vertex of $G - S$ has at least one neighbour in S.[3]

1.3 Focus: Clique-Width

As mentioned, in this survey we focus on clique-width. Despite the usefulness of boundedness of clique-width, our understanding of clique-width itself is still very limited. For

---

[1] MSO_2 refers to the fragment of second order logic where quantified relation symbols must have arity at most 2, which means that, with graphs, one can quantify over both sets of vertices and sets of edges. Many graph problems can be defined using MSO_2, such as deciding whether a graph has a k-colouring (for fixed k) or a Hamiltonian path, but there are also problems that cannot be defined in this way.

[2] MSO_1 is monadic second order logic with the use of quantifiers permitted on relations of arity 1 (such as vertices), but not of arity 2 (such as edges) or more. Hence, MSO_1 is more restricted than MSO_2. We refer to [55] for more information on MSO_1 and MSO_2.

[3] Several other problems, such as LIST COLOURING and PRECOLOURING EXTENSION are polynomial-time solvable on graphs of bounded treewidth [113], but stay NP-complete on graph of bounded clique-width; the latter follows from results of [113] and [20], respectively; see also [88].

example, although computing the clique-width of a graph is known to be NP-hard in general [77],[4] the complexity of computing the clique-width is open even on very restricted graph classes, such as unit interval graphs (see [107] for some partial results). To give another example, the complexity of determining whether a given graph has clique-width at most c is still open for every fixed constant $c \geq 4$. On the positive side, see [49] for a polynomial-time algorithm for $c = 3$ and [75] for a polynomial-time algorithm, for every fixed c, on graphs of bounded treewidth.

To get a better handle on clique-width, many properties of clique-width, and relationships between clique-width and other graph parameters, have been determined over the years. In particular, numerous graph classes of bounded and unbounded clique-width have been identified. This has led to several dichotomies for various families of graph classes, which state exactly which graph classes of the family have bounded or unbounded clique-width. However, determining (un)boundedness of clique-width of a graph class is usually a highly non-trivial task, as it requires a thorough understanding of the structure of graphs in the class. As such, there are still many gaps in our knowledge.

A number of results on clique-width are collected in the surveys on clique-width by Gurski [95] and Kamiński, Lozin and Milanič [116]. Gurski focuses on the behaviour of clique-width (and NLC-width) under graph operations and transformations. Kamiński, Lozin and Milanič also discuss results for special graph classes. We refer to a recent survey of Oum [150] for algorithmic and structural results on the equivalent width parameter rank-width.

1.4 Aims and Outline

In Section 2 we introduce some basic terminology and notation that we use throughout the paper. In Section 3 we formally define clique-width. In the same section we present a number of basic results on clique-width and explain two general techniques for showing that the clique-width of a graph class is bounded or unbounded. For this purpose, in the same section we also list a number of graph operations that preserve (un)boundedness of clique-width for hereditary graph classes.

A graph class is *hereditary* if it is closed under taking induced subgraphs, or equivalently, under vertex deletion. Due to its natural definition, the framework of hereditary graph classes captures many well-known graph classes, such as bipartite, chordal, planar, interval and perfect graphs; we refer to the textbook of Brandstädt, Le and Spinrad [34] for a survey. As we shall see, boundedness of clique-width has been particularly well studied for hereditary graph classes. We discuss the state-of-the-art and other known results on boundedness of clique-width for hereditary graph classes in Section 4. This is all related to our first aim: to update the paper of Kamiński, Lozin and Milanič [116] from 2009 by surveying, in a systematic way, known results and open problems on boundedness of clique-width for hereditary graph classes.

Our second aim is to discuss algorithmic implications of the results from Section 4. We do this in Section 5 by focussing on two well-known problems. We first discuss implications for the COLOURING problem, which is well known to be NP-complete [133]. We focus on (hereditary) graph classes defined by two forbidden induced subgraphs. Afterwards, we consider the algorithmic consequences for the GRAPH ISOMORPHISM problem. This problem can be solved in quasi-polynomial time [7]. It is not known if GRAPH ISOMORPHISM

---

[4]It is also NP-hard to compute treewidth [4] and parameters equivalent to clique-width, such as NLC-width [98], rank-width (see [110, 149]) and boolean-width [159].

can be solved in polynomial time, but it is not NP-complete unless the polynomial hierarchy collapses [160]. As such, we define the complexity class GI, which consists of all problems that can be polynomially reduced to GRAPH ISOMORPHISM and a problem in GI is GI-complete if GRAPH ISOMORPHISM can be polynomially reduced to it. The GRAPH ISOMORPHISM problem is of particular interest, as there are similarities between proving unboundedness of clique-width of some graph class and proving that GRAPH ISOMORPHISM stays GI-complete on this class [161].

Our third aim is to discuss a conjectured relationship between boundedness of clique-width and well-quasi-orderability by the induced subgraph relation. If it can be shown that a graph class is well-quasi-ordered, we can apply several powerful results to prove further properties of the class. This is, for instance, illustrated by the Robertson-Seymour Theorem [157], which states that the set of all finite graphs is well-quasi-ordered by the minor relation. This result makes it possible to test in cubic time whether a graph belongs to some given minor-closed graph class [156] (see [112] for a quadratic algorithm). For the induced subgraph relation, it is easy to construct examples of hereditary graph classes that are not well-quasi-ordered. Take, for instance, the class of graphs of degree at most 2, which contains an infinite anti-chain, namely the set of all cycles.

If every hereditary graph class that is well-quasi-ordered by the induced subgraph relation also has bounded clique-width, then all algorithmic consequences of having bounded clique-width would also hold for being well-quasi-ordered by the induced subgraph relation. However, Lozin, Razgon and Zamaraev [142] gave a negative answer to a question of Daligault, Rao and Thomassé [69] about this implication, by presenting a hereditary graph class of unbounded clique-width that is nevertheless well-quasi-ordered by the induced subgraph relation. Their graph class can be characterized only by infinitely many forbidden induced subgraphs. This led the authors of [142] to conjecture that every finitely defined hereditary graph class that is well-quasi-ordered by the induced subgraph relation has bounded clique-width, which, if true, would still be very useful. All known results agree with this conjecture, and we survey these results in Section 6. In the same section we explain that the graph operations given in Section 3 do not preserve well-quasi-orderability by the induced subgraph relation. However, we also explain that a number of these operations can be used for a stronger property, namely well-quasi-orderability by the labelled induced subgraph relation.

In Section 7 we conclude our survey with a list of other relevant open problems. There, we also discuss some variants of clique-width, including linear clique-width and power-bounded clique-width.

2 Preliminaries

Throughout the paper we consider only finite, undirected graphs without multiple edges or self-loops.

Let $G = (V, E)$ be a graph. The *degree* of a vertex $u \in V$ is the size of its neighbourhood $N(u) = \{v \in V \mid uv \in E\}$. For a subset $S \subseteq V$, the graph $G[S]$ denotes the subgraph of G *induced by* S, which is the graph with vertex set S and an edge between two vertices $u, v \in S$ if and only if $uv \in E$. If F is an induced subgraph of G, then we denote this by $F \subseteq_i G$. Note that $G[S]$ can be obtained from G by deleting the vertices of $V \setminus S$. The *line graph* of G is the graph with vertex set E and an edge between two vertices e_1 and e_2 if and only if e_1 and e_2 share a common end-vertex in G.

An *isomorphism* from a graph G to a graph H is a bijective mapping $f : V(G) \rightarrow V(H)$

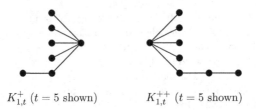

$K_{1,t}^{+}$ $(t = 5$ shown$)$ $K_{1,t}^{++}$ $(t = 5$ shown$)$

Figure 2: The graphs $K_{1,t}^{+}$ and $K_{1,t}^{++}$.

such that there is an edge between two vertices u and v in G if and only if there is an edge between $f(u)$ and $f(v)$ in H. If such an isomorphism exists then G and H are said to be *isomorphic*. We say that G is *H-free* if G contains no induced subgraph isomorphic to H.

Let $G = (V, E)$ be a graph. A set $K \subseteq V$ is a *clique* of G and $G[K]$ is *complete* if there is an edge between every pair of vertices in K. If G is connected, then a vertex $v \in V$ is a *cut-vertex* of G if $G[V \setminus \{v\}]$ is disconnected, and a clique $K \subset V$ is a *clique cut-set* of G if $G[V \setminus K]$ is disconnected. If G is connected and has at least three vertices but no cut-vertices, then G is *2-connected*. A maximal induced subgraph of G that has no cut-vertices is a *block* of G. If G is connected and has no clique cut-set, then G is an *atom*.

The graphs C_n, P_n and K_n denote the cycle, path and complete graph on n vertices, respectively. The *length* of a path or a cycle is the number of its edges. The *distance* between two vertices u and v in a graph G is the length of a shortest path between them. For an integer $r \geq 1$, the *r-th power* of G is the graph with vertex set $V(G)$ and an edge between two vertices u and v if and only if u and v are at distance at most r from each other in G.

If F and G are graphs with disjoint vertex sets, then the *disjoint union* of F and G is the graph $G + F = (V(F) \cup V(G), E(F) \cup E(G))$. The disjoint union of s copies of a graph G is denoted sG. A *forest* is a graph with no cycles, that is, every connected component is a *tree*. A forest is *linear* if it has no vertices of degree at least 3, or equivalently, if it is the disjoint union of paths. A *leaf* in a tree is a vertex of degree 1. In a *complete binary* tree all non-leaf vertices have degree 3.

Let S and T be disjoint vertex subsets of a graph $G = (V, E)$. A vertex v is *(anti-)complete* to T if it is (non-)adjacent to every vertex in T. Similarly, S is *(anti-)complete* to T if every vertex in S is (non-)adjacent to every vertex in T. A set of vertices M is a *module* of G if every vertex of G that is not in M is either complete or anti-complete to M. A module of G is *trivial* if it contains zero, one or all vertices of G, otherwise it is *non-trivial*. We say that G is *prime* if every module of G is trivial.

A graph G is *bipartite* if its vertex set can be partitioned into two (possibly empty) subsets X and Y such that every edge of G has one end-vertex in X and the other one in Y. If X is complete to Y, then G is *complete bipartite*. For two non-negative integers s and t, we denote the complete bipartite graph with partition classes of size s and t, respectively, by $K_{s,t}$. The graph $K_{1,t}$ is also known as the $(t + 1)$-vertex *star*. The *subdivision* of an edge uv in a graph replaces uv by a new vertex w and edges uw and vw. We let $K_{1,t}^{+}$ and $K_{1,t}^{++}$ be the graphs obtained from $K_{1,t}$ by subdividing one of its edges once or twice, respectively.

A graph is *complete r-partite*, for some $r \geq 1$, if its vertex set can be partitioned into r independent sets $V_1, \ldots, V_r$ such that there exists an edge between two vertices u and v

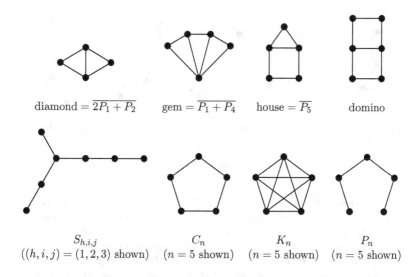

diamond $= \overline{2P_1 + P_2}$ gem $= \overline{P_1 + P_4}$ house $= \overline{P_5}$ domino

$S_{h,i,j}$ C_n K_n P_n
$((h, i, j) = (1, 2, 3)$ shown) $(n = 5$ shown) $(n = 5$ shown) $(n = 5$ shown)

Figure 3: Some common graphs used throughout the paper.

if and only if u and v do not belong to the same set V_i. Note that a non-empty graph is complete r-partite for some $r \geq 1$ if and only if it is $(P_1 + P_2)$-free.

Let $G = (V, E)$ be a graph. Its *complement* $\overline{G}$ is the graph with vertex set V and an edge between two vertices u and v if and only if uv is not an edge of G. We say that G is *self-complementary* if G is isomorphic to $\overline{G}$. The complement of a bipartite graph is a *co-bipartite* graph.

The graphs $K_{1,3}$, $\overline{2P_1 + P_2}$, $\overline{P_1 + P_4}$, and $\overline{P_5}$ are also known as the *claw*, *diamond*, *gem*, and *house*, respectively. The latter three graphs are shown in Figure 3, along with the *domino*. The graph $S_{h,i,j}$, for $1 \leq h \leq i \leq j$, denotes the *subdivided claw*, which is the tree with one vertex x of degree 3 and exactly three leaves, which are of distance h, i and j from x, respectively. Note that $S_{1,1,1} = K_{1,3}$, $S_{1,1,2} = K_{1,3}^+$ and $S_{1,1,3} = K_{1,3}^{++}$. See Figure 3 for an example. We let $\mathcal{S}$ be the class of graphs every connected component of which is either a subdivided claw or a path on at least one vertex. The graph $T_{h,i,j}$ with $0 \leq h \leq i \leq j$ denotes the triangle with pendant paths of length h, i and j, respectively. That is, $T_{h,i,j}$ is the graph with vertices $a_0, \ldots, a_h$, $b_0, \ldots, b_i$ and $c_0, \ldots, c_j$ and edges $a_0 b_0$, $b_0 c_0$, $c_0 a_0$, $a_p a_{p+1}$ for $p \in \{0, \ldots, h-1\}$, $b_p b_{p+1}$ for $p \in \{0, \ldots, i-1\}$ and $c_p c_{p+1}$ for $p \in \{0, \ldots, j-1\}$. Note that $T_{0,0,0} = C_3 = K_3$. The graphs $T_{0,0,1} = \overline{P_1 + P_3}$, $T_{0,1,1}$, $T_{1,1,1}$ and $T_{0,0,2}$ are also known as the *paw*, *bull*, *net* and *hammer*, respectively; see also Figure 4. Also note that $T_{h,i,j}$ is the line graph of $S_{h+1,i+1,j+1}$. We let $\mathcal{T}$ be the class of graphs that are the line graphs of graphs in $\mathcal{S}$. Note that $\mathcal{T}$ contains every graph $T_{h,i,j}$ and every path (as the line graph of P_t is P_{t-1} for $t \geq 2$).

Let $G = (V, E)$ be a graph. For an induced subgraph $F \subseteq_i G$, the *subgraph complementation* operation, which acts on G with respect to F, replaces every edge in F by a non-edge, and vice versa. If we apply this operation on G with respect to G itself, then we obtain the complement $\overline{G}$ of G. For two disjoint vertex subsets S and T in G, the *bipartite complementation* operation, which acts on G with respect to S and T, replaces every edge

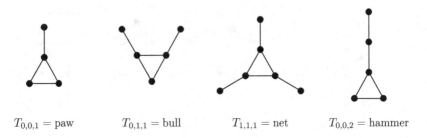

$T_{0,0,1}$ = paw $T_{0,1,1}$ = bull $T_{1,1,1}$ = net $T_{0,0,2}$ = hammer

Figure 4: Examples of graphs $T_{h,i,j}$.

with one end-vertex in S and the other one in T by a non-edge and vice versa. We note that applying a bipartite complementation is equivalent to applying a sequence of three consecutive subgraph complementations, namely on $G[S \cup T]$, $G[S]$ and $G[T]$.

Let $\mathcal{G}$ be a graph class. Denote the number of labelled graphs on n vertices in $\mathcal{G}$ by g_n. Then $\mathcal{G}$ is *superfactorial* if there does not exist a constant c such that $g_n \leq n^{cn}$ for every n.

Recall that a graph class is hereditary if it is closed under taking induced subgraphs. It is not difficult to see that a graph class $\mathcal{G}$ is hereditary if and only if $\mathcal{G}$ can be characterized by a unique set $\mathcal{F}_\mathcal{G}$ of minimal forbidden induced subgraphs. A hereditary graph class $\mathcal{G}$ is *finitely defined* if $\mathcal{F}_\mathcal{G}$ is finite. We note, however, that the set $\mathcal{F}_\mathcal{G}$ may have infinite size. For example, if $\mathcal{G}$ is the class of bipartite graphs, then $\mathcal{F}_\mathcal{G} = \{C_3, C_5, C_7, \ldots\}$. If $\mathcal{F}$ is a set of graphs, we say that a graph G is $\mathcal{F}$-free if G does not contain any graph in $\mathcal{F}$ as an induced subgraph. In particular, this means that if a graph class $\mathcal{G}$ is hereditary, then $\mathcal{G}$ is exactly the class of $\mathcal{F}_\mathcal{G}$-free graphs. If $\mathcal{F} = \{H_1, H_2, \ldots\}$ or $\{H_1, H_2, \ldots, H_p\}$ for some $p \geq 0$, we may also describe a graph G as being $(H_1, H_2, \ldots)$-free or $(H_1, H_2, \ldots, H_p)$-free, respectively, rather than $\mathcal{F}$-free; recall that if $\mathcal{F} = \{H_1\}$ we may write H_1-free instead.

Observation 2.1. *Let $\mathcal{H}$ and $\mathcal{H}^*$ be sets of graphs. The class of $\mathcal{H}$-free graphs is contained in the class of $\mathcal{H}^*$-free graphs if and only if for every graph $H^* \in \mathcal{H}^*$, the set $\mathcal{H}$ contains an induced subgraph of H^*.*

Suppose $\mathcal{H}$ and $\mathcal{H}^*$ are sets of graphs such that for every graph $H^* \in \mathcal{H}^*$, the set $\mathcal{H}$ contains an induced subgraph of H^*. Observation 2.1 implies that any graph problem that is polynomial-time solvable for $\mathcal{H}^*$-free graphs is also polynomial-time solvable for $\mathcal{H}$-free graphs, and any graph problem that is NP-complete for $\mathcal{H}$-free graphs is also NP-complete for $\mathcal{H}^*$-free graphs.

We define the *complement* of a hereditary graph class $\mathcal{G}$ as $\overline{\mathcal{G}} = \{\overline{G} \mid G \in \mathcal{G}\}$. Then $\mathcal{G}$ is *closed under complementation* if $\mathcal{G} = \overline{\mathcal{G}}$. As $\mathcal{F}_\mathcal{G}$ is the unique minimal set of forbidden induced subgraphs for $\mathcal{G}$, we can make the following observation.

Observation 2.2. *A hereditary graph class $\mathcal{G}$ is closed under complementation if and only if $\mathcal{F}_\mathcal{G}$ is closed under complementation.*

Let G be a graph. The *contraction* of an edge uv replaces u and v and their incident edges by a new vertex w and edges wy if and only if either uy or vy was an edge in G (without creating multiple edges or self-loops). Let u be a vertex with exactly two neighbours v, w, which in addition are non-adjacent. The *vertex dissolution* of u removes u, uv and uw, and adds the edge vw. Note that vertex dissolution is a special type of edge contraction, and it is the reverse operation of an edge subdivision (recall that the latter operation replaces an edge uv by a new vertex w with edges uw and vw).

Let G and H be graphs. The graph H is a *subgraph* of G if G can be modified into H by a sequence of vertex deletions and edge deletions. We can define other containment relations using the graph operations defined above. We say that G contains H as a *minor* if G can be modified into H by a sequence of edge contractions, edge deletions and vertex deletions, as a *topological minor* if G can be modified into H by a sequence of vertex dissolutions, edge deletions and vertex deletions, as an *induced minor* if G can be modified into H by a sequence of edge contractions and vertex deletions, and as an *induced topological minor* if G can be modified into H by a sequence of vertex dissolutions and vertex deletions. Let $\{H_1, \ldots, H_p\}$ be a set of graphs. If G does not contain any of the graphs $H_1, \ldots, H_p$ as a subgraph, then G is $(H_1, \ldots, H_p)$-*subgraph-free*. We define the terms $(H_1, \ldots, H_p)$-*minor-free*, $(H_1, \ldots, H_p)$-*topological-minor-free*, $(H_1, \ldots, H_p)$-*induced-minor-free*, and $(H_1, \ldots, H_p)$-*induced-topological-minor-free* analogously. Note that graph classes defined by some set of forbidden subgraphs, minors, topological minors, induced minors, or induced topological minors are hereditary, as they are all closed under vertex deletion.

Example 2.3. A graph is *planar* if it can be embedded in the plane in such a way that any two edges only intersect with each other at their end-vertices. It is well known that the class of planar graphs can be characterized by a set of forbidden minors: Wagner's Theorem [165] states that a graph is planar if and only if it is $(K_{3,3}, K_5)$-minor-free.

We will also need the following folklore observation (see, for example, [90]).

Observation 2.4. *For every $F \in \mathcal{S}$, a graph is F-subgraph-free if and only if it is F-minor-free.*

A k-*colouring* of a graph G is a mapping $c : V \to \{1, \ldots, k\}$ such that $c(u) \neq c(v)$ whenever u and v are adjacent vertices. The *chromatic number* of G is the smallest k such that G has a k-colouring. The *clique number* of G is the size of a largest clique of G.

A graph G is *perfect* if, for every $H \subseteq_i G$, the chromatic number of H is equal to the clique number of H. The Strong Perfect Graph Theorem [45] states that G is perfect if and only if G is $(C_5, C_7, C_9, \ldots)$-free and $(\overline{C_7}, \overline{C_9}, \ldots)$-free. A graph G is *chordal* if it is $(C_4, C_5, C_6, \ldots)$-free and *weakly chordal* if it is $(C_5, C_6, C_7, \ldots)$-free and $(\overline{C_6}, \overline{C_7}, \ldots)$-free. A graph G is a *split graph* if it has a *split partition*, that is, a partition of its vertex set into two (possibly empty) sets K and I, where K is a clique and I is an independent set. It is known that a graph is split if and only if it is $(C_4, C_5, 2P_2)$-free [78]. A graph G is a *permutation graph* if line segments connecting two parallel lines can be associated to its vertices in such a way that two vertices of G are adjacent if and only if the two corresponding line segments intersect. A graph G is a *permutation split graph* if it is both permutation and split, and G is a *permutation bipartite graph* if it is both permutation and bipartite. A graph G is *chordal bipartite* if it is $(C_3, C_5, C_6, C_7, \ldots)$-free. A graph G is *distance-hereditary* if the distance between any two vertices u and v in any connected induced subgraph of G is the same as the distance of u and v in G. Equivalently, a graph is distance-hereditary if and only if it is (domino, gem, house, $C_5, C_6, C_7, \ldots$)-free [9]. A graph is *(unit) interval* if it has a representation in which each vertex u corresponds to an interval I_u (of unit length) of the line such that two vertices u and v are adjacent if and only if $I_u \cap I_v \neq \emptyset$.

We make the following observation. A number of inclusions in Observation 2.5 follow immediately from the definitions and the Strong Perfect Graph Theorem. For the remaining inclusions we refer to [34].

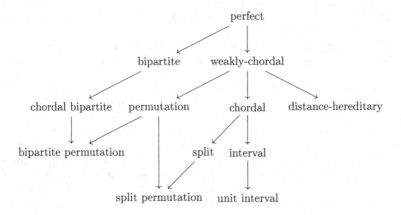

Figure 5: The inclusion relations between well-known classes mentioned in the paper. An arrow from one class to another indicates that the first class contains the second.

Observation 2.5. *The following statements hold:*

1. *every split graph is chordal,*
2. *every (unit) interval graph is chordal,*
3. *every chordal graph is weakly chordal,*
4. *every (bipartite or split) permutation graph is weakly chordal,*
5. *every distance-hereditary graph is weakly chordal,*
6. *every weakly chordal graph is perfect,*
7. *every bipartite permutation graph is chordal bipartite, and*
8. *every (chordal) bipartite graph is perfect.*

The containments listed in Observation 2.5 (and those that follow from them by transitivity) are also displayed Figure 5. It is not difficult to construct counterexamples for the other containments. Indeed, for pairs of classes above for which we have listed the minimal forbidden induced subgraph characterizations, these characterizations immediately provide such counterexamples.

We now introduce the notion of treewidth formally. Recall from Section 1 that treewidth expresses to what extent a graph is "tree-like". A *tree decomposition* of a graph G is a pair $(T, \mathcal{X})$ where T is a tree and $\mathcal{X} = \{X_i \mid i \in V(T)\}$ is a collection of subsets of $V(G)$, such that the following three conditions hold:

(i) $\bigcup_{i \in V(T)} X_i = V(G)$

(ii) for every edge $xy \in E(G)$, there is an $i \in V(T)$ such that $x, y \in X_i$ and

(iii) for every $x \in V(G)$, the set $\{i \in V(T) \mid x \in X_i\}$ induces a connected subtree of T.

The *width* of the tree decomposition $(T, \mathcal{X})$ is $\max\{|X_i| - 1 \mid i \in V(T)\}$, and the *treewidth* $\mathrm{tw}(G)$ of G is the minimum width over all tree decompositions of G. If T is a path, then (X, T) is a *path decomposition* of G. The *path-width* $\mathrm{pw}(G)$ of G is the minimum width over all path decompositions of G.

A *quasi order* $\leq$ on a set X is a reflexive, transitive binary relation. Two elements $x, y \in X$ in $\leq$ are *comparable* if $x \leq y$ or $y \leq x$; otherwise they are *incomparable*. A set of pairwise (in)comparable elements in $\leq$ is called an *(anti)-chain*. A quasi-order $\leq$ is a *well-quasi-order* if every infinite sequence of elements $x_1, x_2, x_3, \ldots$ in X contains a pair (x_i, x_j) with $x_i \leq x_j$ and $i < j$, or equivalently, if $\leq$ has no infinite strictly decreasing sequence and no infinite anti-chain. A *partial order* $\leq$ is a quasi-order which is anti-symmetric, that is, if $x \leq y$ and $y \leq x$ then $x = y$. If we consider two graphs to be "equal" when they are isomorphic, then all quasi orders considered in this paper are in fact partial orders. As such, throughout this paper "quasi order" can be interpreted as "partial order".

For an arbitrary set M, we let M^* denote the set of finite sequences of elements of M. A quasi-order $\leq$ on M defines a quasi-order $\leq^*$ on M^* as follows: $(a_1, \ldots, a_m) \leq^* (b_1, \ldots, b_n)$ if and only if there is a sequence of integers $i_1, \ldots, i_m$ with $1 \leq i_1 < \cdots < i_m \leq n$ such that $a_j \leq b_{i_j}$ for $j \in \{1, \ldots, m\}$. We call $\leq^*$ the *subsequence relation*.

The following lemma is well known and very useful when dealing with quasi-orders.

Lemma 2.6 (Higman's Lemma [109]). *Let $(M, \leq)$ be a well-quasi-order. Then $(M^*, \leq^*)$ is a well-quasi-order.*

3 Clique-Width

In this section we give a number of basic results on clique-width. We begin by giving a formal definition.[5] The *clique-width* of a graph G, denoted by $\mathrm{cw}(G)$, is the minimum number of labels needed to construct G using the following four operations:

1. Create a new graph with a single vertex v with label i. (This operation is written $i(v)$.)
2. Take the disjoint union of two labelled graphs G_1 and G_2 (written $G_1 \oplus G_2$).
3. Add an edge between every vertex with label i and every vertex with label j, $i \neq j$ (written $\eta_{i,j}$).
4. Relabel every vertex with label i to have label j (written $\rho_{i \rightarrow j}$).

We say that a construction of a graph G with the four operations is a k-*expression* if it uses at most k labels. Thus the clique-width of G is the minimum k for which G has a k-expression. We refer to [57, 106, 108] for a number of characterizations of clique-width and to [115] for a compact representation of graphs of clique-width k.

Example 3.1. We first note that $\mathrm{cw}(P_1) = 1$ and $\mathrm{cw}(P_2) = \mathrm{cw}(P_3) = 2$. Now consider a path on four vertices v_1, v_2, v_3, v_4, in that order. Then this path can be constructed using the four operations (using only three labels) as follows:

$$\eta_{3,2}(3(v_4) \oplus \rho_{3 \rightarrow 2}(\rho_{2 \rightarrow 1}(\eta_{3,2}(3(v_3) \oplus \eta_{2,1}(2(v_2) \oplus 1(v_1)))))).$$

Note that at the end of this construction, only v_4 has label 3. It is easy to see that a construction using only two labels is not possible. Hence, we deduce that $\mathrm{cw}(P_4) = 3$. This construction can readily be generalized to longer paths: for $n \geq 5$ let E be a 3-expression for the path P_{n-1} on vertices $v_1, \ldots, v_{n-1}$, with only the vertex v_{n-1} having label 3, then

---

[5]The term *clique-width* and the definition in essentially same form we give here were introduced by Courcelle and Olariu [59] based on operations and related decompositions from Courcelle, Engelfriet and Rozenberg [56]; see also [55]. Although we consider only undirected graphs, the definitions of [59] also covered the case of directed graphs. Other equivalent width parameters have also been studied for directed graphs. For example, Kanté and Rao [118] considered the rank-width of directed graphs.

$\eta_{3,2}(3(v_n) \oplus \rho_{3\to2}(\rho_{2\to1}(E)))$ is a 3-expression for the path P_n on vertices $v_1, \ldots, v_n$, with only the vertex v_n having label 3. Therefore $\mathrm{cw}(P_n) = 3$ for all $n \geq 4$. Moreover, by changing the construction to give the first vertex v_1 on a path P_n ($n \geq 3$) a unique fourth label, we can connect it to the last constructed vertex v_n of P_n (the only vertex with label 3) via an edge-adding operation to obtain C_n. Hence, we find that $\mathrm{cw}(C_n) \leq 4$ for every $n \geq 3$. In fact $\mathrm{cw}(C_n) = 4$ holds for every $n \geq 7$ [145].

A class of graphs $\mathcal{G}$ has *bounded* clique-width if there is a constant c such that the clique-width of every graph in $\mathcal{G}$ is at most c. If such a constant c does not exist, we say that the clique-width of $\mathcal{G}$ is *unbounded*. A hereditary graph class $\mathcal{G}$ is a *minimal* class of unbounded clique-width if it has unbounded clique-width and every proper hereditary subclass of $\mathcal{G}$ has bounded clique-width.

The following two observations, which are both well known and readily seen, give two graph classes of small clique-width. In particular, Proposition 3.3 follows from Example 3.1 after observing that a graph of maximum degree at most 2 is the disjoint union of paths and cycles. For more examples of graph classes of small width, see, for instance, [26, 27].

Proposition 3.2. *Every forest has clique-width at most 3.*

Proof. Let T be a tree with a root vertex v. We claim that there is a 3-expression which creates T such that, in the resulting labelled tree, only v has label 3. We prove this by induction on $|V(T)|$. Clearly this holds when $|V(T)| = 1$. Otherwise, let $v_1, \ldots, v_k$ be the children of v and let $T_1, \ldots, T_k$ be the subtrees of T rooted at $v_1, \ldots, v_k$, respectively. By the induction hypothesis, for each i there is a 3-expression which creates T_i such that, in the resulting labelled tree, only v_i has label 3. We take the disjoint union $\oplus$ of these expressions and let E be the resulting 3-expression. Then $\eta_{3,2}(3(v) \oplus \rho_{3\to2}(\rho_{2\to1}(E)))$ is a 3-expression which creates T such that, in the resulting labelled tree, only v has label 3. Therefore for every tree T, there is a 3-expression that constructs T. Since a forest is a disjoint union of trees, we can then use the $\oplus$ operation to extend this to a 3-expression for any forest. The proposition follows. $\square$

Proposition 3.3. *Every graph of maximum degree at most 2 has clique-width at most 4.*

Recall that for general graphs, the complexity of computing the clique-width of a graph was open for a number of years, until Fellows, Rosamund, Rotics and Szeider [77] proved that this is NP-hard. However, Proposition 3.2 implies that we can determine the clique-width of a forest F in polynomial time: if F contains an induced P_4, then $\mathrm{cw}(F) = 3$; if F is P_4-free but has an edge, then $\mathrm{cw}(F) = 2$; and if $F = sP_1$ for some $s \geq 1$, then $\mathrm{cw}(F) = 1$.

In contrast to Proposition 3.3, graphs of maximum degree at most 3 may have arbitrarily large clique-width. An example of this is a *wall* of arbitrary height, which can be thought of as a hexagonal grid. We do not formally define the wall, but instead we refer to Figure 6, in which three examples of walls of different heights are depicted; see, for example, [46] for a formal definition. Note that walls of height at least 2 have maximum degree 3. The following result is well known; see for example [116].

Theorem 3.4. *The class of walls has unbounded clique-width.*

As mentioned, clique-width is more general than treewidth. Courcelle and Olariu [59] proved that $\mathrm{cw}(G) \leq 4 \cdot 2^{\mathrm{tw}(G)-1} + 1$ for every graph G (see [87] for an alternative proof). Corneil and Rotics [50] improved this bound by showing that $\mathrm{cw}(G) \leq 3 \cdot 2^{\mathrm{tw}(G)-1}$ for every graph G. They also proved that for every k, there is a graph G with $\mathrm{tw}(G) = k$

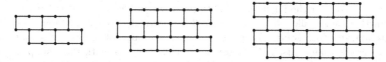

Figure 6: Walls of height 2, 3 and 4, respectively.

and $\mathrm{cw}(G) \geq 2^{\left\lfloor \frac{\mathrm{tw}(G)}{2} \right\rfloor - 1}$. The following result shows that for restricted graph classes the two parameters may be equivalent (see [53, 54] for graph classes for which treewidth and clique-width are even linearly related).

Theorem 3.5 ([97]). *For $t \geq 1$, every class of $K_{t,t}$-subgraph-free graphs of bounded clique-width has bounded treewidth.*

Corollary 3.6. *A class of graphs of bounded maximum degree has bounded clique-width if and only if it has bounded treewidth.*

Gurski and Wanke gave another connection between treewidth and clique-width.

Theorem 3.7 ([100]). *A class of graphs $\mathcal{G}$ has bounded treewidth if and only if the class of line graphs of graphs in $\mathcal{G}$ has bounded clique-width.*

As mentioned in Section 1, boundedness of clique-width has been determined for many hereditary graph classes. However, using the definition of clique-width directly to prove that a certain hereditary graph class $\mathcal{G}$ has bounded clique-width is often difficult. An alternative way to show that a hereditary graph class $\mathcal{G}$ has bounded clique-width is to prove that for infinitely many values of n, the number of labelled graphs in $\mathcal{G}$ on n vertices is at most the Bell number B_n [3], but this has limited applicability. The following BCW Method is more commonly used:

Bounding Clique-Width (BCW Method)

1. If possible, consider only graphs in $\mathcal{G}$ that have some suitable property π.

2. Take a graph class $\mathcal{G}'$ for which it is known that its clique-width is bounded.

3. For every graph $G \in \mathcal{G}$ (possibly with property π), reduce G to a graph in $\mathcal{G}'$ by using a constant number of graph operations that do not change the clique-width of G by "too much".

Note that the subclass of graphs in $\mathcal{G}$ that have some property π in Step 1 need not be hereditary. For example, it is known [18, 139] that we may choose the property π to be that of being 2-connected and that we can delete some constant number k of vertices from a graph without affecting the clique-width by more than some bounded amount. Then we could try to prove that $\mathcal{G}$ has bounded clique-width by showing that for every 2-connected graph in $\mathcal{G}$, we can delete no more than k vertices to obtain a graph in some class $\mathcal{G}'$ that we know to have bounded clique-width. We give some concrete examples of this method in the next section.

The power of the method depends on both the graph property π in Step 1 and the graph operations that we are allowed to use in Step 3. In particular we will use graph operations to modify a graph G of some class $\mathcal{G}$ into the disjoint union of some graphs that have a simpler structure than G itself. As a result, we can then deal with these simpler graphs separately. This approach is particularly useful if $\mathcal{G}$ is hereditary: if the simpler graphs are induced subgraphs of the original graph G, then we can still make use of earlier deduced properties for $\mathcal{G}$ when dealing with the simpler induced subgraphs of G. Before giving important examples of these operations and properties, we first formalize our approach.

Let $k \geq 0$ be a constant and let γ be some graph operation. We say that a graph class $\mathcal{G}'$ is (k, γ)-*obtained* from a graph class $\mathcal{G}$ if the following two conditions hold:

1. every graph in $\mathcal{G}'$ can be obtained from a graph in $\mathcal{G}$ by performing γ at most k times, and

2. for every graph $G \in \mathcal{G}$ there exists at least one graph in $\mathcal{G}'$ that can be obtained from G by performing γ at most k times (note that $\mathcal{G}$ is not necessarily a subclass of $\mathcal{G}'$).

A graph operation γ *preserves* boundedness of clique-width if, for every finite constant k and every graph class $\mathcal{G}$, every graph class $\mathcal{G}'$ that is (k, γ)-obtained from $\mathcal{G}$ has bounded clique-width if and only if $\mathcal{G}$ has bounded clique-width. We note that Condition 1 is necessary for this definition to be meaningful; without this condition the class of all graphs (which has unbounded clique-width) would be (k, γ)-obtained from every other graph class. Similarly, we also need Condition 2, as otherwise every graph class would be (k, γ)-obtained from the class of all graphs. If $k = \infty$ is allowed, then γ preserves boundedness of clique-width *ad infinitum*. Similarly, a graph property π *preserves* boundedness of clique-width if, for every graph class $\mathcal{G}$, the subclass of $\mathcal{G}$ with property π has bounded clique-width if and only if $\mathcal{G}$ has bounded clique-width. If necessary, we may restrict these definitions to only be valid for some specific types of graph classes.

We refer to the survey of Gurski [95] for a detailed overview of graph operations that preserve boundedness of clique-width and for bounds that tell us more precisely by how much the clique-width can change when applying various operations.[6] Here, we only state the most important graph operations, together with two well-known properties that preserve boundedness of clique-width.

Facts about clique-width:

Fact 1. Vertex deletion preserves boundedness of clique-width [139].

Fact 2. Subgraph complementation preserves boundedness of clique-width [116].

Fact 3. Bipartite complementation preserves boundedness of clique-width [116].

Fact 4. Being prime preserves boundedness of clique-width for hereditary graph classes [59].

Fact 5. Being 2-connected preserves boundedness of clique-width for hereditary graph classes [18, 139].

Fact 6. Edge subdivision preserves boundedness of clique-width ad infinitum for graph classes of bounded maximum degree [116].

---

[6]We note that some of these graph operations may exponentially increase the upper bound of the clique-width.

We note that Fact 3 follows from Fact 2, as bipartite complementations can be mimicked by three subgraph complementations. Moreover, an edge deletion is a special case of subgraph complementation, whereas an edge contraction is a vertex deletion and a bipartite complementation. Finally, recall that an edge subdivision is the reverse operation of a vertex dissolution, which can be seen as a type of edge contraction. Hence, from Facts 1–3 it follows that edge deletion, edge contraction and edge subdivision each preserve boundedness of clique-width.

Vertex deletions, edge deletions and edge contractions do not preserve boundedness of clique-width ad infinitum: one can take any graph class of unbounded clique-width and apply one of these operations until one obtains the empty graph or an edgeless graph. Hence, Facts 1–3 do not preserve boundedness of clique-width ad infinitum. This holds even for graphs of maximum degree at most 3, as the class of walls and their induced subgraphs has unbounded clique-width by Theorem 3.4.

In contrast, Fact 6 says that edge subdivisions applied on graphs of bounded maximum degree do preserve boundedness of clique-width ad infinitum. We note that Fact 6 follows from Corollary 3.6 and the fact that an edge subdivision does not change the treewidth of a graph (see, for example, [140]). However, the condition on the maximum degree is necessary for the "only if" direction of Fact 6. Otherwise, as discussed in [67], one could start with a clique K on at least two vertices (which has clique-width 2) and then apply an edge subdivision on an edge uv in K if and only if uv is not an edge in some graph G of arbitrarily large clique-width with $|V(G)| = |V(K)|$. This yields a graph G' that contains G as an induced subgraph, implying that $\mathrm{cw}(G') \geq \mathrm{cw}(G)$, which is arbitrarily larger than $\mathrm{cw}(K) = 2$.

As an aside, note that edge contractions do not increase the clique-width of graphs of bounded maximum degree either. We can apply Corollary 3.6 again after observing from the definition of treewidth that edge contractions do not increase treewidth. However, the condition on the maximum degree is necessary here as well; a (non-trivial) counterexample is given by Courcelle [52], who proved that the class of graphs that are obtained by edge contractions from the class of graphs of clique-width 3 has unbounded clique-width.

For the BCW Method, operations that preserve boundedness of clique-width may be combined, but these operations may not always be used in combination with some property π that preserves boundedness of clique-width. This is because applying a graph operation may result in a graph that does not have property π. Moreover, it is not always clear whether two or more properties that preserve boundedness of clique-width may be unified into one property. For instance, every non-empty class of 2-connected graphs is not hereditary and every class of prime graphs containing a graph on more than two vertices is not hereditary. As such, it is unknown whether Facts 4 and 5, which may only be applied on hereditary graph classes, can be combined. That is, the following problem is open.

Open Problem 3.8. *Let $\mathcal{G}$ be a hereditary class of graphs and let $\mathcal{F}$ be the class of 2-connected prime graphs in $\mathcal{G}$. If $\mathcal{F}$ has bounded clique-width, does this imply that $\mathcal{G}$ has bounded clique-width?*

To prove that a graph class $\mathcal{G}$ has unbounded clique-width, a similar method to the BCW Method can be used.

Unbounding Clique-Width (UCW Method)

1. Take a graph class $\mathcal{G}'$ known to have unbounded clique-width.

2. For every graph $G' \in \mathcal{G}'$, reduce G' to a graph in $\mathcal{G}$ by using a constant number of graph operations that do not change the clique-width of G' by "too much".

By Theorem 3.4, we can consider the class of walls as a starting point for the graph class $\mathcal{G}'$. A *k-subdivided wall* is a graph obtained from a wall after subdividing each edge exactly k times for some constant $k \geq 0$. Combining Fact 6 with Theorem 3.4 and the observation that walls of height at least 2 have maximum degree 3 leads to the following result.

Corollary 3.9 ([140]). *For any constant $k \geq 0$, the class of k-subdivided walls has unbounded clique-width.*

Corollary 3.9 has proven to be very useful. For instance, it can be used to obtain the following result (recall that $\mathcal{S}$ is the class of graphs each connected component of which is either a subdivided claw or a path).

Corollary 3.10 ([140]). *Let $\{H_1, \ldots, H_p\}$ be a finite set of graphs. If $H_i \notin \mathcal{S}$ for all $i \in \{1, \ldots, p\}$, then the class of $(H_1, \ldots, H_p)$-free graphs has unbounded clique-width.*

As a side note, we remark that "limit classes" of hereditary graph classes of unbounded clique-width may have bounded clique-width. For instance, the class of $(C_k, \ldots, C_\ell)$-subgraph-free graphs has unbounded clique-width for any two integers $k \geq 3$ and $\ell \geq k$ due to Corollary 3.9. However, for every $k \geq 3$, the class of $(C_k, C_{k+1}, \ldots)$-subgraph-free graphs has bounded clique-width [137]. We refer to [137] for more details on limit classes.

Corollary 3.9 is further generalized by the following theorem.

Theorem 3.11 ([67]). *For $m \geq 0$ and $n > m + 1$ the clique-width of a graph G is at least $\lfloor \frac{n-1}{m+1} \rfloor + 1$ if $V(G)$ has a partition into sets $V_{i,j}$ ($i, j \in \{0, \ldots, n\}$) with the following properties:*

1. *$|V_{i,0}| \leq 1$ for all $i \geq 1$,*
2. *$|V_{0,j}| \leq 1$ for all $j \geq 1$,*
3. *$|V_{i,j}| \geq 1$ for all $i, j \geq 1$,*
4. *$G[\cup_{j=0}^n V_{i,j}]$ is connected for all $i \geq 1$,*
5. *$G[\cup_{i=0}^n V_{i,j}]$ is connected for all $j \geq 1$,*
6. *for $i, j, k \geq 1$, if a vertex of $V_{k,0}$ is adjacent to a vertex of $V_{i,j}$ then $i \leq k$,*
7. *for $i, j, k \geq 1$, if a vertex of $V_{0,k}$ is adjacent to a vertex of $V_{i,j}$ then $j \leq k$, and*
8. *for $i, j, k, \ell \geq 1$, if a vertex of $V_{i,j}$ is adjacent to a vertex of $V_{k,\ell}$ then $|k - i| \leq m$ and $|\ell - j| \leq m$.*

Many other constructions of graphs of large clique-width follow from Theorem 3.11 using the UCW Method (possibly by applying Facts 1–3). For instance, this is the case for square grids [145], whose exact clique-width was determined by Golumbic and Rotics [91]. This is also the case for the constructions of Brandstädt, Engelfriet, Le and Lozin [27], Lozin and Volz [143], Korpelainen, Lozin and Mayhill [124] and Kwon, Pilipczuk and Siebertz [128] for proving that the classes of K_4-free co-chordal graphs, $2P_3$-free bipartite graphs, split permutation graphs and twisted chain graphs, respectively, have unbounded clique-width.

Constructions of graphs of arbitrarily large clique-width not covered by Theorem 3.11 can be found in [91] and [35], which prove that unit interval graphs and bipartite permutation graphs, respectively, have unbounded clique-width. We discuss these results in more detail in the next section, but we note the following.

First, the classes of split permutation graphs (and the analogous bipartite class of bichain graphs) [5], unit interval graphs [136] and bipartite permutation graphs [136] are even minimal hereditary graph classes of unbounded clique-width. Collins, Foniok, Korpelainen, Lozin and Zamaraev [48] proved that the number of minimal hereditary graphs of unbounded clique-width is infinite. Second, for classes, such as split graphs, bipartite graphs, co-bipartite graphs and $(K_{1,3}, 2K_2)$-free graphs, unboundedness of clique-width also follows from the fact that these classes are superfactorial [18] and an application of the following result.

Theorem 3.12 ([18]). *Every superfactorial graph class has unbounded clique-width.*

4 Results on Clique-Width for Hereditary Graph Classes

In this section we survey known results on (un)boundedness of clique-width for hereditary graph classes in a systematic way.[7] The proofs of these results often use the BCW Method or UCW Method. As mentioned earlier, many well-studied graph classes are hereditary. From the point of view of clique-width, these are also natural classes to consider, as the definition of clique-width implies that if a graph G contains a graph H as an induced subgraph, then $cw(H) \leq cw(G)$.

Recall that a graph class $\mathcal{G}$ is hereditary if and only if it can be characterized by a (possibly infinite) set of forbidden induced subgraphs $\mathcal{F}_\mathcal{G}$. We start by giving a dichotomy for the case when $\mathcal{F}_\mathcal{G}$ consists of a single graph H. This result is folklore: observe that P_4 has clique-width 3 and see [59] for a proof that P_4-free graphs have clique-width at most 2 and [67] for a proof of the other claims of Theorem 4.1.

Theorem 4.1. *Let H be a graph. The class of H-free graphs has bounded clique-width if and only if H is an induced subgraph of P_4. Furthermore, a graph has clique-width at most 2 if and only if it is P_4-free.*

Note that by Theorem 4.1 we can test whether a graph G has clique-width at most 2 in polynomial time by checking whether G is P_4-free. We recall that deciding whether a graph has clique-width at most c is known to be polynomial-time solvable for $c = 3$ [49], but open for $c \geq 4$.

As discussed in Section 1, an important reason for studying boundedness of clique-width for special graph classes is to obtain more classes of graphs for which a wide range of classical NP-complete problems become polynomial-time solvable. Theorem 4.1 shows that this cannot be done for (most) classes of H-free graphs. In order to find more graph classes of bounded clique-width, we can follow several approaches that try to extend Theorem 4.1.

To give an example, Vanherpe [163] considered the class of partner-limited graphs, which were introduced by Roussel, Rusu and Thuillier in [158]. A vertex u in a graph G is a *partner* of an induced subgraph H isomorphic to P_4 of G if $V(H) \cup \{u\}$ induces at least

---

[7]The Information System on Graph Classes and their Inclusions [71] also keeps a record of many graph classes for which boundedness or unboundedness of clique-width is known.

two P_4s in G. A graph G is said to be *partner-limited* if every induced P_4 has at most two partners. Vanherpe proved that the clique-width of partner-limited graphs is at most 4. This result generalized a corresponding result of Courcelle, Makowsky and Rotics [58] for P_4-*tidy* graphs, which are graphs in which every induced P_4 has at most one partner.

To give another example, Makowsky and Rotics [145] considered the classes of (q, t)-graphs, which were introduced by Babel and Olariu in [8]. For two integers q and t, a graph is a (q, t)-*graph* if every subset of q vertices induces a subgraph that has at most t distinct induced P_4s. Note that P_4-free graphs are the $(4, 0)$-graphs, whereas $(5, 1)$-graphs are also known as P_4-*sparse* graphs; note that the latter class of graphs is a subclass of the class of P_4-tidy graphs. Makowsky and Rotics proved the following result.

Theorem 4.2 ([145]). *Let $q \geq 4$ and $t \geq 0$. Then the class of (q, t)-graphs has bounded clique-width if*
- $q \leq 6$ and $t \leq q - 4$, or
- $q \geq 7$ and $t \leq q - 3$

and it has unbounded clique-width if
- $q \leq 6$ and $t \geq q - 3$
- $q = 7$ and $t \geq q - 2$, or
- $q \geq 8$ and $t \geq q - 1$.

Theorem 4.2 covers all cases except where $q \geq 8$ and $t = q - 2$. Makowsky and Rotics [145] therefore posed the following open problem (see also [116]).

Open Problem 4.3. *Is the clique-width of $(q, q - 2)$-graphs bounded if $q \geq 8$?*

Below we list five other systematic approaches, which we discuss in detail in the remainder of this section. First, we can try to replace "H-free graphs" by "H-free graphs in some hereditary graph class $\mathcal{X}$" in Theorem 4.1. We discuss this line of research in Section 4.1.

Second, we may try to determine boundedness of clique-width of hereditary graph classes $\mathcal{G}$ for which $\mathcal{F}_\mathcal{G}$ is small. However, even the classification for (H_1, H_2)-free graphs is not straightforward and is still incomplete. We discuss the state-of-the-art for (H_1, H_2)-free graphs in Section 4.2. There, we also explain how results in Section 4.1 are helpful for proving results for (H_1, H_2)-free graphs.[8]

Third, we may try to determine boundedness of clique-width for hereditary graph classes $\mathcal{G}$ for which $\mathcal{F}_\mathcal{G}$ only contains graphs of small size. For instance, Brandstädt, Dragan, Le and Mosca [26] classified boundedness of clique-width for those hereditary graph classes for which $\mathcal{F}_\mathcal{G}$ consists of 1-vertex extensions of P_4. We discuss their result, together with other results in this direction, in Section 4.3.

Fourth, we observe that P_4 is self-complementary. As such we can try to extend Theorem 4.1 to graph classes closed under complementation. Determining boundedness of clique-width for such graph classes is also natural to consider due to Fact 2. We present the current state-of-the-art in this direction in Section 4.4.

Fifth, we may consider hereditary graph classes that can be described not only in terms of forbidden induced subgraphs but also using some other forbidden subgraph containment. For instance, we can consider hereditary graph classes characterized by some set $\mathcal{F}$ of forbidden minors. We survey the known results in this direction in Section 4.5.

---

[8]We emphasize that the underlying research goal is not to start classifying the case of three forbidden induced subgraphs H_1, H_2 and H_3 after the classification for two graphs H_1 and H_2 has been completed. Instead the aim is to develop new techniques through a systematic study, by looking at hereditary graph classes from different angles in order to increase our understanding of clique-width.

4.1 Considering H-Free Graphs Contained in Some Hereditary Graph Class

Theorem 4.1 shows that the class of H-free graphs has bounded clique-width only if H is an induced subgraph of P_4. In this section we survey the effect on boundedness of clique-width of restricting the class of H-free graphs to just those graphs that belong to some hereditary graph class $\mathcal{X}$. Initially we do not want to make the hereditary graph class $\mathcal{X}$, in which we look for these H-free graphs, too narrow. However, if we let $\mathcal{X}$ be too large, the classification might remain the same as the one for general H-free graphs in Theorem 4.1. This is the case if we let $\mathcal{X}$ be the class of perfect graphs, or even the class of weakly chordal graphs, which form a proper subclass of perfect graphs by Observation 2.5.

Theorem 4.4 ([25]). *Let H be a graph. The class of H-free weakly chordal graphs has bounded clique-width if and only if H is an induced subgraph of P_4.*

If we restrict $\mathcal{X}$ further, then there are several potential classes of graphs to consider, such as chordal graphs, permutation graphs and distance-hereditary graphs (see also Figure 5). However, distance-hereditary graphs are known to have clique-width at most 3 [91] (and hence their clique-width can be computed in polynomial time using the algorithm of [49]). On the other hand, the classes of chordal graphs and permutation graphs have unbounded clique-width. This follows from combining Observation 2.5 with one of the following three theorems.

Theorem 4.5 ([91]). *The class of unit interval graphs has unbounded clique-width.*

Theorem 4.6 ([124]). *The class of split permutation graphs has unbounded clique-width.*

Theorem 4.7 ([35]). *The class of bipartite permutation graphs has unbounded clique-width.*

The case when $\mathcal{X}$ is the class of chordal graphs has received particular attention, as we now discuss. Brandstädt, Engelfriet, Le and Lozin [27] proved that the class of $4P_1$-free chordal graphs has unbounded clique-width. However, there are many graphs H besides P_4 for which the class of H-free chordal graphs has bounded clique-width. A result of [50] implies that K_r-free chordal graphs have bounded clique-width for every integer $r \geq 1$. Brandstädt, Le and Mosca [32] showed that $(P_1 + P_4)$-free chordal graphs have clique-width at most 8 and that $\overline{P_1 + P_4}$-free chordal graphs are distance-hereditary graphs and thus have clique-width at most 3. Brandstädt, Dabrowski, Huang and Paulusma [25] proved that bull-free chordal graphs have clique-width at most 3, improving a known bound of 8 [132]. The same authors also proved that $\overline{S_{1,1,2}}$-free chordal graphs have clique-width at most 4, and that the classes of $\overline{K_{1,3} + 2P_1}$-free chordal graphs, $(P_1 + \overline{P_1 + P_3})$-free chordal graphs and $(P_1 + \overline{2P_1 + P_2})$-free chordal graphs each have bounded clique-width.

Combining all the above results [25,27,32,50,91,145] leads to the following summary for H-free chordal graphs; see Figure 7 for definitions of the graphs F_1 and F_2 and Figure 8 for pictures of all (maximal) graphs H for which the class of H-free chordal graphs is known to have bounded clique-width.

Theorem 4.8 ([25]). *Let H be a graph with $H \notin \{F_1, F_2\}$. The class of H-free chordal graphs has bounded clique-width if and only if:*
  (i) $H = K_r$ for some $r \geq 1$,
  (ii) $H \subseteq_i$ bull,
  (iii) $H \subseteq_i P_1 + P_4$,
  (iv) $H \subseteq_i$ gem,

F_1 F_2

Figure 7: The two graphs H for which the boundedness of clique-width of the class of H-free chordal graphs is open.

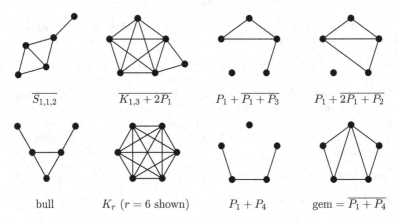

$\overline{S_{1,1,2}}$ $\overline{K_{1,3}+2P_1}$ $P_1+\overline{P_1+P_3}$ $P_1+\overline{2P_1+P_2}$

bull K_r ($r=6$ shown) P_1+P_4 gem $=\overline{P_1+P_4}$

Figure 8: The graphs H listed in Theorem 4.8, for which the class of H-free chordal graphs has bounded clique-width.

(v) $H \subseteq_i \overline{K_{1,3}+2P_1}$,
(vi) $H \subseteq_i P_1+\overline{P_1+P_3}$,
(vii) $H \subseteq_i P_1+\overline{2P_1+P_2}$, or
(viii) $H \subseteq_i \overline{S_{1,1,2}}$.

As can be seen from its statement, Theorem 4.8 leaves only two cases open, namely F_1 and F_2; see also [25].

Open Problem 4.9. *Determine whether the class of H-free chordal graphs has bounded or unbounded clique-width when $H = F_1$ or $H = F_2$.*

Recall that split graphs are chordal by Observation 2.5 and have been shown to have unbounded clique-width [145] (this also follows from Theorem 4.6). We now let $\mathcal{X}$ be the class of split graphs, that is, we consider classes of H-free split graphs, and find graphs H for which the class of H-free split graphs has bounded clique-width. We first note that as the class of split graphs is the class of $(C_4, C_5, 2P_2)$-free graphs [78], the complement of a split graph is also a split graph by Observation 2.2. By Fact 2 this implies the following observation, which we discuss in more depth in Section 4.4.

Observation 4.10. *For a graph H, the class of H-free split graphs has bounded clique-width if and only if the class of $\overline{H}$-free split graphs has bounded clique-width.*

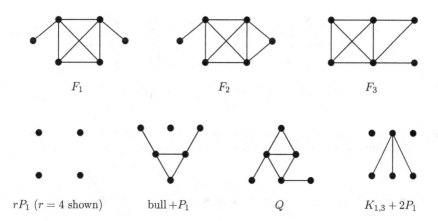

F_1 F_2 F_3

rP_1 ($r = 4$ shown) bull $+P_1$ Q $K_{1,3} + 2P_1$

Figure 9: The graphs H from Theorem 4.11 for which the classes of H-free split graphs and $\overline{H}$-free split graphs have bounded clique-width.

Brandstädt, Dabrowski, Huang and Paulusma considered H-free split graphs in [24]. They considered the two cases $H = F_1$ and $H = F_2$ that are open for H-free chordal graphs (Open Problem 4.9) and proved that the classes of F_1-free split graphs and F_2-free split graphs have bounded clique-width. They showed the same result for (bull $+P_1$)-free split graphs, Q-free split graphs, $(K_{1,3} + 2P_1)$-free split graphs and F_3-free split graphs; see Figure 9 for a description of each of these graphs. They also proved that for every integer $r \geq 1$, the clique-width of rP_1-free split graphs is at most $r + 1$. Moreover, they showed the following: if H is a graph with at least one edge and at least one non-edge that is not an induced subgraph of a graph in $\{F_4, \overline{F_4}, F_5, \overline{F_5}\}$ (see Figure 10), then the class of H-free split graphs has unbounded clique-width. Note that both F_4 and F_5 have seven vertices. The 6-vertex induced subgraphs of F_4 are: bull $+P_1, \overline{F_1}, \overline{F_3}$ and $K_{1,3} + 2P_1$. The 6-vertex induced subgraphs of F_5 are: bull $+P_1, F_1, F_2, \overline{F_2}, F_3, \overline{F_3}$ and Q. The above results lead to the following theorem.

Theorem 4.11 ([24]). *Let H be a graph not in $\{F_4, \overline{F_4}, F_5, \overline{F_5}\}$. The class of H-free split graphs has bounded clique-width if and only if:*
(i) $H = rP_1$ for some $r \geq 1$,
(ii) $H = K_r$ for some $r \geq 1$, or
(iii) H is an induced subgraph of a graph in $\{F_4, \overline{F_4}, F_5, \overline{F_5}\}$.

Theorem 4.11, combined with Observation 4.10, leaves two open cases: F_4 (or equivalently $\overline{F_4}$) and F_5 (or equivalently $\overline{F_5}$); see also [24].

Open Problem 4.12. *Determine whether the class of H-free split graphs has bounded or unbounded clique-width when $H = F_4$ or $H = F_5$.*

Note that a split graph with split partition (K, I) can be changed into a bipartite graph with bipartition classes K and I by applying a subgraph complementation on K. Hence, due to Fact 2, there is a close relationship between boundedness of clique-width for subclasses of split graphs and for subclasses of bipartite graphs. As such, it is natural to also consider the class of bipartite graphs as our class $\mathcal{X}$. We note that the relationship between

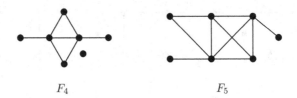

$$F_4 \qquad\qquad\qquad\qquad F_5$$

Figure 10: The (only) two graphs for which it is not known whether or not the classes of H-free split graphs and $\overline{H}$-free split graphs have bounded clique-width.

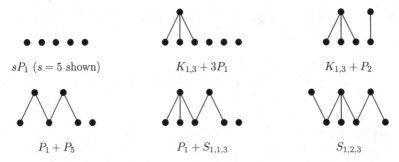

sP_1 ($s = 5$ shown) $K_{1,3} + 3P_1$ $K_{1,3} + P_2$

$P_1 + P_5$ $P_1 + S_{1,1,3}$ $S_{1,2,3}$

Figure 11: The graphs H for which the class of H-free bipartite graphs has bounded clique-width.

split graphs and bipartite graphs involves some subtleties as a split graph can have two non-isomorphic split partitions and a (disconnected) bipartite graph may have more than one bipartition (see [24] for a precise explanation). Nevertheless, results on boundedness of clique-width for H-free bipartite graphs, which we discuss below, have proved useful in proving Theorem 4.11.

Lozin [135] proved that the clique-width of $S_{1,2,3}$-free bipartite graphs is at most 5. He previously proved this bound in [134] for $(\text{sun}_4, S_{1,2,3})$-free bipartite graphs where sun_4 is the graph obtained from a 4-vertex cycle on vertices $u_1, \ldots, u_4$ by adding four new vertices $v_1, \ldots, v_4$ with edges $u_i v_i$ for $i \in \{1, \ldots, 4\}$. Fouquet, Giakoumakis and Vanherpe [81] proved that $(P_7, S_{1,2,3})$-free bipartite graphs have clique-width at most 4.

Lozin and Volz [143] used the above results to continue the study of [140] into boundedness of clique-width of H-free bipartite graphs. They fully classified the boundedness of clique-width for a variant of H-free bipartite graphs called strongly H^ℓ-free graphs, where H is forbidden with respect to a specified bipartition given by some labelling ℓ (which is unique if H is connected). Dabrowski and Paulusma [66] proved a similar (but different) dichotomy for a relaxation of this variant called weakly H^ℓ-free graphs, which is the variant used for proving some of the cases in Theorem 4.11. We refer to [66] for an explanation of strongly and weakly H^ℓ-free bipartite graphs. Using the above results Dabrowski and Paulusma [66] also gave a full classification for H-free bipartite graphs, that is, with H forbidden as an induced subgraph, as before; see also Figure 11.

Theorem 4.13 ([66]). *Let H be a graph. The class of H-free bipartite graphs has bounded clique-width if and only if:*

(i) $H = sP_1$ for some $s \geq 1$,

(ii) $H \subseteq_i K_{1,3} + 3P_1$,

(iii) $H \subseteq_i K_{1,3} + P_2$,

(iv) $H \subseteq_i P_1 + S_{1,1,3}$, or

(v) $H \subseteq_i S_{1,2,3}$.

We refer to [23] for some specific bounds on the clique-width of subclasses of H-free split graphs, bipartite graphs and co-bipartite graphs obtained from a decomposition property of 1-Sperner hypergraphs.

We continue our discussion on finding suitable graph classes $\mathcal{X}$ for which the classification of boundedness of the clique-width of its H-free subclasses differs from the (general) classification for H-free graphs in Theorem 4.1. Theorem 4.5 states that the class of unit interval graphs has unbounded clique-width. Unit interval graphs are contained in the class of interval graphs, which are contained in the class of chordal graphs by Observation 2.5. Hence, as well as narrowing the class of chordal graphs to split graphs, it is also natural to consider unit interval graphs and interval graphs to be the class $\mathcal{X}$. We recall that the class of unit interval graphs is a minimal hereditary graph class of unbounded clique-width [136]. Hence the clique-width of H-free unit interval graphs is bounded if and only if H is a unit interval graph. We refer to [147] for bounds on the clique-width of certain subclasses of unit interval graphs and pose the following open problem.

Open Problem 4.14. *Determine for which graphs H the class of H-free interval graphs has bounded clique-width.*

As mentioned earlier, instead of chordal graphs we can consider other subclasses of weakly chordal graphs as our class $\mathcal{X}$, such as permutation graphs (the containment follows from Observation 2.5). Recall that even the classes of split permutation graphs and bipartite permutation graphs have unbounded clique-width, as stated in Theorems 4.6 and 4.7, respectively. Hence, we could also take each of these three graph classes as the class $\mathcal{X}$. However, we recall that the classes of split permutation graphs [5] and bipartite permutation graphs [136] are minimal hereditary graph classes of unbounded clique-width. Hence, the clique-width of H-free split permutation graphs is bounded if and only if H is a split permutation graph, and similarly, the clique-width of H-free bipartite permutation graphs is bounded if and only if H is a bipartite permutation graph. Recall that Theorem 4.1 states that the class of H-free graphs has bounded clique-width if and only if H is an induced subgraph of P_4 and that Theorem 4.4 states that the same classification holds if we restrict to H-free weakly chordal graphs. Brignall and Vatter proved that the same classification also holds if we further restrict to H-free permutation graphs.

Theorem 4.15 ([40]). *Let H be a graph. The class of H-free permutation graphs has bounded clique-width if and only if H is an induced subgraph of P_4.*

Proof. Let H be a graph and note that if H is not a permutation graph, then the class of H-free permutation graphs equals the class of permutation graphs, which has unbounded clique-width by Theorem 4.1. We may therefore assume that H is a permutation graph. If H is an induced subgraph of P_4 then the class of H-free permutation graphs is a subclass of the class of P_4-free graphs and in this case Theorem 4.1 completes the proof.

The class of C_3-free permutation graphs is equal to the class of bipartite permutation graphs, which has unbounded clique-width by Theorem 4.7. Since the class of permutation graphs is closed under complementation (in the definition of permutation graphs, reverse the order of intersections of the line segments with one of the parallel lines), Fact 2 implies that $3P_1$-free permutation graphs also have unbounded clique-width. It therefore remains to consider the case when H is a $(C_3, 3P_1)$-free graph that is not an induced subgraph of P_4.

It is easy to verify that the only $(C_3, 3P_1)$-free graph on more than four vertices is C_5. Since C_5 is not a permutation graph, we may assume that H has at most four vertices. By inspection, the only $(C_3, 3P_1)$-free graphs H on at most four vertices that are not induced subgraphs of P_4 are C_4 and $2P_2$. As C_5 is not a permutation graph, the class of $(C_4, 2P_2)$-free permutation graphs is equal to the class of split permutation graphs, which has unbounded clique-width by Theorem 4.6. Therefore the class of H-free permutation graphs has unbounded clique-width if $H \in \{C_4, 2P_2\}$. This completes the proof. □

Recall from Observation 2.5 that bipartite permutation graphs are chordal bipartite, and that by Theorem 4.7 the class of bipartite permutation graphs has unbounded clique-width. From these two facts it follows that the class of chordal bipartite graphs has unbounded clique-width. In contrast, Lozin and Rautenbach [138] proved that $K_{1,t}^+$-free chordal bipartite graphs have bounded clique-width (recall that $K_{1,t}^+$ is the graph obtained from the star $K_{1,t}$ by subdividing one of its edges). Subdividing all three edges of the claw $K_{1,3}$ yields the graph $S_{2,2,2}$. As every bipartite permutation graph is $S_{2,2,2}$-free chordal bipartite, the class of $S_{2,2,2}$-free chordal bipartite graphs has unbounded clique-width, again due to Theorem 4.7.

The above discussion leads to the following open problems. Let E_t denote the graph obtained from the star $K_{1,t+1}$ after subdividing exactly two of its edges. Kamiński, Lozin and Milanič [116] asked the question: for which t, does the class of E_t-free chordal bipartite graphs have bounded clique-width? For $t \leq 2$, the class of E_t-free graphs has bounded clique-width by Theorem 4.13, as $E_2 = S_{1,2,2}$. Hence $t = 3$ is the first open case. By taking the class of chordal bipartite graphs as the class $\mathcal{X}$, we can pose a more general open problem.

Open Problem 4.16. *Determine for which graphs H the class of H-free chordal bipartite graphs has bounded clique-width.*

Boliac and Lozin [18] proved that for a graph H, the class of H-free claw-free graphs has bounded clique-width if and only if $H \subseteq_i P_4$, $H \subseteq_i$ paw or $H \subseteq_i K_3 + P_1$ (see also the more general Theorem 4.18 in Section 4.2). Line graphs form a subclass of the class of claw-free graphs. Gurski and Wanke [101] proved that if a line graph has a vertex whose non-neighbours induce a subgraph of clique-width k, then it has clique-width at most $8k+4$, which would imply, for instance, that (P_1+P_4)-free line graphs have clique-width at most 18 (they then improved this bound to 14). In fact we can show the following classification for the boundedness of clique-width of $(H_1, \ldots, H_p)$-free line graphs. Recall that $\mathcal{S}$ is the class of graphs every connected component of which is either a subdivided claw or a path on at least one vertex, whereas $\mathcal{T}$ consists of all line graphs of graphs in $\mathcal{S}$.

Theorem 4.17. *Let $\{H_1, \ldots, H_p\}$ be a finite set of graphs. Then the class of $(H_1, \ldots, H_p)$-free line graphs has bounded clique-width if and only if $H_i \in \mathcal{T}$ for some $i \in \{1, \ldots, p\}$.*

Proof. First suppose that $H_i \in \mathcal{T}$ for some $i \in \{1, \ldots, p\}$. By definition of $\mathcal{T}$, it follows that H_i is the line graph of some graph $F \in \mathcal{S}$. Because F is in $\mathcal{S}$, forbidding F as a (not necessarily induced) subgraph of G is the same as forbidding F as a minor by Observation 2.4. Moreover, F is planar. By a result of Bienstock, Robertson, Seymour and Thomas [13], every graph that does not contain some fixed planar graph as a minor has bounded path-width. Hence, the class of F-subgraph-free graphs has bounded path-width and consequently, bounded treewidth. Then, by Theorem 3.7, the class of H_i-free graphs, and thus the class of $(H_1, \ldots, H_p)$-free graphs, has bounded clique-width.

Now suppose that $H_i \notin \mathcal{T}$ for every $i \in \{1, \ldots, p\}$. Then every H_i has a connected component $H_i' \notin \mathcal{T}$. We may assume without loss of generality that each H_i is a line graph (otherwise forbidding it does not affect the class defined; if no H_i is a line graph, then the class of $(H_1, \ldots, H_p)$-free line graphs is the class of all line graphs, which has unbounded clique-width [18]). Since every $H_i' \notin \mathcal{T}$, every H_i' is not isomorphic to K_3. Hence, for every H_i' there exists a unique graph F_i' such that H_i' is the line graph of F_i' (see, for example, [103]). Since $H_i' \notin \mathcal{T}$, it follows that $F_i' \notin \mathcal{S}$, which means that there exists a positive integer k_i, such that the class of F_i'-subgraph-free graphs contains the class of k_i-subdivided walls. We let $k = \max\{k_i \mid 1 \leq i \leq p\}$. Then the class of $(F_1', \ldots, F_p')$-subgraph-free graphs contains the class of k-subdivided walls. As the class of k-subdivided walls has unbounded clique-width by Corollary 3.9, it follows that the class of $(F_1', \ldots, F_p')$-subgraph-free graphs has unbounded clique-width and hence unbounded treewidth [59]. Then, by Theorem 3.7, the class of $(H_1', \ldots, H_p')$-free line graphs has unbounded clique-width. Since the class of $(H_1, \ldots, H_p)$-free line graphs contains the class of $(H_1', \ldots, H_p')$-free line graphs, it follows that the class of $(H_1, \ldots, H_p)$-free line graphs also has unbounded clique-width. $\square$

4.2   Forbidding A Small Number of Graphs

As discussed, even the case when only two induced subgraphs H_1 and H_2 are forbidden has not yet been fully classified, and there are only partial results for the cases where three or four induced subgraphs are forbidden. Besides the class of $(C_4, C_5, 2P_2)$-free graphs (split graphs) [145], it is, for example, known that the classes of $(C_4, K_{1,3}, K_4, \text{diamond})$-free graphs [18,27] and $(3P_2, P_2 + P_4, P_6, \text{gem})$-free graphs have unbounded clique-width [67]. Recall that the gem is the graph $\overline{P_1 + P_4}$ (see Figure 3) and that the hammer is the graph $T_{0,0,2}$ (see Figure 4). It is known that the clique-width of (hammer, gem, $S_{1,1,2}$)-free graphs is at most 7 [33]. However, unlike the case for two forbidden induced subgraphs, no large-scale systematic study has been initiated for finitely defined hereditary graphs classes with more than two forbidden induced subgraphs; in Sections 4.3 and 4.4, respectively, we discuss two studies [14,26] with partial results in this direction. In this section, we focus only on (H_1, H_2)-free graphs.

Despite the classification for H-free graphs (Theorem 4.1) and many existing results for (un)boundedness of clique-width for (H_1, H_2)-free graphs [18,24,27,30–32,36,61,62,65] over the years, the number of open cases (H_1, H_2) was only recently proven to be finite, in [67]. This was done by combining the existing known results together with a number of new results for (H_1, H_2)-free graphs, and led to a classification that left 13 non-equivalent open cases.[9] This number has been reduced to five non-equivalent open cases by four later papers [14,19,60,63], and the current state-of-the-art is as follows (recall that $\mathcal{S}$ is the class of graphs each connected component of which is either a subdivided claw or a path and see also Figures 3, 4 and 11 in which a number of the graphs mentioned below are displayed).

---

[9]Given four graphs H_1, H_2, H_3, H_4, the classes of (H_1, H_2)-free graphs and (H_3, H_4)-free graphs are said to be equivalent if the unordered pair H_3, H_4 can be obtained from the unordered pair H_1, H_2 by some combination of the operations: (i) complementing both graphs in the pair, and (ii) if one of the graphs in the pair is $3P_1$, replacing it with $P_1 + P_3$ or vice versa. If two classes are equivalent, then one of them has bounded clique-width if and only if the other one does [67].

Theorem 4.18 ([19]). *Let $\mathcal{G}$ be a class of graphs defined by two forbidden induced subgraphs. Then:*

1. *$\mathcal{G}$ has bounded clique-width if it is equivalent to a class of (H_1, H_2)-free graphs such that one of the following holds:*
   - *(i)* H_1 *or* $H_2 \subseteq_i P_4$,
   - *(ii)* $H_1 = K_s$ *and* $H_2 = tP_1$ *for some* $s, t \geq 1$,
   - *(iii)* $H_1 \subseteq_i$ *paw and* $H_2 \subseteq_i K_{1,3} + 3P_1$, $K_{1,3} + P_2$, $P_1 + P_2 + P_3$, $P_1 + P_5$, $P_1 + S_{1,1,2}$, $P_2 + P_4$, P_6, $S_{1,1,3}$ *or* $S_{1,2,2}$,
   - *(iv)* $H_1 \subseteq_i$ *diamond and* $H_2 \subseteq_i P_1 + 2P_2$, $3P_1 + P_2$ *or* $P_2 + P_3$,
   - *(v)* $H_1 \subseteq_i$ *gem and* $H_2 \subseteq_i P_1 + P_4$ *or* P_5,
   - *(vi)* $H_1 \subseteq_i K_3 + P_1$ *and* $H_2 \subseteq_i K_{1,3}$, *or*
   - *(vii)* $H_1 \subseteq_i \overline{2P_1 + P_3}$ *and* $H_2 \subseteq_i 2P_1 + P_3$.
2. *$\mathcal{G}$ has unbounded clique-width if it is equivalent to a class of (H_1, H_2)-free graphs such that one of the following holds:*
   - *(i)* $H_1 \not\subseteq \mathcal{S}$ *and* $H_2 \not\subseteq \mathcal{S}$,
   - *(ii)* $H_1 \not\subseteq \overline{\mathcal{S}}$ *and* $H_2 \not\subseteq \overline{\mathcal{S}}$,
   - *(iii)* $H_1 \supseteq_i K_3 + P_1$ *or* C_4 *and* $H_2 \supseteq_i 4P_1$ *or* $2P_2$,
   - *(iv)* $H_1 \supseteq_i$ *diamond and* $H_2 \supseteq_i K_{1,3}$, $5P_1$, $P_2 + P_4$ *or* P_6,
   - *(v)* $H_1 \supseteq_i K_3$ *and* $H_2 \supseteq_i 2P_1 + 2P_2$, $2P_1 + P_4$, $4P_1 + P_2$, $3P_2$ *or* $2P_3$,
   - *(vi)* $H_1 \supseteq_i K_4$ *and* $H_2 \supseteq_i P_1 + P_4$ *or* $3P_1 + P_2$, *or*
   - *(vii)* $H_1 \supseteq_i$ *gem and* $H_2 \supseteq_i P_1 + 2P_2$.

Example 4.19. As an example of how results from Section 4.1 were useful in proving Theorem 4.18, consider the case when $(H_1, H_2) = (K_4, 2P_1 + P_3)$. In [25], it was shown that $(K_4, 2P_1 + P_3)$-free graphs have bounded clique-width. This was proven as follows. First, Theorem 4.8 was applied to solve the case when the given $(K_4, 2P_1 + P_3)$-free graph G is chordal. If G is not chordal, then G must contain a cycle C of length at least 4. As G is $(2P_1 + P_3)$-free, C can have length at most 7. This leads to a case distinction depending on the length of C. In each case, the set of vertices of G not on C is partitioned according to the intersection of their set of neighbours with C. This partition is then analysed and the facts from Section 3 are used to modify G into a graph belonging to a class known to have bounded clique-width.

As mentioned earlier, Theorem 4.18 does not cover five (non-equivalent) cases; see also [19].

Open Problem 4.20. *Does the class of (H_1, H_2)-free graphs have bounded or unbounded clique-width when:*

(i) $H_1 = K_3$ *and* $H_2 \in \{P_1 + S_{1,1,3}, S_{1,2,3}\}$,

(ii) $H_1 = $ *diamond and* $H_2 \in \{P_1 + P_2 + P_3, P_1 + P_5\}$

(iii) $H_1 = $ *gem and* $H_2 = P_2 + P_3$.

As discussed in [63], it would be interesting to find out if H-free bipartite graphs and H-free triangle-free graphs have the same classification with respect to the boundedness of their clique-width. It follows from Theorems 4.13 and 4.18 that the evidence so far is affirmative. Nevertheless, Open Problem 4.20.(i) shows that two remaining cases still need to be solved, namely $H = P_1 + S_{1,1,2}$ and $H = S_{1,2,3}$.

We will prove two partial results for the two cases in Open Problem 4.20.(i). These results also illustrate some of the previously discussed techniques. Namely, we show that the class of prime $(K_3, C_5, S_{1,2,3})$-free graphs has bounded clique-width (Proposition 4.22) and that the class of $(K_3, C_5, P_1 + S_{1,1,3})$-free graphs has bounded clique-width (Proposition 4.23). Combining Propositions 4.22 and 4.23 with Fact 4 implies that in both cases of Open Problem 4.20.(i) we need only consider prime graphs that contain C_5 as an induced subgraph.

For Proposition 4.22 we need the following lemma, which follows from [60, Lemma 8].[10] Proposition 4.23 is a new result.

Lemma 4.21 ([60]). *If G is a prime $(K_3, C_5, S_{1,2,3})$-free graph, then G is either bipartite or a cycle.*

Proposition 4.22. *The class of prime $(K_3, C_5, S_{1,2,3})$-free graphs has bounded clique-width.*

Proof. If a $(K_3, C_5, S_{1,2,3})$-free graph is bipartite, then it is an $S_{1,2,3}$-free bipartite graph and we are done by Theorem 4.13. If it is a cycle then it has maximum degree 2, and we are done by Proposition 3.3. By Lemma 4.21 this completes the proof. □

Proposition 4.23. *The class of $(K_3, C_5, P_1 + S_{1,1,3})$-free graphs has bounded clique-width.*

Proof. Let G be a $(K_3, C_5, P_1 + S_{1,1,3})$-free graph. Since the clique-width of a graph equals the maximum of the clique-width of its components, we may assume that G is connected. We may assume that G is not bipartite, otherwise it is a $(P_1 + S_{1,1,3})$-free bipartite graph, in which case it has bounded clique-width by Theorem 4.13. As G is (C_3, C_5)-free (since $C_3 = K_3$), it contains an induced odd cycle C on k vertices, say $v_1, v_2, \ldots, v_k$ in that order, where $k \geq 7$. We may assume without loss of generality that C is an odd cycle of minimum length in G.

If $V(G) = V(C)$, then G has maximum degree 2 and we can use Proposition 3.3. From now on we assume that G contains at least one vertex not on C. Suppose that there is a vertex v that is adjacent to at least two vertices of C. As C has minimal length and G is (C_3, C_5)-free, v must be adjacent to precisely two vertices of C, which must be at distance 2 from each other on C.

For $i \in \{1, \ldots, k\}$, let V_i be the set of vertices outside C that are adjacent to v_{i-1} and v_{i+1} (subscripts on vertices and vertex sets are interpreted modulo k throughout the proof), and let W_i be the set of vertices whose unique neighbour in C is v_i. Finally, let U be the set of vertices that have no neighbour in C. Thus every vertex in G is in C, U or in some set V_i or W_i for some $i \in \{1, \ldots, k\}$. Moreover, as G is connected, there must be at least one set of the form V_i or W_i that is non-empty. We may assume without loss of generality that there is a vertex $v \in V_1 \cup W_2$. If $k \geq 9$ then $G[v_7, v_2, v, v_1, v_3, v_4, v_5]$ is a $P_1 + S_{1,1,3}$, a contradiction. We conclude that $k = 7$.

We now prove five claims, the first of which follows immediately from the fact that G is K_3-free.

Claim 1. For $i \in \{1, \ldots, 7\}$, V_i and W_i are independent sets.

Claim 2. For every $i \in \{1, \ldots, 7\}$, V_i and W_i are complete to U, and $|U| \leq 1$.

Suppose, for contradiction, that a vertex $x \in V_1 \cup W_2$ is non-adjacent to $y \in U$. Then $G[y, v_2, x, v_1, v_3, v_4, v_5]$ is a $P_1 + S_{1,1,3}$, a contradiction. By symmetry, this proves the first

---

[10] [60, Lemma 8] is about $(K_3, C_5, S_{1,2,3})$-free graphs without *false twins*, that is, without pairs of non-adjacent vertices which have the same set of neighbours. Prime graphs have no false twins by definition.

part of the claim. Now suppose that U contains at least two vertices y and y'. Then $v \in V_1 \cup W_2$ is adjacent to both y and y'. Since G is K_3-free, it follows that y and y' are not adjacent. Then $G[v_6, v, y, y', v_2, v_3, v_4]$ is a $P_1 + S_{1,1,3}$, a contradiction. This proves the second part of the claim.

Claim 3. For $i \in \{1, \ldots, 7\}$, $|W_i| \leq 1$.
Suppose that $x, y \in W_1$. By Claim 1, we find that x is non-adjacent to y. Then $G[v_6, v_1, x, y, v_2, v_3, v_4]$ is a $P_1 + S_{1,1,3}$, a contradiction. The claim follows by symmetry.

A set of vertices is *large* if it contains at least two vertices and *small* otherwise.

Claim 4. For $i, j \in \{1, \ldots, 7\}$, if v_i is adjacent to v_j and at least one of V_i and V_j is large, then V_i is complete to V_j.
Suppose that there are vertices $x, x' \in V_2$ and $y \in V_3$ such that y is non-adjacent to x'. By Claim 1, x is non-adjacent to x'. Then $G[x', y, x, v_2, v_4, v_5, v_6]$ or $G[y, v_1, x, x', v_7, v_6, v_5]$ is a $P_1 + S_{1,1,3}$ if y is adjacent or non-adjacent to x, respectively, a contradiction. The claim follows by symmetry.

Claim 5. For distinct $i, j \in \{1, \ldots, 7\}$, if a vertex of V_i has a neighbour in V_j, then v_i is adjacent to v_j.
Since G is K_3-free, for every i the set V_i is anti-complete to the set V_{i+2}. Moreover, if i and j are such that the vertices v_i and v_j are at distance more than 2 on the cycle, then V_i and V_j must be anti-complete, as otherwise there would be a smaller odd cycle than C in G, contradicting the minimality of k. This proves Claim 5.

Let G' be the graph obtained from G by deleting all vertices in small sets V_i, W_i or U (note that in doing this we delete at most $7 + 7 + 1 = 15$ vertices). By Fact 1, it is sufficient to show that G' has bounded clique-width. Let V_i' be V_i if V_i is large and $\emptyset$ otherwise. By Claims 2 and 3, G' only contains vertices in C and the sets V_i'. By Claim 1, each set V_i' is independent. Furthermore, by Claim 4, if v_i and v_j are adjacent vertices of C then V_i' is complete to V_j'. By Claim 5, for all other choices of i and j, the set V_i' is anti-complete to V_j'. This implies that for every $i \in \{1, \ldots, 7\}$, the set $V_i' \cup \{v_i\}$ is a module that is an independent set. We apply seven bipartite complementations, namely between V_i and V_{i+1} for $i \in \{1, \ldots, 7\}$. This yields an edgeless graph, which has clique-width 1. By Fact 3, it follows that G' has bounded clique-width. Hence G has bounded clique-width. This completes the proof.   □

4.3 Forbidding Small Induced Subgraphs

Theorem 4.1 states that a class of H-free graphs has bounded clique-width if and only if H is an induced subgraph of P_4. As discussed, one way to obtain more graph classes of bounded clique-width is to extend P_4 by one extra vertex, but then we need to forbid at least one other graph as an induced subgraph besides this 1-vertex extension of P_4. In this context, Brandstädt and Mosca [37] classified the boundedness of clique-width for $\mathcal{H}$-free graphs, where $\mathcal{H}$ is a subset of the set of P_4-sparse graphs with five vertices. Brandstädt, Hoàng and Le [29] proved that (bull, $S_{1,1,2}$, $\overline{S_{1,1,2}}$)-free graphs have bounded clique-width. Brandstädt, Dragan, Le and Mosca proved the following more general dichotomy containing the results of [29, 37]; see also Figure 12.

Theorem 4.24 ([26]). *Let $\mathcal{H}$ be a set of 1-vertex extensions of P_4. The class of $\mathcal{H}$-free graphs has bounded clique-width if and only if $\mathcal{H}$ is not a subset of any of the following sets:*
(i) $\{P_1 + P_4, P_5, S_{1,1,2}, \overline{\text{banner}}, C_5, \overline{S_{1,1,2}}\}$,

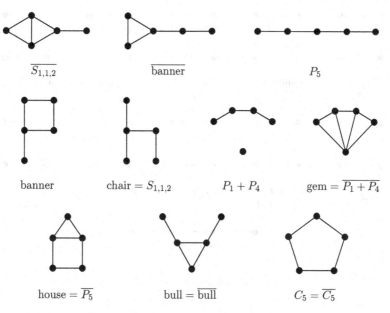

Figure 12: The 1-vertex extensions of P_4.

(ii) $\{\overline{P_1 + P_4}, \overline{P_5}, S_{1,1,2}, \text{banner}, C_5, \overline{S_{1,1,2}}\}$,
(iii) $\{P_1 + P_4, P_5, S_{1,1,2}, \overline{\text{banner}}, \text{banner}, C_5, \text{bull}\}$,
(iv) $\{\overline{P_1 + P_4}, \overline{P_5}, \overline{S_{1,1,2}}, \overline{\text{banner}}, \text{banner}, C_5, \text{bull}\}$ or
(v) $\{P_5, \overline{\text{banner}}, \text{banner}, C_5, \overline{P_5}\}$.

Brandstädt, Engelfriet, Le and Lozin [27] considered all sets $\mathcal{H}$ of graphs on at most four vertices and determined for which such sets $\mathcal{H}$ the class of $\mathcal{H}$-free graphs has bounded clique-width. They proved the following dichotomy for sets $\mathcal{H}$ of 4-vertex graphs and showed that all cases involving at least one graph with fewer than four vertices follow from known cases (see also Theorems 4.1 and 4.18); the graphs in Theorem 4.25 are displayed in Figure 13.

Theorem 4.25 ([27]). *Let $\mathcal{H}$ be a set of 4-vertex graphs. The class of $\mathcal{H}$-free graphs has bounded clique-width if and only if $\mathcal{H}$ is not a subset of any of the following sets:*
   (i) $\{C_4, 2P_2\}$,
   (ii) $\{K_4, 2P_2\}$,
   (iii) $\{C_4, 4P_1\}$,
   (iv) $\{K_4, \text{diamond}, C_4, \text{claw}\}$,
   (v) $\{4P_1, 2P_1 + P_2, 2P_2, K_3 + P_1\}$,
   (vi) $\{K_4, \text{diamond}, C_4, \text{paw}, K_3 + P_1\}$, or
   (vii) $\{4P_1, 2P_1 + P_2, 2P_2, P_1 + P_3, \text{claw}\}$.

4.4   Considering Hereditary Graph Classes Closed Under Complementation

Recall that subgraph complementation preserves boundedness of clique-width by Fact 2. It is therefore natural to consider hereditary classes of graphs $\mathcal{G}$ that are closed under complementation. In this section we survey the known results for these graph classes.

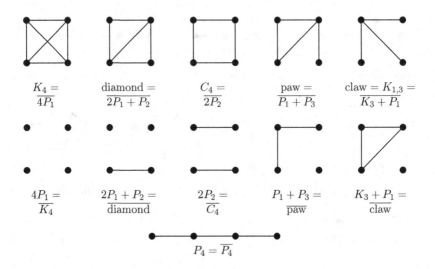

Figure 13: The graphs on four vertices.

Recall that by Observation 2.2 a hereditary graph class $\mathcal{G}$ is closed under complementation if and only if $\mathcal{H} = \mathcal{F}_{\mathcal{G}}$ is closed under complementation. We start by considering the cases where $|\mathcal{H}|$ is small.

The only two non-empty self-complementary induced subgraphs of P_4 are P_1 and P_4. Hence, from Theorem 4.1 it follows that the only self-complementary graphs H for which the class of H-free graphs has bounded clique-width are $H = P_1$ and $H = P_4$. This result settles the $|\mathcal{H}| = 1$ case and was generalized as follows.

Theorem 4.26 ([14]). *For any set $\mathcal{H}$ of non-empty self-complementary graphs, the class of $\mathcal{H}$-free graphs has bounded clique-width if and only if either $P_1 \in \mathcal{H}$ or $P_4 \in \mathcal{H}$.*

We now discuss the $|\mathcal{H}| = 2$ case. By Theorem 4.26, it remains to consider the case when $\mathcal{H} = \{H_1, H_2\}$ with $H_2 = \overline{H_1}$ and H_1 is not self-complementary. This leads to the following classification, which also follows from Theorem 4.18. The graphs in this classification are displayed in Figure 14.

Theorem 4.27 ([14]). *For a graph H, the class of $(H, \overline{H})$-free graphs has bounded clique-width if and only if H or $\overline{H}$ is an induced subgraph of $K_{1,3}$, $P_1 + P_4$, $2P_1 + P_3$ or sP_1 for some $s \geq 1$.*

As we will see, the $|\mathcal{H}| = 3$ case has not yet been fully settled. Up to permutations of the graphs H_1, H_2, H_3, a class of (H_1, H_2, H_3)-free graphs is closed under complementation if and only if H_i is self-complementary for all $i \in \{1, 2, 3\}$, or $H_1 = \overline{H_2}$ and H_3 is self-complementary (note that we may assume that $\mathcal{H}$ is minimal). By Theorem 4.26, we only need to consider the second case. By Theorem 4.1, we may exclude the case when $H_3 = P_1$ or $H_3 = P_4$. The next two smallest self-complementary graphs H_3 are the C_5 and the bull.

Blanché, Dabrowski, Johnson, Lozin, Paulusma and Zamaraev [14] proved that the classification of boundedness of clique-width for $(H, \overline{H}, C_5)$-free graphs coincides with the one of Theorem 4.27. This raised the question of whether the same is true for other sets

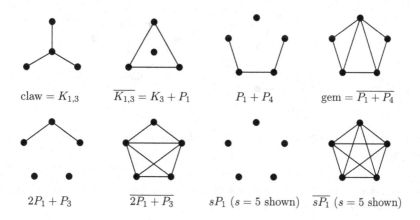

Figure 14: Graphs H for which the clique-width of $(H, \overline{H})$-free graphs is bounded.

of self-complementary graphs $\mathcal{F} \neq \{C_5\}$. However, the bull is self-complementary, and if $\mathcal{F}$ contains the bull, then the answer is negative, which can be seen as follows. By Theorem 4.27, both the class of $(S_{1,1,2}, \overline{S_{1,1,2}})$-free graphs and the class of $(2P_2, C_4)$-free graphs have unbounded clique-width. In contrast, by Theorem 4.24, both the class of $(S_{1,1,2}, \overline{S_{1,1,2}}, \text{bull})$-free graphs and even the class of $(P_5, \overline{P_5}, \text{bull})$-free graphs have bounded clique-width. However, as shown in the next theorem, the bull turned out to be the *only* exception if we exclude the "trivial" cases $H_3 = P_1$ and $H_3 = P_4$, which are the only non-empty self-complementary graphs on fewer than five vertices.

Theorem 4.28 ([14]). *Let $\mathcal{F}$ be a set of self-complementary graphs on at least five vertices not equal to the bull. For a graph H, the class of $(\{H, \overline{H}\} \cup \mathcal{F})$-free graphs has bounded clique-width if and only if H or $\overline{H}$ is an induced subgraph of $K_{1,3}$, $P_1 + P_4$, $2P_1 + P_3$ or sP_1 for some $s \geq 1$.*

By Theorems 4.26 and 4.28 the case $|\mathcal{H}| = 3$ is settled except when $H_1 = \overline{H_2}$ and H_3 is the bull; see also [14].

Open Problem 4.29. *For which graphs H does the class of $(H, \overline{H}, \text{bull})$-free graphs have bounded clique-width?*

In light of Theorem 4.28, Open Problem 4.29 can also be extended to sets $\mathcal{F}$ of self-complementary graphs containing the bull.

4.5 Forbidding with Respect to Other Graph Containment Relations

In this section we survey results on (un)boundedness of clique-width for hereditary graph classes that can alternatively be characterized by some other graph containment relation. In particular, when we forbid a finite collection of either subgraphs, minors or topological minors, it is possible to completely characterize those graph classes that have bounded clique-width.

Theorem 4.30 ([67, 116]). *Let $\{H_1, \ldots, H_p\}$ be a finite set of graphs. Then the following statements hold:*

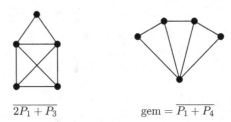

$$\overline{2P_1 + P_3} \qquad\qquad \text{gem} = \overline{P_1 + P_4}$$

Figure 15: The graphs H such that the class of H-induced-minor-free graphs has bounded clique-width.

(i) The class of $(H_1, \ldots, H_p)$-subgraph-free graphs has bounded clique-width if and only if $H_i \in \mathcal{S}$ for some $i \in \{1, \ldots, p\}$.

(ii) The class of $(H_1, \ldots, H_p)$-minor-free graphs has bounded clique-width if and only if H_i is planar for some $i \in \{1, \ldots, p\}$.

(iii) The class of $(H_1, \ldots, H_p)$-topological-minor-free graphs has bounded clique-width if and only if H_i is planar and has maximum degree at most 3 for some $i \in \{1, \ldots, p\}$.

The graph classes in Theorem 4.30 have in common that the corresponding containment relation allows edge deletions. If edge deletions are not permitted, then the situation becomes less clear, as we already saw for the induced subgraph relation. This is also true for the induced minor relation, for which only the following (non-trivial) result is known. We refer to Figure 15 for a picture of the graphs $\overline{2P_1 + P_3}$ and $\overline{P_1 + P_4}$ (recall that the latter graph is also known as the gem).

Theorem 4.31 ([10]). *Let H be a graph. The class of H-induced-minor-free graphs has bounded clique-width if and only if $H \subseteq_i \overline{2P_1 + P_3}$ or $H \subseteq_i \overline{P_1 + P_4}$.*

With an eye on Theorem 4.18, Theorem 4.31 leads to the following open problem.

Open Problem 4.32. *Determine for which pairs of graphs (H_1, H_2) the class of (H_1, H_2)-induced-minor-free graphs has bounded clique-width.*

We end this section with two more open problems; we note that a class of H-contraction-free graphs need not be hereditary and that Open Problem 4.33 is trivial if we allow disconnected graphs, since edge contractions preserve the number of components in a graph.

Open Problem 4.33. *Determine for which graphs H the class of connected H-contraction-free graphs has bounded clique-width.*

Open Problem 4.34. *Determine for which graphs H the class of H-induced-topological-minor-free graphs has bounded clique-width.*

5   Algorithmic Consequences

In this section we illustrate how bounding clique-width (or one of its equivalent parameters) can be used to find polynomial-time algorithms to solve problems on special graph classes, even when these problems are NP-hard on general graphs. In Section 5.1 we discuss meta-theorems, and in Section 5.2 we show how they can be used as part of a general strategy for solving problems. In Section 5.3 we focus on atoms, which are often used as a specific ingredient for the general strategy. Finally, in Sections 5.4 and 5.5 we look at two problems in particular, namely COLOURING and GRAPH ISOMORPHISM, respectively. For other graph problems where boundedness of clique-width is used to classify their computational complexity on hereditary graph classes, see, for example, [28, 44]. We refer to [11, 41, 79, 80, 89] for parameterized complexity results on clique-width.

5.1   Meta-Theorems

We observed in Section 1 that one of the advantages of showing that a graph class has bounded clique-width is that one can apply meta-theorems that say that any problem definable within certain constraints can be solved in polynomial time on the class. We mentioned such a theorem concerning any problem that can be defined in MSO_1 [58]. The result of [58] has been extended by others to address graph problems that cannot be defined in MSO_1. An important example of such a problem is the $\mathcal{F}$-PARTITION problem, which asks, for a graph G and an integer k, whether $V(G)$ can be partitioned into (possibly empty) sets $V_1, \ldots, V_k$ such that every V_i induces a graph in $\mathcal{F}$. In particular, if $\mathcal{F}$ consists of the edgeless graphs, then the $\mathcal{F}$-PARTITION problem is equivalent to the COLOURING problem.

Espelage, Gurski and Wanke [74] gave a general method to show that on graphs of bounded clique-width, $\mathcal{F}$-PARTITION is polynomial-time solvable for a number of graph classes $\mathcal{F}$ including complete graphs, edgeless graphs, forests and triangles. Their method can also be applied to other problems, such as HAMILTON CYCLE (see also [166] and see [12] for a faster algorithm) and CUBIC SUBGRAPH. Later, Kobler and Rotics [121] proved that a variety of other NP-complete graph partition problems (where either the set of vertices or the set of edges is partitioned) can be solved in polynomial time for graphs of bounded clique-width. Again, their set of problems includes COLOURING (see [131] for the fastest known algorithm, parameterized by clique-width, for finding a k-colouring if k is constant). However, their work also captures other graph partition problems, such as LIST k-COLOURING and EDGE-DOMINATING SET.

Gerber and Kobler [86] gave a framework of vertex partition problems with respect to a fixed interval degree constraint matrix. They showed that these problems, which include INDUCED BOUNDED DEGREE SUBGRAPH, INDUCED k-REGULAR SUBGRAPH, H-COLOURING and H-COVERING, are all solvable in polynomial time on graphs of bounded clique-width. In the same paper, they extended their framework to include more general problems, such as SATISFACTORY GRAPH PARTITIONING and MAJORITY DOMINATION NUMBER. Rao [153] gave another family of vertex partitioning problems that can be solved in polynomial time for graphs of bounded clique-width. Besides COLOURING, this family also includes DOMATIC NUMBER, HAMILTON CYCLE and $\mathcal{F}$-PARTITION where $\mathcal{F}$ consists of complete and edgeless graphs; perfect graphs; or H-free graphs, for an arbitrary fixed graph H.

The algorithms in [58, 74, 86, 92, 121] all require a c-expression of the input graph G for some constant c. Recall that computing the clique-width of a graph is NP-hard [77] (and that the complexity of deciding whether a graph has clique-width at most c is still open for every constant $c \geq 4$). Of course, this suggests we cannot hope to compute a $\mathrm{cw}(G)$-expression in polynomial time. However, it is sufficient to use the algorithm of Seymour and Oum [151], which returns a c-expression for some $c \leq 2^{3\,\mathrm{cw}(G)+2} - 1$ in $O(n^9 \log n)$ time, or the later improvements of Oum [149] and Hliněný and Oum [110] that provide cubic-time algorithms which yield a c-expression for some $c \leq 8^{\mathrm{cw}(G)} - 1$ and $c \leq 2^{\mathrm{cw}\,G+1} - 1$, respectively.

We note that there exist problems that are polynomial-time solvable for graphs of clique-width c, but NP-complete for graphs of clique-width d for constants c and d with $c < d$. For example, this holds for the DISJOINT PATHS problem, which is linear-time solvable for graphs of clique-width at most 2, but NP-complete for graphs of clique-width at most 6 [99].

5.2 A General Strategy for Finding Algorithms

Below we describe an approach that has often been used as a general strategy when we want to solve a problem Π on a graph class $\mathcal{G}$. We suppose that there exists some meta-algorithm A that can be used to solve Π on classes of bounded clique-width. We say that the graph class $\mathcal{G}$ is *reducible* to some subclass $\mathcal{G}' \subseteq \mathcal{G}$ with respect to Π if the following holds: if Π can be solved in polynomial time on $\mathcal{G}'$, then Π can also be solved in polynomial time on $\mathcal{G}$. We can now state the following general approach.

Clique-Width Method

1. Check if $\mathcal{G}$ has bounded clique-width (for instance, by using the BCW Method).

2. If so, then apply A. Otherwise choose between 3a and 3b.

3a. Reduce $\mathcal{G}$ to some subclass $\mathcal{G}'$ of bounded clique-width and apply A.

3b. Partition $\mathcal{G}$ into two classes $\mathcal{G}_1$ and $\mathcal{G}_2$, such that $\mathcal{G}_1$ has bounded clique-width and is as large as possible. Apply A to solve Π on $\mathcal{G}_1$. Use some problem-specific algorithm to solve Π on $\mathcal{G}_2$.

To give an example where Step 3a of this method is used, we can let $\mathcal{G}'$ be the class that consists of all atoms in $\mathcal{G}$. Recall that a connected graph is an atom if it has no clique cut-set. Dirac [73] introduced the notion of a clique cut-set and proved that every chordal graph is either complete or has a clique cut-set. As complete graphs have clique-width 2, this means that chordal graphs that are atoms have clique-width at most 2, whereas the class of chordal graphs has unbounded clique-width (see, for example, Theorem 4.8). Over the years, decomposition into atoms has become a widely used tool for solving decision problems on hereditary graph classes. For instance, a classical result of Tarjan [162] implies that COLOURING and other problems, such as those of determining the size of a largest independent set (INDEPENDENT SET) or a largest clique (CLIQUE), are polynomial-time solvable on a hereditary graph class $\mathcal{G}$ if and only if they are polynomial-time solvable on the atoms of $\mathcal{G}$. We will discuss atoms in more detail in Section 5.3.

To give an example where Step 3b of this method is used, Fraser, Hamel, Hoàng, Holmes and LaMantia [82] proved that COLOURING can be solved in polynomial time for $(C_4, C_5, 4P_1)$-free graphs by proving that the non-perfect graphs from this class have bounded clique-width and by recalling that COLOURING can be solved in polynomial time on perfect graphs [93].

5.3 Atoms

As mentioned, atoms are an important example for Step 3a in the Clique-Width Method. To determine new polynomial-time results for COLOURING, Gaspers, Huang and

Paulusma [85] investigated whether there exist graph classes of unbounded clique-width whose atoms have bounded clique-width. They found that this is not the case for the classes of H-free graphs. That is, the classification for H-free atoms coincides with the classification for H-free graphs in Theorem 4.1.

Theorem 5.1 ([85]). *Let H be a graph. The class of H-free atoms has bounded clique-width if and only if H is an induced subgraph of P_4.*

As split graphs are chordal by Observation 2.5, it follows that split atoms (split graphs that are atoms) are complete graphs, and thus have clique-width at most 2, whereas the class of general split graphs has unbounded clique-width [145]. As the class of split graphs coincides with the class of $(C_4, C_5, 2P_2)$-free graphs [78], Gaspers, Huang and Paulusma [85] asked whether there exists a class of (H_1, H_2)-free graphs of unbounded clique-width whose atoms form a class of bounded clique-width. They proved that this is indeed the case by showing a constant bound on the clique-width of atoms in the class of (C_4, P_6)-free graphs, which form a superclass of split graphs (they used this to prove that COLOURING is polynomial-time solvable for (C_4, P_6)-free graphs).

Theorem 5.2 ([85]). *Every (C_4, P_6)-free atom has clique-width at most 18.*

We are not aware of any other examples, which leads us to ask the following open problem (see also [85]).

Open Problem 5.3. *Determine all pairs of graphs H_1, H_2 such that the class of (H_1, H_2)-free graphs has unbounded clique-width, but the class of (H_1, H_2)-free atoms has bounded clique-width.*

Recall from Open Problem 4.20 that there are still five non-equivalent pairs H_1, H_2 for which we do not know whether the clique-width of (H_1, H_2)-free graphs is bounded or unbounded. Due to the algorithmic implications mentioned above, the following problem is therefore also of interest.

Open Problem 5.4. *Does the class of (H_1, H_2)-free atoms have bounded clique-width when:*

(i) $H_1 = K_3$ and $H_2 \in \{P_1 + S_{1,1,3}, S_{1,2,3}\}$

(ii) $H_1 = $ diamond and $H_2 \in \{P_1 + P_2 + P_3, P_1 + P_5\}$

(iii) $H_1 = $ gem and $H_2 = P_2 + P_3$.

5.4 Graph Colouring

Král', Kratochvíl, Tuza, and Woeginger completely classified the complexity of COLOURING for H-free graphs.

Theorem 5.5 ([126]). *Let H be a graph. If $H \subseteq_i P_4$ or $H \subseteq_i P_1 + P_3$, then COLOURING restricted to H-free graphs is polynomial-time solvable, otherwise it is NP-complete.*

For (H_1, H_2)-free graphs, the classification of COLOURING is open for many pairs of graphs H_1, H_2. A summary of the known results can be found in [88], but several other results have since appeared [15, 43, 68, 85, 119, 120, 146]; see [68] for further details. In relation to boundedness of clique-width, the following is of importance. There still exist ten classes of (H_1, H_2)-free graphs, for which COLOURING could potentially be solved in polynomial time by showing that their clique-width is bounded. That is, for these classes, the complexity of COLOURING is not resolved, and it is not known whether the clique-width is bounded. This list is obtained by updating the list of [60], which contains 13 cases, with the result of [15] for $(H_1, H_2) = (2P_1 + P_3, \overline{2P_1 + P_3})$ and the results of [19] for $(H_1, H_2) = (\text{gem}, P_1 + 2P_2)$ and $(H_1, H_2) = (P_1 + P_4, \overline{P_1 + 2P_2})$.

Open Problem 5.6. *Can the* COLOURING *problem be solved in polynomial time on* (H_1, H_2)-*free graphs when:*

  (i) $H_1 \in \{K_3, \text{paw}\}$ *and* $H_2 \in \{P_1 + S_{1,1,3}, S_{1,2,3}\}$,

  (ii) $H_1 = 2P_1 + P_2$ *and* $H_2 \in \{\overline{P_1 + P_2 + P_3}, \overline{P_1 + P_5}\}$,

  (iii) $H_1 = \text{diamond}$ *and* $H_2 \in \{P_1 + P_2 + P_3, P_1 + P_5\}$,

  (iv) $H_1 = P_1 + P_4$ *and* $H_2 = \overline{P_2 + P_3}$,

  (v) $H_1 = \text{gem}$ *and* $H_2 = P_2 + P_3$.

5.5 Graph Isomorphism

Grohe and Schweitzer [92] proved that GRAPH ISOMORPHISM is polynomial-time solvable for graphs of bounded clique-width. Hence, identifying graph classes of bounded clique-width is of importance for the GRAPH ISOMORPHISM problem.

The classification for the computational complexity of GRAPH ISOMORPHISM for H-free graphs can be found in a technical report of Booth and Colbourn [22], who credited the result to an unpublished manuscript of Colbourn and Colbourn. Another proof of this result appears in a paper of Kratsch and Schweitzer [127].

Theorem 5.7 ([22]). *Let H be a graph. If $H \subseteq_i P_4$, then* GRAPH ISOMORPHISM *for H-free graphs can be solved in polynomial time, otherwise it is* GI-*complete.*

Note that GRAPH ISOMORPHISM is polynomial-time solvable even for the class of permutation graphs [47], which contains the class of P_4-free graphs.

Schweitzer [161] observed great similarities between the techniques used for classifying boundedness of clique-width and classifying the complexity of GRAPH ISOMORPHISM for hereditary graph classes. He proved that GRAPH ISOMORPHISM is GI-complete for any graph class $\mathcal{G}$ that allows a so-called simple path encoding and also showed that every such graph class $\mathcal{G}$ has unbounded clique-width. Indeed, the UCW Method relies on some clique-width-boundedness-preserving transformations of an arbitrary graph from some known graph class $\mathcal{G}'$ of unbounded clique-width, such as the class of walls, to a graph of the unknown class $\mathcal{G}$. One way to do this is to show that the graphs in $\mathcal{G}$ contain a simple path encoding of graphs from $\mathcal{G}'$.

Kratsch and Schweitzer [127] initiated a complexity classification for GRAPH ISOMORPHISM for (H_1, H_2)-free graphs. Schweitzer [161] extended the results of [127] and proved that the number of unknown cases is finite, but did not explicitly list what these cases were. As mentioned earlier, GRAPH ISOMORPHISM is polynomial-time solvable for graphs of bounded clique-width [92]. Bonamy, Dabrowski, Johnson and Paulusma [19] therefore combined the known results for boundedness of clique-width for bigenic classes (Theorem 4.18) with the results of [127] and [161] to obtain an explicit list of only 14 cases, for which the complexity of GRAPH ISOMORPHISM was unknown. In the same paper they reduced this number to 7 and gave the following state-of-the-art summary; recall that $K_{1,t}^+$ and $K_{1,t}^{++}$ are the graphs obtained from $K_{1,t}$ by subdividing one edge once or twice, respectively.

Theorem 5.8 ([19]). *For a class $\mathcal{G}$ of graphs defined by two forbidden induced subgraphs, the following holds:*

1. GRAPH ISOMORPHISM *is solvable in polynomial time on $\mathcal{G}$ if $\mathcal{G}$ is equivalent[11] to a class of (H_1, H_2)-free graphs such that one of the following holds:*
   - *(i)* $\underline{H_1}$ *or* $H_2 \subseteq_i P_4$
   - *(ii)* $\underline{H_1}$ *and* $H_2 \subseteq_i K_{1,t} + P_1$ *for some $t \geq 1$*
   - *(iii)* $\overline{H_1}$ *and* $H_2 \subseteq_i tP_1 + P_3$ *for some $t \geq 1$*
   - *(iv)* $H_1 \subseteq_i K_t$ *and* $H_2 \subseteq_i 2K_{1,t}, K_t^+$ *or* P_5 *for some $t \geq 1$*
   - *(v)* $H_1 \subseteq_i$ paw *and* $H_2 \subseteq_i P_2 + P_4, P_6, S_{1,2,2}$ *or* $K_{1,t}^{++} + P_1$ *for some $t \geq 1$*
   - *(vi)* $H_1 \subseteq_i$ diamond *and* $H_2 \subseteq_i P_1 + 2P_2$
   - *(vii)* $H_1 \subseteq_i$ gem *and* $H_2 \subseteq_i P_1 + P_4$ *or* P_5 *or*
   - *(viii)* $H_1 \subseteq_i 2P_1 + P_3$ *and* $H_2 \subseteq_i P_2 + P_3$.

2. GRAPH ISOMORPHISM *is GI-complete on $\mathcal{G}$ if $\mathcal{G}$ is equivalent to a class of (H_1, H_2)-free graphs such that one of the following holds:*
   - *(i)* *neither* $\underline{H_1}$ *nor* H_2 *is a path star forest*
   - *(ii)* *neither* $\overline{H_1}$ *nor* $\overline{H_2}$ *is a path star forest*
   - *(iii)* $H_1 \supseteq_i K_3$ *and* $H_2 \supseteq_i 2P_1 + 2P_2, P_1 + 2P_3, 2P_1 + P_4$ *or* $3P_2$
   - *(iv)* $H_1 \supseteq_i K_4$ *and* $H_2 \supseteq_i K_{1,4}^{++}, P_1 + 2P_2$ *or* $P_1 + P_4$
   - *(v)* $H_1 \supseteq_i K_5$ *and* $H_2 \supseteq_i K_{1,3}$
   - *(vi)* $H_1 \supseteq_i C_4$ *and* $H_2 \supseteq_i K_{1,3}, 3P_1 + P_2$ *or* $2P_2$
   - *(vii)* $H_1 \supseteq_i$ diamond *and* $H_2 \supseteq_i K_{1,3}, P_2 + P_4, 2P_3$ *or* P_6 *or*
   - *(viii)* $H_1 \supseteq_i P_1 + P_4$ *and* $H_2 \supseteq_i P_1 + 2P_2$.

As shown in [19], Theorem 5.8 leads to the following open problem.

Open Problem 5.9. *What is the complexity of* GRAPH ISOMORPHISM *on (H_1, H_2)-free graphs in the following seven cases?*
- *(i)* $H_1 = K_3$ *and* $H_2 \in \{P_7, S_{1,2,3}\}$,
- *(ii)* $H_1 = K_4$ *and* $H_2 = S_{1,1,3}$,
- *(iii)* $H_1 = $ diamond *and* $H_2 \in \{P_1 + P_2 + P_3, P_1 + P_5\}$,
- *(iv)* $H_1 = $ gem *and* $H_2 = P_2 + P_3$,
- *(v)* $H_1 = \overline{2P_1 + P_3}$ *and* $H_2 = P_5$.

For H-induced-minor-free graphs the classification for the complexity of GRAPH ISOMORPHISM is given in Theorem 5.10. Note that the second and third tractable cases follow from Theorem 4.31 and the fact that GRAPH ISOMORPHISM is polynomial-time solvable on graphs of bounded clique-width [92]. We refer to Figure 15 for a picture of the graphs $\overline{2P_1 + P_3}$ and $\overline{P_1 + P_4}$.

Theorem 5.10 ([10]). *Let H be a graph. The* GRAPH ISOMORPHISM *problem on H-induced-minor-free graphs is polynomial-time solvable if:*
- *(i)* H *is a complete graph,*
- *(ii)* $H \subseteq_i \overline{2P_1 + P_3}$ *or*
- *(iii)* $H \subseteq_i \overline{P_1 + P_4}$

and GI-*complete otherwise.*

---

[11]Equivalence is defined in the same way as for clique-width (see Footnote 9). If two classes are equivalent, then the complexity of GRAPH ISOMORPHISM is the same on both of them. [19]

6 Well-Quasi-Orderability

We recall that the Robertson-Seymour Theorem [157] states that the set of all finite graphs is well-quasi-ordered by the minor relation. This result, combined with the cubic-time algorithm of [156] for testing if a graph G contains some fixed graph H as a minor, gives a cubic-time algorithm for testing whether a graph belongs to some minor-closed graph class. Other known results on well-quasi-orderability include a result of Ding [72], which implies that every class of graphs with bounded vertex cover number is well-quasi-ordered by the induced subgraph relation and a result of Mader [144], who showed that every class of graphs with bounded feedback vertex number is well-quasi-ordered by the topological minor relation. Fellows, Hermelin and Rosamund [76] simplified the proofs of Ding and Mader. They also showed that every class of graphs of bounded circumference is well-quasi-ordered by the induced minor relation. As applications they gave linear-time algorithms for recognizing graphs from any topological-minor-closed graph class with bounded feedback vertex number; any induced-minor-closed graph class of bounded circumference; and any induced-subgraph-closed graph class with bounded vertex cover number.

The Robertson-Seymour Theorem also implies that there exist graph classes of un-bounded clique-width that are well-quasi-ordered by the minor relation. For hereditary graph classes, the notion of well-quasi-orderability by the induced subgraph relation is closely related to boundedness of clique-width, but the exact relationship between the two notions is not yet fully understood. In this section we survey results on well-quasi-orderability by the induced subgraph relation for hereditary classes, together with some more results for other containment relations.

In Section 1, we noted that Daligault, Rao and Thomassé [69] asked if every hereditary graph class that is well-quasi-ordered by the induced subgraph relation has bounded clique-width. Lozin, Razgon and Zamaraev [142] gave a negative answer to this question. That is, they found an example of a hereditary graph class that is well-quasi-ordered by the induced subgraph relation but has unbounded clique-width. As the hereditary graph class in their example is not finitely defined (that is, this graph class is defined by infinitely many forbidden induced subgraphs), they conjectured the following.

Conjecture 6.1 ([142]). *If a finitely defined hereditary class of graphs $\mathcal{G}$ is well-quasi-ordered by the induced subgraph relation, then $\mathcal{G}$ has bounded clique-width.*

We note that the reverse implication of the statement in Conjecture 6.1 is not true. We can take the (hereditary) class of graphs of maximum degree at most 2, which have clique-width at most 4 by Proposition 3.3. However, the class of graphs of maximum degree at most 2 contains all cycles, which form an infinite anti-chain. Furthermore, the class of graphs of maximum degree at most 2, is finitely defined: it is the class of $(\text{claw}, \text{paw}, \text{diamond}, K_4)$-free graphs.

6.1 Well-Quasi-Orderability Preserving Operations

In order to prove that some class of graphs is well-quasi-ordered by the induced subgraph relation or not, we would like to use similar facts to those used to prove boundedness or unboundedness of clique-width. This is not straightforward, as there is no analogue of Facts 1–6 for well-quasi-orderability by the induced subgraph relation. We show this in the three examples below, but first we recall that these facts concern, respectively, vertex deletion (Fact 1), subgraph complementation (Fact 2), bipartite complementation

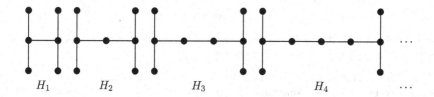

Figure 16: The graphs H_i from Example 6.3.

(Fact 3), being prime (Fact 4), being 2-connected (Fact 5), and edge subdivision for graphs of bounded maximum degree (Fact 6).

Example 6.2. A counterexample for analogues of Facts 1–3 is formed by the class of cycles [64]: deleting a vertex of a cycle, complementing the subgraph induced by two adjacent vertices, or applying a bipartite complementation between two adjacent vertices yields a path. The set of cycles is an infinite anti-chain with respect to the induced subgraph relation, but the set of paths is well-quasi-ordered.

Example 6.3. A counterexample for analogues of Facts 4–5 is formed by the following class of graphs. For $i \geq 1$, take a path of length i with end-vertices u and v and add vertices u', u'', v', v'' with edges uu', uu'', vv' and vv''. Call the resulting graph H_i (see also Figure 16) and let $\mathcal{H}$ be the class of graphs H_i (and their induced subgraphs). If $i \neq j$, then H_i is not an induced subgraph of H_j, which implies that $\mathcal{H}$ is not well-quasi-ordered by the induced subgraph relation. However, the prime graphs of $\mathcal{H}$ are paths, which are well-quasi-ordered by the induced subgraph relation. This shows that the analogue to Fact 4 does not hold for well-quasi-orderability by the induced subgraph relation. The analogue to Fact 5 does not hold either, as $\mathcal{H}$ contains no 2-connected graphs.

Example 6.4. To obtain a counterexample for the analogue of Fact 6 we consider the class of graphs $\mathcal{H}_1$ consisting of the graph H_1 from Example 6.3 only. This class is well-quasi-ordered by the induced subgraph relation. However, we can obtain the class $\mathcal{H}$ in Example 6.3, which is not well-quasi-ordered, from $\mathcal{H}_1$ via edge subdivisions. That is, for $i \geq 1$, the graph H_{i+1} is obtained from H_i by the subdivision of an edge of the path of length i.

As these examples suggest, we need a stronger variant of well-quasi-orderability by the induced subgraph relation. To define this variant, consider an arbitrary quasi-order $(W, \leq)$. Then a graph G is a *labelled* graph if each vertex v of G is equipped with a *label* $l_G(v) \in W$. A graph F with labelling l_F is a *labelled induced subgraph* of G if F is isomorphic to an induced subgraph G' of G such that there is an isomorphism which maps each vertex v of F to a vertex w of G' with $l_F(v) \leq l_G(w)$. If $(W, \leq)$ is a well-quasi-order, then it is not possible for a graph class $\mathcal{G}$ to contain an infinite sequence of labelled graphs that is strictly-decreasing with respect to the labelled induced subgraph relation. We say that $\mathcal{G}$ is well-quasi-ordered by the *labelled* induced subgraph relation if for *every* well-quasi-order $(W, \leq)$ the class $\mathcal{G}$ contains no infinite anti-chains of labelled graphs.

Observation 6.5. *Every graph class that is well-quasi-ordered by the labelled induced subgraph relation is well-quasi-ordered by the induced subgraph relation.*

Daligault, Rao and Thomassé proved the following result.

Theorem 6.6 ([69]). *Every hereditary class of graphs that is well-quasi-ordered by the labelled induced subgraph relation is finitely defined.*

By Theorem 6.6 it is easy to prove that there exist hereditary graph classes that are well-quasi-ordered by the induced subgraph relation but not by the labelled induced subgraph relation. Korpelainen, Lozin and Razgon [125] gave the class of linear forests as an example (see also Example 6.9 below). The same authors conjectured that if a hereditary class of graphs $\mathcal{G}$ is defined by a finite set of forbidden induced subgraphs, then $\mathcal{G}$ is well-quasi-ordered by the induced subgraph relation if and only if it is well-quasi-ordered by the labelled induced subgraph relation. However, Brignall, Engen and Vatter [38] recently found a counterexample for this conjecture.

Theorem 6.7 ([38]). *There exists a graph class $\mathcal{G}^*$ with $|\mathcal{F}_{\mathcal{G}^*}| = 14$ that is well-quasi-ordered by the induced subgraph relation but not by the labelled induced subgraph relation.*

Theorem 6.7 leads to the following open problem.

Open Problem 6.8. *Does there exist a hereditary graph class $\mathcal{G}$ with $|\mathcal{F}_G| \leq 13$ that is well-quasi-ordered by the induced subgraph relation but not by the labelled induced subgraph relation?*

We consider an approach similar to one used for boundedness of clique-width. A graph operation γ *preserves* well-quasi-orderability by the labelled induced subgraph relation if, for every finite constant k and every graph class $\mathcal{G}$, every graph class $\mathcal{G}'$ that is (k, γ)-obtained from $\mathcal{G}$ is well-quasi-ordered by this relation if and only if $\mathcal{G}$ is. We also say that a graph property π *preserves* well-quasi-orderability by the labelled induced subgraph relation if for every graph class $\mathcal{G}$, the subclass of $\mathcal{G}$ with property π is well-quasi-ordered by the labelled induced subgraph relation if and only if this is the case for $\mathcal{G}$.

Facts about well-quasi orderability:

Fact 1. Vertex deletion preserves well-quasi-orderability by the labelled induced subgraph relation [64].

Fact 2. Subgraph complementation preserves well-quasi-orderability by the labelled induced subgraph relation [64].

Fact 3. Bipartite complementation preserves well-quasi-orderability by the labelled induced subgraph relation [64].

Fact 4. Being prime preserves well-quasi-orderability by the labelled induced subgraph relation for hereditary classes [6].

For labelled well-quasi-orders, there is no analogue to Fact 5 (on 2-connectivity) and Fact 6 (on edge subdivision) as illustrated by the following counterexample.

Example 6.9. Let $\mathcal{F}$ be the (hereditary) class of linear forests. The class $\mathcal{F}$ contains the class $\mathcal{P}$ of all paths on at least two vertices. If we label the end-vertices of every path in $\mathcal{P}$ with one label and all other vertices with a second label incomparable with the first, we obtain an infinite anti-chain with respect to the labelled induced subgraph relation

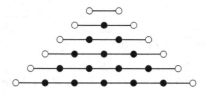

Figure 17: An anti-chain of paths under the labelled induced subgraph relation. The two colours are incomparable.

(see also Figure 17). Hence $\mathcal{F}$ is not well-quasi-ordered by the labelled induced subgraph relation. However, the restriction of $\mathcal{F}$ to 2-connected graphs is the empty class, which is well-quasi-ordered by the labelled induced subgraph relation. Moreover, every graph of $\mathcal{F}$ has maximum degree at most 2. However, every path of $\mathcal{P}$ can be obtained by repeatedly subdividing P_2, and the class $\{P_2\}$ is well-quasi-ordered by the labelled induced subgraph relation. We conclude that Facts 5 and 6 for clique-width do not have a counterpart for well-quasi-orderability by the labelled induced subgraph relation.

As a final remark in this section, we note that it is easy to verify that graph classes of bounded neighbourhood diversity (introduced in [130]) have bounded clique-width and are well-quasi-ordered by the labelled induced subgraph relation. Moreover, the same property also holds for graph classes of bounded uniformicity (introduced in [123]) or bounded lettericity (introduced in [152]); uniformicity and lettericity are more general than neighbourhood diversity.

6.2   Results for Hereditary Graph Classes

We now survey known results for well-quasi-orderability of hereditary graphs by the induced subgraph relation. As we shall see, all known results agree with Conjecture 6.1. We start with a result of Damaschke.

Theorem 6.10 ([70]). *Let H be a graph. The class of H-free graphs is well-quasi-ordered by the induced subgraph relation if and only if $H \subseteq_i P_4$.*

In fact, the same classification holds for the labelled induced subgraph relation [6], which means that if there is a hereditary class $\mathcal{G}$ which gives a positive answer to Open Problem 6.8, then $|\mathcal{F}_\mathcal{G}| \geq 2$. We also note that the classification of Theorem 6.10 coincides with the one of Theorem 4.1 for boundedness of clique-width. In order to increase our understanding of well-quasi-orderability by the induced subgraph relation we can follow the same approaches as done in Section 4 for clique-width. However, considerably less work has been done on this subject.

Just as in Section 4, we can first restrict ourselves to H-free graphs contained in some other hereditary graph class. In particular, results for H-free bipartite graphs, such as those in [122], have shown to be useful. For instance, they have been used to prove results on well-quasi-orderability for (H_1, H_2)-free graphs [123]. Combining the results for H-free bipartite and H-free triangle-free graphs of [122, 123] with the results of [6, 63, 72] and Ramsey's Theorem for the case when $H = sP_1$ $(s \geq 1)$ yields the following two classifications (see [63] for further explanation).

Theorem 6.11 ([63]). *Let H be a graph. The class of H-free bipartite graphs is well-quasi-ordered by the induced subgraph relation if and only if $H = sP_1$ for some $s \geq 1$ or H is an induced subgraph of $P_1 + P_5$, $P_2 + P_4$ or P_6.*

Theorem 6.12 ([63]). *Let H be a graph. The class of (K_3, H)-free graphs is well-quasi-ordered by the induced subgraph relation if and only if $H = sP_1$ for some $s \geq 1$ or H is an induced subgraph of $P_1 + P_5$, $P_2 + P_4$, or P_6.*

We note that the classifications of Theorem 6.11 and 6.12 coincide. In contrast, we recall that it is not yet clear if the classifications for boundedness of clique-width on H-free bipartite graphs and (K_3, H)-free graphs also coincide; see Open Problem 4.20.

We now present the state-of-the-art summary for well-quasi-orderability for classes on (H_1, H_2)-free graphs, which is obtained by combining results from [6, 63, 64, 72, 122, 123]. Note that Theorem 6.13 implies Theorem 6.12.

Theorem 6.13 ([63]). *Let $\mathcal{G}$ be a class of graphs defined by two forbidden induced subgraphs. Then:*

1. *$\mathcal{G}$ is well-quasi-ordered by the labelled induced subgraph relation if it is equivalent[12] to a class of (H_1, H_2)-free graphs such that one of the following holds:*
   - *(i) H_1 or $H_2 \subseteq_i P_4$,*
   - *(ii) $H_1 = K_s$ and $H_2 = tP_1$ for some $s, t \geq 1$,*
   - *(iii) $H_1 \subseteq_i$ paw and $H_2 \subseteq_i P_1 + P_5, P_2 + P_4$ or P_6,*
   - *(iv) $H_1 \subseteq_i$ diamond and $H_2 \subseteq_i P_2 + P_3$ or P_5.*
2. *$\mathcal{G}$ is not well-quasi-ordered by the induced subgraph relation if it is equivalent to a class of (H_1, H_2)-free graphs such that one of the following holds:*
   - *(i) neither H_1 nor H_2 is a linear forest,*
   - *(ii) $H_1 \supseteq_i K_3$ and $H_2 \supseteq_i 3P_1 + P_2, 3P_2$ or $2P_3$,*
   - *(iii) $H_1 \supseteq_i C_4$ and $H_2 \supseteq_i 4P_1$ or $2P_2$,*
   - *(iv) $H_1 \supseteq_i$ diamond and $H_2 \supseteq_i 4P_1, P_2 + P_4$ or P_6,*
   - *(v) $H_1 \supseteq_i$ gem and $H_2 \supseteq_i P_1 + 2P_2$.*

Theorem 6.13 does not cover six cases, which are all still open (see also [63]).

Open Problem 6.14. *Is the class of (H_1, H_2)-free graphs well-quasi-ordered by the induced subgraph relation when:*
   (i) $H_1 =$ diamond and $H_2 \in \{P_1 + 2P_2, P_1 + P_4\}$,
   (ii) $H_1 =$ gem and $H_2 \in \{P_1 + P_4, 2P_2, P_2 + P_3, P_5\}$?

It follows from Theorems 4.18 and 6.13 that the class of $(\overline{P_1 + P_4}, P_2 + P_3)$-free graphs is the only class of (H_1, H_2)-free graphs left for which Conjecture 6.1 still needs to be verified (see also [64]).

Open Problem 6.15. *Is Conjecture 6.1 true for the class of (H_1, H_2)-free graphs when $H_1 = \overline{P_1 + P_4}$ and $H_2 = P_2 + P_3$?*

Finally, instead of the induced subgraph relation or the minor relation, one can also consider other containment relations. Ding [72] proved that for a graph H, the class of H-subgraph-free graphs is well-quasi-ordered by the subgraph relation if and only if H is a linear forest. This result can be readily generalized.

---

[12]Equivalence is defined in the same way as for clique-width (see Footnote 9). If two classes are equivalent, then one of them is well-quasi-ordered by the induced subgraph relation if and only if the other one is [123].

Theorem 6.16. *Let* $\{H_1, \ldots, H_p\}$ *be a finite set of graphs. The class of* $(H_1, \ldots, H_p)$*-subgraph-free graphs is well-quasi-ordered by the subgraph relation if and only if* H_i *is a linear forest for some* $i \in \{1, \ldots, p\}$*.*

Proof. If H_i is a linear forest for some $i \in \{1, \ldots, p\}$, then we can apply the result of Ding [72]. Now suppose that every H_i either has a cycle or an induced claw. We let g be the maximum girth over all H_i that contain a cycle. Then the set of cycles of length at least $g + 1$ is an infinite antichain of $(H_1, \ldots, H_p)$-free graphs with respect to the subgraph relation. □

Kamiński, Raymond and Trunck [117] and Błasiok, Kamiński, Raymond and Trunck [16] gave classifications for the contraction relation and induced minor relation, respectively. We note that the connectivity condition in Theorem 6.17 is natural, as the edgeless graphs form an antichain under the contraction relation. We refer to Figure 15 for pictures of the graphs $\overline{2P_1 + P_3}$ and $\overline{P_1 + P_4}$ in Theorem 6.18.

Theorem 6.17 ([117]). *Let H be a graph. The class of connected H-contraction-free graphs is well-quasi-ordered by the contraction relation if and only if* $H \in \{C_3, \text{diamond}, P_1, P_2, P_3\}$*.*

Theorem 6.18 ([16]). *Let H be a graph. The class of H-induced-minor-free graphs is well-quasi-ordered by the induced-minor relation if and only if* $H \subseteq_i \overline{2P_1 + P_3}$ *or* $H \subseteq_i \overline{P_1 + P_4}$*.*

We pose the following two open problems.

Open Problem 6.19. *Determine for which pairs of graphs (H_1, H_2) the class of connected (H_1, H_2)-contraction-free graphs is well-quasi-ordered by the contraction relation.*

Open Problem 6.20. *Determine for which pairs of graphs (H_1, H_2) the class of (H_1, H_2)-induced-minor-free graphs is well-quasi-ordered by the induced minor relation.*

For containment relations other than the induced subgraph relation we can ask the following question: does there exist a containment-closed graph class of unbounded clique-width that is well-quasi-ordered by the same containment relation? The Robertson-Seymour Theorem [157] tells us that that the class of all (finite) graphs is well-quasi-ordered by the minor relation. Hence, if we forbid minors, we can consider the class of all finite graphs, which has unbounded clique-width. By Theorems 4.30.(i) and 6.16, we would need to forbid an infinite set of graphs for the subgraph relation to find a positive answer to this question. A *clique-cactus graph* is a graph in which each block is either a complete graph or a cycle (these graphs are also known as cactus block graphs). The class of diamond-contraction-free graphs coincides with the class of clique-cactus graphs [117]. As complete graphs and cycles have clique-width at most 2 and 4, respectively, clique-cactus graphs have bounded clique-width due to Fact 4. Hence, by Theorem 6.17 we would need to forbid a set of at least two graphs for the contraction relation (when considering connected graphs). The classification in Theorem 6.18 coincides with the classification in Theorem 4.31 for boundedness of the clique-width of H-induced-minor-free graphs. Hence we would also need to forbid a set of at least two graphs for the induced minor relation. We note that the hereditary graph class given in [142] (that is well-quasi-ordered by the induced subgraph relation, but has unbounded clique-width) is not closed under contractions, subgraphs or induced minors. Hence, this class does not give a positive answer to the question for contractions, subgraphs or induced minors.

7 Variants of Clique-Width

We have surveyed results and techniques for proving (un)boundedness of clique-width for various families of hereditary graph classes and stated a number of open problems. We conclude our paper with a brief discussion of some other variants of clique-width. Lozin and Rautenbach [141] introduced the notion of relative clique-width, whose definition is more consistent with the definition of treewidth. Computing relative clique-width is NP-hard, as shown by Müller and Urner [148], but the concept has not been studied for hereditary graph classes. Courcelle [52] and Fürer [83] defined symmetric clique-width and multi-clique-width, respectively. Both these width parameters are equivalent to clique-width [52, 83]. As this survey focuses on boundedness of clique-width, we therefore do not discuss these parameters any further here. Instead we focus on two other variants, namely linear clique-width (Section 7.1) and power-bounded clique-width (Section 7.2).

7.1 Linear Clique-Width

Linear clique-width [98, 141], also called *sequential clique-width*, is defined in the same way as clique-width except that in Operation 2 (the disjoint union operation) of the definition of clique-width, at least one of the two graphs must consist of a single vertex. Just as clique-width is equivalent to NLC-width and rank-width, linear clique-width is equivalent to linear NLC-width [98] and linear rank-width (see, for example, [150]).[13] Moreover, just as is the case for clique-width, the notion of linear clique-width is also not well understood, and similar approaches to those for clique-width have been followed. To illustrate this, the following analogous results to those for clique-width are known. Computing linear clique-width is NP-hard [77] for general graphs, but it is polynomial-time solvable for forests [1] and distance-hereditary graphs [2]. Moreover, graphs of linear clique-width at most 3 can be recognized in polynomial time [104], but the computational complexity of recognizing graphs of linear clique-width at most c is unknown for $c \geq 4$ (see [105] for some partial results for $c = 4$). Another analogous result is due to Gurski and Wanke who proved the following theorem (compare to Theorem 3.7).

Theorem 7.1 ([100]). *A class of graphs $\mathcal{G}$ has bounded path-width if and only if the class of the line graphs of graphs in $\mathcal{G}$ has bounded linear clique-width.*

By definition, every graph class of bounded linear clique-width has bounded clique-width, but the reverse implication does not hold. For example, recall that every P_4-free graph has clique-width at most 2 by Theorem 4.1 and that every tree has clique-width at most 3 by Proposition 3.2. In contrast, Gurski and Wanke [98] proved that the class of P_4-free graphs and even the class of complete binary trees have unbounded linear clique-width. This led Brignall, Korpelainen and Vatter to consider hereditary subclasses of P_4-free graphs. They proved the following dichotomy result.

Theorem 7.2 ([39]). *A hereditary subclass of P_4-free graphs has bounded linear clique-width if and only if it contains neither the class of (C_4, P_4)-free graphs nor the class of $(2P_2, P_4)$-free graphs.*

We note that (C_4, P_4)-free graphs are also known as the *trivially perfect* or *quasi-threshold* graphs.

---

[13]We note that the corresponding variants for directed graphs were recently introduced by Gurski and Rehs [96].

To obtain an analogous result to Theorem 4.1, we state the following two results. The first one is due to Gurski. The second can be easily derived from known results.

Theorem 7.3 ([94]). *A graph has linear clique-width at most 2 if and only if it is* $(2P_2, \overline{2P_3}, P_4)$-*free.*

Theorem 7.4. *Let H be a graph. The class of H-free graphs has bounded linear clique-width if and only if H is an induced subgraph of $P_1 + P_2$ or P_3. Furthermore, $(P_1 + P_2)$-free graphs and P_3-free graphs have linear clique-width at most 2 and 3, respectively.*

Proof. By Theorem 4.1 it suffices to consider the case when H is an induced subgraph of P_4 and by Theorem 7.2 we may assume that $H \neq P_4$. Let G be an H-free graph. If $H \subseteq_i P_3$, then every connected component of G is a complete graph. Complete graphs are readily seen to have linear clique-width at most 2. Hence, G has linear clique-width at most 3 (after creating each connected component, we relabel all of its vertices to a third label). The only remaining case is $H = P_1 + P_2$. Since $P_1 + P_2$ is an induced subgraph of $2P_2$, $\overline{2P_3}$ and P_4, Theorem 7.3 implies that G has linear clique-width at most 2. $\quad\square$

Theorem 7.4 leads to the following open problem.

Open Problem 7.5. *Determine for which pairs of graphs (H_1, H_2) the class of (H_1, H_2)-free graphs has bounded linear clique-width.*

Just as is the case for clique-width, we expect that results on boundedness of linear clique-width for H-free bipartite graphs would be useful for solving Open Problem 7.5. We therefore also pose the following open problem.

Open Problem 7.6. *Determine for which graphs H the class of H-free bipartite graphs has bounded linear clique-width.*

Finally, we can also prove an analogous result to Theorem 4.17.

Theorem 7.7. *Let $\{H_1, \ldots, H_p\}$ be a finite set of graphs. Then the class of $(H_1, \ldots, H_p)$-free line graphs has bounded linear clique-width if and only if $H_i \in \mathcal{T}$ for some $i \in \{1, \ldots, p\}$.*

Proof. First suppose that $H_i \in \mathcal{T}$ for some $i \in \{1, \ldots, p\}$. By definition of $\mathcal{T}$, it follows that H_i is the line graph of some graph $F_i \in \mathcal{S}$. We repeat the arguments of the proof of Theorem 4.17 to find that the class of F_i-subgraph-free graphs has bounded path-width. Then, by Theorem 7.1, the class of H_i-free graphs, and thus the class of $(H_1, \ldots, H_p)$-free graphs, has bounded linear clique-width. Now suppose that $H_i \notin \mathcal{T}$ holds for every $i \in \{1, \ldots, p\}$. By Theorem 4.17, the class of $(H_1, \ldots, H_p)$-free line graphs has unbounded clique-width, and thus unbounded linear clique-width. $\quad\square$

7.2　Power-Bounded Clique-Width

Recall that the r-th power of G ($r \geq 1$) is the graph with vertex set $V(G)$ and an edge between two vertices u and v if and only if u and v are at distance at most r from each other in G. Gurski and Wanke [102] proved that the clique-width of the r-th power of a tree is at most $r + 2 + \max\{\lfloor \frac{r}{2} \rfloor - 1, 0\}$ and that the r-th power of a graph G has clique-width at most $2(r + 1)^{\mathrm{tw}(G)+1} - 1$.

A graph class $\mathcal{G}$ has *power-bounded clique-width* if there is a constant r such that the graph class consisting of all r-th powers of all graphs from $\mathcal{G}$ has bounded clique-width;

otherwise $\mathcal{G}$ has *power-unbounded clique-width*. Hence, if a graph class has bounded clique-width, it has power-bounded clique-width (we can take $r = 1$). The reverse implication does not hold. This follows, for example, from a comparison of Theorem 4.1 with the following classification for H-free graphs of Bonomo, Grippo, Milanič and Safe.

Theorem 7.8 ([21]). *Let H be a graph. Then the class of H-free graphs has power-bounded clique-width if and only if H is a linear forest.*

Bonomo, Grippo, Milanič and Safe also proved the following classification for (H_1, H_2)-free graphs when both H_1 and H_2 are connected.

Theorem 7.9 ([21]). *Let H_1 and H_2 be two connected graphs. Then the class of (H_1, H_2)-free graphs has power-bounded clique-width if and only if*
  (i) at least one of H_1 and H_2 is a path, or
  (ii) $H_1 = S_{1,i,j}$ for some $i, j \geq 1$ and $H_2 = T_{0,i',j'}$ for some $i', j' \geq 0$.

The case when H_1 or H_2 is disconnected has not yet been settled.

Open Problem 7.10. *Determine for which pairs of graphs (H_1, H_2) the class of (H_1, H_2)-free graphs has power-bounded clique-width.*

We note that analogous results to Theorems 4.5 and 4.7 exist for power-bounded clique-width; that is, the classes of bipartite permutation graphs and unit interval graphs have power-unbounded clique-width [21]. For more open problems on power-bounded clique-width, we refer to [21].

Acknowledgements

We thank Robert Brignall, Vincent Vatter and an anonymous reviewer for helpful comments. The work was supported by the Leverhulme Trust Research Project Grant RPG-2016-258.

References

[1] Isolde Adler and Mamadou Moustapha Kanté, *Linear rank-width and linear clique-width of trees*, Theoretical Computer Science **589** (2015), 87–98.

[2] Isolde Adler, Mamadou Moustapha Kanté, and O-joung Kwon, *Linear rank-width of distance-hereditary graphs I. A polynomial-time algorithm*, Algorithmica **78** (2017), no. 1, 342–377.

[3] Peter Allen, Vadim V. Lozin, and Michaël Rao, *Clique-width and the speed of hereditary properties*, The Electronic Journal of Combinatorics **16** (2009), no. 1, Research Paper 35, pp. 11.

[4] Stefan Arnborg, Derek G. Corneil, and Andrzej Proskurowski, *Complexity of finding embeddings in a k-tree*, SIAM Journal on Algebraic Discrete Methods **8** (1987), no. 2, 277–284.

[5] Aistis Atminas, Robert Brignall, Vadim V. Lozin, and Juraj Stacho, *Minimal classes of graphs of unbounded clique-width defined by finitely many forbidden induced subgraphs*, CoRR abs/1503.01628 (2015).

[6] Aistis Atminas and Vadim V. Lozin, *Labelled induced subgraphs and well-quasi-ordering*, Order **32** (2015), no. 3, 313–328.

[7] László Babai, *Graph isomorphism in quasipolynomial time [extended abstract]*, Proc. STOC 2016 (2016), 684–697.

[8] Luitpold Babel and Stephan Olariu, *On the structure of graphs with few P_4s*, Discrete Applied Mathematics **84** (1998), no. 1–3, 1–13.

[9] Hans-Jürgen Bandelt and Henry Martyn Mulder, *Distance-hereditary graphs*, Journal of Combinatorial Theory, Series B **41** (1986), no. 2, 182–208.

[10] Rémy Belmonte, Yota Otachi, and Pascal Schweitzer, *Induced minor free graphs: Isomorphism and clique-width*, Algorithmica **80** (2018), no. 1, 29–47.

[11] Benjamin Bergougnoux and Mamadou Moustapha Kanté, *Fast exact algorithms for some connectivity problems parametrized by clique-width*, CoRR abs/1707.03584 (2017).

[12] Benjamin Bergougnoux, Mamadou Moustapha Kanté, and O-joung Kwon, *An optimal XP algorithm for Hamiltonian cycle on graphs of bounded clique-width*, Proc. WADS 2017, LNCS **10389** (2017), 121–132.

[13] Dan Bienstock, Neil Robertson, Paul D. Seymour, and Robin Thomas, *Quickly excluding a forest*, Journal of Combinatorial Theory, Series B **52** (1991), no. 2, 274–283.

[14] Alexandre Blanché, Konrad K. Dabrowski, Matthew Johnson, Vadim V. Lozin, Daniël Paulusma, and Viktor Zamaraev, *Clique-width for graph classes closed under complementation*, Proc. MFCS 2017, LIPIcs **83** (2017), 73:1–73:14.

[15] Alexandre Blanché, Konrad K. Dabrowski, Matthew Johnson, and Daniël Paulusma, *Hereditary graph classes: When the complexities of* COLORING *and* CLIQUE COVER *coincide*, Journal of Graph Theory (in press).

[16] Jarosław Błasiok, Marcin Kamiński, Jean-Florent Raymond, and Théophile Trunck, *Induced minors and well-quasi-ordering*, Journal of Combinatorial Theory, Series B **134** (2019), 110–142.

[17] Hans L. Bodlaender, *A linear-time algorithm for finding tree-decompositions of small treewidth*, SIAM Journal on Computing **25** (1996), no. 6, 1305–1317.

[18] Rodica Boliac and Vadim V. Lozin, *On the clique-width of graphs in hereditary classes*, Proc. ISAAC 2002, LNCS **2518** (2002), 44–54.

[19] Marthe Bonamy, Konrad K. Dabrowski, Matthew Johnson, and Daniël Paulusma, *Graph isomorphism for (H_1, H_2)-free graphs: an almost complete dichotomy*, CoRR abs/1811.12252 (2018).

[20] Flavia Bonomo, Guillermo Durán, and Javier Marenco, *Exploring the complexity boundary between coloring and list-coloring*, Annals of Operations Research **169** (2009), no. 1, 3–16.

[21] Flavia Bonomo, Luciano N. Grippo, Martin Milanič, and Martín D. Safe, *Graph classes with and without powers of bounded clique-width*, Discrete Applied Mathematics **199** (2016), 3–15.

[22] Kellogg Speed Booth and Charles Joseph Colbourn, *Problems polynomially equivalent to graph isomorphism*, Tech. Report CS-77-04, Department of Computer Science, University of Waterloo, 1979.

[23] Endre Boros, Vladimir Gurvich, and Martin Milanič, *Characterizing and decomposing classes of threshold, split, and bipartite graphs via 1-Sperner hypergraphs*, CoRR abs/1805.03405 (2018).

[24] Andreas Brandstädt, Konrad K. Dabrowski, Shenwei Huang, and Daniël Paulusma, *Bounding the clique-width of H-free split graphs*, Discrete Applied Mathematics **211** (2016), 30–39.

[25] Andreas Brandstädt, Konrad K. Dabrowski, Shenwei Huang, and Daniël Paulusma, *Bounding the clique-width of H-free chordal graphs*, Journal of Graph Theory **86** (2017), no. 1, 42–77.

[26] Andreas Brandstädt, Feodor F. Dragan, Hoàng-Oanh Le, and Raffaele Mosca, *New graph classes of bounded clique-width*, Theory of Computing Systems **38** (2005), no. 5, 623–645.

[27] Andreas Brandstädt, Joost Engelfriet, Hoàng-Oanh Le, and Vadim V. Lozin, *Clique-width for 4-vertex forbidden subgraphs*, Theory of Computing Systems **39** (2006), no. 4, 561–590.

[28] Andreas Brandstädt, Vassilis Giakoumakis, and Martin Milanič, *Weighted efficient domination for some classes of H-free and of (H_1, H_2)-free graphs*, Discrete Applied Mathematics **250** (2018), 130–144.

[29] Andreas Brandstädt, Chính T. Hoàng, and Van Bang Le, *Stability number of bull- and chair-free graphs revisited*, Discrete Applied Mathematics **131** (2003), no. 1, 39–50.

[30] Andreas Brandstädt, Tilo Klembt, and Suhail Mahfud, *P_6- and triangle-free graphs revisited: structure and bounded clique-width*, Discrete Mathematics and Theoretical Computer Science **8** (2006), no. 1, 173–188.

[31] Andreas Brandstädt, Hoàng-Oanh Le, and Raffaele Mosca, *Gem- and co-gem-free graphs have bounded clique-width*, International Journal of Foundations of Computer Science **15** (2004), no. 1, 163–185.

[32] Andreas Brandstädt, Hoàng-Oanh Le, and Raffaele Mosca, *Chordal co-gem-free and (P_5,gem)-free graphs have bounded clique-width*, Discrete Applied Mathematics **145** (2005), no. 2, 232–241.

[33] Andreas Brandstädt, Hoàng-Oanh Le, and Jean-Marie Vanherpe, *Structure and stability number of chair-, co-P- and gem-free graphs revisited*, Information Processing Letters **86** (2003), no. 3, 161–167.

[34] Andreas Brandstädt, Van Bang Le, and Jeremy P. Spinrad, *Graph classes: A survey*, SIAM Monographs on Discrete Mathematics and Applications, vol. 3, SIAM, 1999.

[35] Andreas Brandstädt and Vadim V. Lozin, *On the linear structure and clique-width of bipartite permutation graphs*, Ars Combinatoria **67** (2003), 273–281.

[36] Andreas Brandstädt and Suhail Mahfud, *Maximum weight stable set on graphs without claw and co-claw (and similar graph classes) can be solved in linear time*, Information Processing Letters **84** (2002), no. 5, 251–259.

[37] Andreas Brandstädt and Raffaele Mosca, *On variations of P_4-sparse graphs*, Discrete Applied Mathematics **129** (2003), no. 2–3, 521–532.

[38] Robert Brignall, Michael Engen, and Vincent Vatter, *A counterexample regarding labelled well-quasi-ordering*, Graphs and Combinatorics **34** (2018), no. 6, 1395–1409.

[39] Robert Brignall, Nicholas Korpelainen, and Vincent Vatter, *Linear clique-width for hereditary classes of cographs*, Journal of Graph Theory **84** (2017), no. 4, 501–511.

[40] Robert Brignall and Vincent Vatter, *(private communication)*, 2019.

[41] Hajo Broersma, Petr A. Golovach, and Viresh Patel, *Tight complexity bounds for FPT subgraph problems parameterized by the clique-width*, Theoretical Computer Science **485** (2013), 69–84.

[42] Binh-Minh Bui-Xuan, Jan Arne Telle, and Martin Vatshelle, *Boolean-width of graphs*, Theoretical Computer Science **412** (2011), no. 39, 5187–5204.

[43] Kathie Cameron and Chính T. Hoàng, *Solving the clique cover problem on (bull,C_4)-free graphs*, CoRR **abs/1704.00316** (2017).

[44] Domingos Moreira Cardoso, Nicholas Korpelainen, and Vadim V. Lozin, *On the complexity of the dominating induced matching problem in hereditary classes of graphs*, Discrete Applied Mathematics **159** (2011), no. 7, 521–531.

[45] Maria Chudnovsky, Neil Robertson, Paul D. Seymour, and Robin Thomas, *The strong perfect graph theorem*, Annals of Mathematics **164** (2006), no. 1, 51–229.

[46] Julia Chuzhoy, *Improved bounds for the flat wall theorem*, Proc. SODA 2015 (2015), 256–275.

[47] Charles Joseph Colbourn, *On testing isomorphism of permutation graphs*, Networks **11** (1981), no. 1, 13–21.

[48] Andrew Collins, Jan Foniok, Nicholas Korpelainen, Vadim V. Lozin, and Viktor Zamaraev, *Infinitely many minimal classes of graphs of unbounded clique-width*, Discrete Applied Mathematics **248** (2018), 145–152.

[49] Derek G. Corneil, Michel Habib, Jean-Marc Lanlignel, Bruce A. Reed, and Udi Rotics, *Polynomial-time recognition of clique-width ≤ 3 graphs*, Discrete Applied Mathematics **160** (2012), no. 6, 834–865.

[50] Derek G. Corneil and Udi Rotics, *On the relationship between clique-width and treewidth*, SIAM Journal on Computing **34** (2005), 825–847.

[51] Bruno Courcelle, *The monadic second-order logic of graphs. I. Recognizable sets of finite graphs*, Information and Computation **85** (1990), no. 1, 12–75.

[52] Bruno Courcelle, *Clique-width and edge contraction*, Information Processing Letters **114** (2014), no. 1–2, 42–44.

[53] Bruno Courcelle, *From tree-decompositions to clique-width terms*, Discrete Applied Mathematics **248** (2018), 125–144.

[54] Bruno Courcelle, *On quasi-planar graphs: Clique-width and logical description*, Discrete Applied Mathematics (in press).

[55] Bruno Courcelle and Joost Engelfriet, *Graph structure and monadic second-order logic: A language-theoretic approach*, Encyclopedia of Mathematics and its Applications, vol. 138, Cambridge University Press, 2012.

[56] Bruno Courcelle, Joost Engelfriet, and Grzegorz Rozenberg, *Handle-rewriting hypergraph grammars*, Journal of Computer and System Sciences **46** (1993), no. 2, 218–270.

[57] Bruno Courcelle, Pinar Heggernes, Daniel Meister, Charis Papadopoulos, and Udi Rotics, *A characterisation of clique-width through nested partitions*, Discrete Applied Mathematics **187** (2015), 70–81.

[58] Bruno Courcelle, Johann A. Makowsky, and Udi Rotics, *Linear time solvable optimization problems on graphs of bounded clique-width*, Theory of Computing Systems **33** (2000), no. 2, 125–150.

[59] Bruno Courcelle and Stephan Olariu, *Upper bounds to the clique width of graphs*, Discrete Applied Mathematics **101** (2000), no. 1–3, 77–114.

[60] Konrad K. Dabrowski, François Dross, and Daniël Paulusma, *Colouring diamond-free graphs*, Journal of Computer and System Sciences **89** (2017), 410–431.

[61] Konrad K. Dabrowski, Petr A. Golovach, and Daniël Paulusma, *Colouring of graphs with Ramsey-type forbidden subgraphs*, Theoretical Computer Science **522** (2014), 34–43.

[62] Konrad K. Dabrowski, Shenwei Huang, and Daniël Paulusma, *Bounding clique-width via perfect graphs*, Journal of Computer and System Sciences (in press).

[63] Konrad K. Dabrowski, Vadim V. Lozin, and Daniël Paulusma, *Clique-width and well-quasi ordering of triangle-free graph classes*, Proc. WG 2017, LNCS **10520** (2017), 220–233.

[64] Konrad K. Dabrowski, Vadim V. Lozin, and Daniël Paulusma, *Well-quasi-ordering versus clique-width: New results on bigenic classes*, Order **35** (2018), no. 2, 253–274.

[65] Konrad K. Dabrowski, Vadim V. Lozin, Rajiv Raman, and Bernard Ries, *Colouring vertices of triangle-free graphs without forests*, Discrete Mathematics **312** (2012), no. 7, 1372–1385.

[66] Konrad K. Dabrowski and Daniël Paulusma, *Classifying the clique-width of H-free bipartite graphs*, Discrete Applied Mathematics **200** (2016), 43–51.

[67] Konrad K. Dabrowski and Daniël Paulusma, *Clique-width of graph classes defined by two forbidden induced subgraphs*, The Computer Journal **59** (2016), no. 5, 650–666.

[68] Konrad K. Dabrowski and Daniël Paulusma, *On colouring $(2P_2, H)$-free and (P_5, H)-free graphs*, Information Processing Letters **134** (2018), 35–41.

[69] Jean Daligault, Michaël Rao, and Stéphan Thomassé, *Well-quasi-order of relabel functions*, Order **27** (2010), no. 3, 301–315.

[70] Peter Damaschke, *Induced subgraphs and well-quasi-ordering*, Journal of Graph Theory **14** (1990), no. 4, 427–435.

[71] H.N. de Ridder et al., *Information system on graph classes and their inclusions*, 2001–2019, http://www.graphclasses.org.

[72] Guoli Ding, *Subgraphs and well-quasi-ordering*, Journal of Graph Theory **16** (1992), no. 5, 489–502.

[73] Gabriel Andrew Dirac, *On rigid circuit graphs*, Abhandlungen aus dem Mathematischen Seminar der Universität Hamburg **25** (1961), no. 1–2, 71–76.

[74] Wolfgang Espelage, Frank Gurski, and Egon Wanke, *How to solve NP-hard graph problems on clique-width bounded graphs in polynomial time*, Proc. WG 2001, LNCS **2204** (2001), 117–128.

[75] Wolfgang Espelage, Frank Gurski, and Egon Wanke, *Deciding clique-width for graphs of bounded tree-width*, Journal of Graph Algorithms and Applications **7** (2003), no. 2, 141–180.

[76] Michael R. Fellows, Danny Hermelin, and Frances A. Rosamond, *Well quasi orders in subclasses of bounded treewidth graphs and their algorithmic applications*, Algorithmica **64** (2012), no. 1, 3–18.

[77] Michael R. Fellows, Frances A. Rosamond, Udi Rotics, and Stefan Szeider, *Clique-width is NP-Complete*, SIAM Journal on Discrete Mathematics **23** (2009), no. 2, 909–939.

[78] Stéphane Földes and Peter Ladislaw Hammer, *Split graphs*, Congressus Numerantium **XIX** (1977), 311–315.

[79] Fedor V. Fomin, Petr A. Golovach, Daniel Lokshtanov, and Saket Saurabh, *Intractability of clique-width parameterizations*, SIAM Journal on Computing **39** (2010), no. 5, 1941–1956.

[80] Fedor V. Fomin, Petr A. Golovach, Daniel Lokshtanov, and Saket Saurabh, *Almost optimal lower bounds for problems parameterized by clique-width*, SIAM Journal on Computing **43** (2014), no. 5, 1541–1563.

[81] Jean-Luc Fouquet, Vassilis Giakoumakis, and Jean-Marie Vanherpe, *Bipartite graphs totally decomposable by canonical decomposition*, International Journal of Foundations of Computer Science **10** (1999), no. 4, 513–533.

[82] Dallas J. Fraser, Angèle M. Hamel, Chính T. Hoàng, Kevin Holmes, and Tom P. LaMantia, *Characterizations of $(4K_1, C_4, C_5)$-free graphs*, Discrete Applied Mathematics **231** (2017), 166–174.

[83] Martin Fürer, *Multi-clique-width*, Proc. ITCS 2017, LIPIcs **67** (2017), 14:1–14:13.

[84] Michael Randolph Garey and David S. Johnson, *Computers and intractability: A guide to the theory of NP-Completeness*, W.H. Freeman & Co., New York, NY, USA, 1979.

[85] Serge Gaspers, Shenwei Huang, and Daniël Paulusma, *Colouring square-free graphs without long induced paths*, Proc. STACS 2018, LIPIcs **96** (2018), 35:1–35:15.

[86] Michael U. Gerber and Daniel Kobler, *Algorithms for vertex-partitioning problems on graphs with fixed clique-width*, Theoretical Computer Science **299** (2003), no. 1, 719–734.

[87] Alexander Glikson and Johann A. Makowsky, *NCE graph grammars and clique-width*, Proc. WG 2003 **2880** (2003), 237–248.

[88] Petr A. Golovach, Matthew Johnson, Daniël Paulusma, and Jian Song, *A survey on the computational complexity of colouring graphs with forbidden subgraphs*, Journal of Graph Theory **84** (2017), no. 4, 331–363.

[89] Petr A. Golovach, Daniel Lokshtanov, Saket Saurabh, and Meirav Zehavi, *Cliquewidth III: the odd case of graph coloring parameterized by cliquewidth*, Proc. SODA 2018 (2018), 262–273.

[90] Petr A. Golovach, Daniël Paulusma, and Bernard Ries, *Coloring graphs characterized by a forbidden subgraph*, Discrete Applied Mathematics **180** (2015), 101–110.

[91] Martin C. Golumbic and Udi Rotics, *On the clique-width of some perfect graph classes*, International Journal of Foundations of Computer Science **11** (2000), no. 3, 423–443.

[92] Martin Grohe and Pascal Schweitzer, *Isomorphism testing for graphs of bounded rank width*, Proc. FOCS 2015 (2015), 1010–1029.

[93] Martin Grötschel, László Lovász, and Alexander Schrijver, *Polynomial algorithms for perfect graphs*, Annals of Discrete Mathematics **21** (1984), 325–356.

[94] Frank Gurski, *Characterizations for co-graphs defined by restricted NLC-width or clique-width operations*, Discrete Mathematics **306** (2006), no. 2, 271–277.

[95] Frank Gurski, *The behavior of clique-width under graph operations and graph transformations*, Theory of Computing Systems **60** (2017), no. 2, 346–376.

[96] Frank Gurski and Carolin Rehs, *Comparing linear width parameters for directed graphs*, CoRR abs/1812.06653 (2018).

[97] Frank Gurski and Egon Wanke, *The tree-width of clique-width bounded graphs without $K_{n,n}$*, Proc. WG 2000, LNCS **1928** (2000), 196–205.

[98] Frank Gurski and Egon Wanke, *On the relationship between NLC-width and linear NLC-width*, Theoretical Computer Science **347** (2005), no. 1–2, 76–89.

[99] Frank Gurski and Egon Wanke, *Vertex disjoint paths on clique-width bounded graphs*, Theoretical Computer Science **359** (2006), no. 1–3, 188–199.

[100] Frank Gurski and Egon Wanke, *Line graphs of bounded clique-width*, Discrete Mathematics **307** (2007), no. 22, 2734–2754.

[101] Frank Gurski and Egon Wanke, *A local characterization of bounded clique-width for line graphs*, Discrete Mathematics **307** (2007), no. 6, 756–759.

[102] Frank Gurski and Egon Wanke, *The NLC-width and clique-width for powers of graphs of bounded tree-width*, Discrete Applied Mathematics **157** (2009), no. 4, 583–595.

[103] Frank Harary, *Graph theory*, Addison-Wesley Series in Mathematics, Addison-Wesley, 1969.

[104] Pinar Heggernes, Daniel Meister, and Charis Papadopoulos, *Graphs of linear clique-width at most 3*, Theoretical Computer Science **412** (2011), no. 39, 5466–5486.

[105] Pinar Heggernes, Daniel Meister, and Charis Papadopoulos, *Characterising the linear clique-width of a class of graphs by forbidden induced subgraphs*, Discrete Applied Mathematics **160** (2012), no. 6, 888–901.

[106] Pinar Heggernes, Daniel Meister, Charis Papadopoulos, and Udi Rotics, *Clique-width of path powers*, Discrete Applied Mathematics **205** (2016), 62–72.

[107] Pinar Heggernes, Daniel Meister, and Udi Rotics, *Exploiting restricted linear structure to cope with the hardness of clique-width*, Proc. TAMC 2010, LNCS **6108** (2010), 284–295.

[108] Marijn J.H. Heule and Stefan Szeider, *A SAT approach to clique-width*, ACM Transactions on Computational Logic **16** (2015), no. 3, 24:1–24:27.

[109] Graham Higman, *Ordering by divisibility in abstract algebras*, Proceedings of the London Mathematical Society **s3-2** (1952), no. 1, 326–336.

[110] Petr Hliněný and Sang-il Oum, *Finding branch-decompositions and rank-decompositions*, SIAM Journal on Computing **38** (2008), no. 3, 1012–1032.

[111] Petr Hliněný, Sang-il Oum, Detlef Seese, and Georg Gottlob, *Width parameters beyond tree-width and their applications*, The Computer Journal **51** (2008), no. 3, 326–362.

[112] Ken-ichi Kawarabayashi, Yusuke Kobayashi, and Bruce Reed, *The disjoint paths problem in quadratic time*, Journal of Combinatorial Theory, Series B **102** (2012), no. 2, 424–435.

[113] Klaus Jansen and Petra Scheffler, *Generalized coloring for tree-like graphs*, Discrete Applied Mathematics **75** (1997), no. 2, 135–155.

[114] Öjvind Johansson, *Clique-decomposition, NLC-decomposition, and modular decomposition - relationships and results for random graphs*, Congressus Numerantium **132** (1998), 39–60.

[115] Shahin Kamali, *Compact representation of graphs of small clique-width*, Algorithmica **80** (2018), no. 7, 2106–2131.

[116] Marcin Kamiński, Vadim V. Lozin, and Martin Milanič, *Recent developments on graphs of bounded clique-width*, Discrete Applied Mathematics **157** (2009), no. 12, 2747–2761.

[117] Marcin Kamński, Jean-Florent Raymond, and Théophile Trunck, *Well-quasi-ordering H-contraction-free graphs*, Discrete Applied Mathematics **248** (2018), 18–27.

[118] Mamadou Moustapha Kanté and Michaël Rao, *The rank-width of edge-coloured graphs*, Theory of Computing Systems **52** (2013), no. 4, 599–644.

[119] T. Karthick and Frédéric Maffray, *Coloring (gem, co-gem)-free graphs*, Journal of Graph Theory **89** (2018), no. 3, 288–303.

[120] T. Karthick, Frédéric Maffray, and Lucas Pastor, *Polynomial cases for the vertex coloring problem*, Algorithmica **81** (2019), no. 3, 1053–1074.

[121] Daniel Kobler and Udi Rotics, *Edge dominating set and colorings on graphs with fixed clique-width*, Discrete Applied Mathematics **126** (2003), no. 2–3, 197–221.

[122] Nicholas Korpelainen and Vadim V. Lozin, *Bipartite induced subgraphs and well-quasi-ordering*, Journal of Graph Theory **67** (2011), no. 3, 235–249.

[123] Nicholas Korpelainen and Vadim V. Lozin, *Two forbidden induced subgraphs and well-quasi-ordering*, Discrete Mathematics **311** (2011), no. 16, 1813–1822.

[124] Nicholas Korpelainen, Vadim V. Lozin, and Colin Mayhill, *Split permutation graphs*, Graphs and Combinatorics **30** (2014), no. 3, 633–646.

[125] Nicholas Korpelainen, Vadim V. Lozin, and Igor Razgon, *Boundary properties of well-quasi-ordered sets of graphs*, Order **30** (2013), no. 3, 723–735.

[126] Daniel Král', Jan Kratochvíl, Zsolt Tuza, and Gerhard J. Woeginger, *Complexity of coloring graphs without forbidden induced subgraphs*, Proc. WG 2001, LNCS **2204** (2001), 254–262.

[127] Stefan Kratsch and Pascal Schweitzer, *Graph isomorphism for graph classes characterized by two forbidden induced subgraphs*, Discrete Applied Mathematics **216, Part 1** (2017), 240–253.

[128] O-joung Kwon, Michał Pilipczuk, and Sebastian Siebertz, *On low rank-width colorings*, Proc. WG 2017, LNCS **10520** (2017), 372–385.

[129] Richard E. Ladner, *On the structure of polynomial time reducibility*, Journal of the ACM **22** (1975), no. 1, 155–171.

[130] Michael Lampis, *Algorithmic meta-theorems for restrictions of treewidth*, Algorithmica **64** (2012), no. 1, 19–37.

[131] Michael Lampis, *Finer tight bounds for coloring on clique-width*, Proc. ICALP 2018, LIPIcs **107** (2018), 86:1–86:14.

[132] Hoàng-Oanh Le, *Contributions to clique-width of graphs*, Ph.D. thesis, University of Rostock, 2003, Cuvillier Verlag Göttingen, 2004.

[133] László Lovász, *Coverings and coloring of hypergraphs*, Congressus Numerantium **VIII** (1973), 3–12.

[134] Vadim V. Lozin, *On a generalization of bi-complement reducible graphs*, Proc. MFCS 2000, LNCS **1893** (2000), 528–538.

[135] Vadim V. Lozin, *Bipartite graphs without a skew star*, Discrete Mathematics **257** (2002), no. 1, 83–100.

[136] Vadim V. Lozin, *Minimal classes of graphs of unbounded clique-width*, Annals of Combinatorics **15** (2011), no. 4, 707–722.

[137] Vadim V. Lozin and Martin Milanič, *Critical properties of graphs of bounded clique-width*, Discrete Mathematics **313** (2013), no. 9, 1035–1044.

[138] Vadim V. Lozin and Dieter Rautenbach, *Chordal bipartite graphs of bounded tree- and clique-width*, Discrete Mathematics **283** (2004), no. 1–3, 151–158.

[139] Vadim V. Lozin and Dieter Rautenbach, *On the band-, tree-, and clique-width of graphs with bounded vertex degree*, SIAM Journal on Discrete Mathematics **18** (2004), no. 1, 195–206.

[140] Vadim V. Lozin and Dieter Rautenbach, *The tree- and clique-width of bipartite graphs in special classes*, Australasian Journal of Combinatorics **34** (2006), 57–67.

[141] Vadim V. Lozin and Dieter Rautenbach, *The relative clique-width of a graph*, Journal of Combinatorial Theory, Series B **97** (2007), no. 5, 846–858.

[142] Vadim V. Lozin, Igor Razgon, and Viktor Zamaraev, *Well-quasi-ordering versus clique-width*, Journal of Combinatorial Theory, Series B **130** (2018), 1–18.

[143] Vadim V. Lozin and Jordan Volz, *The clique-width of bipartite graphs in monogenic classes*, International Journal of Foundations of Computer Science **19** (2008), no. 2, 477–494.

[144] Wolfgang Mader, *Wohlquasigeordnete Klassen endlicher Graphen*, Journal of Combinatorial Theory, Series B **12** (1972), no. 2, 105–122.

[145] Johann A. Makowsky and Udi Rotics, *On the clique-width of graphs with few P_4's*, International Journal of Foundations of Computer Science **10** (1999), no. 3, 329–348.

[146] Dmitriy S. Malyshev and O.O. Lobanova, *Two complexity results for the vertex coloring problem*, Discrete Applied Mathematics **219** (2017), 158–166.

[147] Daniel Meister and Udi Rotics, *Clique-width of full bubble model graphs*, Discrete Applied Mathematics **185** (2015), 138–167.

[148] Haiko Müller and Ruth Urner, *On a disparity between relative cliquewidth and relative NLC-width*, Discrete Applied Mathematics **158** (2010), no. 7, 828–840.

[149] Sang-il Oum, *Approximating rank-width and clique-width quickly*, ACM Transactions on Algorithms **5** (2008), no. 1, 10:1–10:20.

[150] Sang-il Oum, *Rank-width: Algorithmic and structural results*, Discrete Applied Mathematics **231** (2017), 15–24.

[151] Sang-il Oum and Paul D. Seymour, *Approximating clique-width and branch-width*, Journal of Combinatorial Theory, Series B **96** (2006), no. 4, 514–528.

[152] Marko Petkovšek, *Letter graphs and well-quasi-order by induced subgraphs*, Discrete Mathematics **244** (2002), no. 1–3, 375–388.

[153] Michaël Rao, *MSOL partitioning problems on graphs of bounded treewidth and clique-width*, Theoretical Computer Science **377** (2007), no. 1–3, 260–267.

[154] Michaël Rao, *Clique-width of graphs defined by one-vertex extensions*, Discrete Mathematics **308** (2008), no. 24, 6157–6165.

[155] Neil Robertson and Paul D. Seymour, *Graph minors. X. Obstructions to tree-decomposition*, Journal of Combinatorial Theory, Series B **52** (1991), no. 2, 153–190.

[156] Neil Robertson and Paul D. Seymour, *Graph minors. XIII. The disjoint paths problem*, Journal of Combinatorial Theory, Series B **63** (1995), no. 1, 65–110.

[157] Neil Robertson and Paul D. Seymour, *Graph minors. XX. Wagner's conjecture*, Journal of Combinatorial Theory, Series B **92** (2004), no. 2, 325–357.

[158] Florian Roussel, Irena Rusu, and Henri Thuillier, *On graphs with limited number of P_4-partners*, International Journal of Foundations of Computer Science **10** (1999), no. 1, 103–121.

[159] Sigve Hortemo Sæther and Martin Vatshelle, *Hardness of computing width parameters based on branch decompositions over the vertex set*, Theoretical Computer Science **615** (2016), 120–125.

[160] Uwe Schöning, *Graph isomorphism is in the low hierarchy*, Journal of Computer and System Sciences **37** (1988), no. 3, 312–323.

[161] Pascal Schweitzer, *Towards an isomorphism dichotomy for hereditary graph classes*, Theory of Computing Systems **61** (2017), no. 4, 1084–1127.

[162] Robert Endre Tarjan, *Decomposition by clique separators*, Discrete Mathematics **55** (1985), no. 2, 221–232.

[163] Jean-Marie Vanherpe, *Clique-width of partner-limited graphs*, Discrete Mathematics **276** (2004), no. 1–3, 363–374.

[164] Martin Vatshelle, *New width parameters of graphs*, Ph.D. thesis, University of Bergen, 2012.

[165] Klaus Wagner, *Über eine Eigenschaft der ebenen Komplexe*, Mathematische Annalen **114** (1937), no. 1, 570–590.

[166] Egon Wanke, *k-NLC graphs and polynomial algorithms*, Discrete Applied Mathematics **54** (1994), no. 2–3, 251–266.

Department of Computer Science
Durham University
Durham, UK
{konrad.dabrowski,matthew.johnson2,daniel.paulusma}@durham.ac.uk

Analytic representations of large graphs

Andrzej Grzesik and Daniel Král'[1]

Abstract

The recently emerged theory of graph limits provides analytic tools to represent and analyse large graphs, which appear in various scenarios in mathematics and computer science. We survey basic concepts concerning dense graph limits and then focus on recent results on finitely forcible graph limits. We conclude by presenting some of the existing notions concerning sparse graph limits and discussing their mutual relation.

1 Introduction

Large graphs appear as representations of huge networks in many different areas of life. One should mention in particular the internet network of hyperlinks, acquaintance graphs of social networks, etc. Since such graphs are often too huge to be examined by standard graph theoretic or algorithmic approaches, there has been a need for developing tools specifically for large graphs that can be used to gain some information from local sampling, studying global properties, or observing the behaviour of various processes on the graph through a longer time interval. In this short survey, we present analytic tools for representing and analysing large graphs provided by the theory of graph limits as a response to these new challenges.

The theory of graph limits has highlighted new exciting links between analysis, combinatorics, computer science, ergodic theory, group theory and probability theory. The techniques have been developed to some extent independently for dense graphs and sparse graphs, which is also reflected in the way that this survey is structured. For many applications, the concept of a convergence of a sequence of graphs, without explicitly defining an analytic object representing its limit, could be sufficient. However, a better understanding can often be gained if an analytic object that properly captures the interplay of local and global parameters is available. In our exposition, we will be concerned with the convergence and limit representations of graphs. However, many of the results presented further can be translated to other discrete objects, e.g., permutations [37,53,54,63] or partial orders [51,55]. We also refer the reader to a recent monograph by Lovász [67], where the theory of graph limits is treated in a more detailed and thorough way.

In this survey, we are primarily concerned with results on limits of dense graphs, i.e., graphs where the number of edges is quadratic in its number of vertices. The foundations of the theory of dense graph limits were laid in a series of papers by Borgs, Chayes, Lovász, Sós, Szegedy and Vesztergombi [15–17,69,70]. In Section 3, we survey basic concepts concerning limits of dense graphs, and we then focus on the uniqueness of the limit structures in Section 4. Limits of dense graphs turned out to be very useful with respect to applications in extremal combinatorics. In particular, the closely related flag algebra method, which was introduced by Razborov [84], enables the use of semidefinite programming to search for bounds on problems studied in extremal graph theory. Using this method, Razborov [85] solved the famous problem, which dates back to the work of Rademacher in the 1940's

---

[1]The work of both authors has received funding from the European Research Council (ERC) under the European Union's Horizon 2020 research and innovation programme (grant agreement No 648509). This publication reflects only its authors' view; the European Research Council Executive Agency is not responsible for any use that may be made of the information it contains.

and Erdős in the 1950's, on the minimum possible density of triangles in a graph with a given edge density. This result was later generalized by Nikiforov [81] and by Reiher [87] using similar but finer techniques from triangles to larger complete graphs. It should be emphasized that the flag algebra method can also be used in relation to other combinatorial objects such as directed graphs, hypergraphs, permutations, etc. The method has seen many profound applications and resulted in substantial progress on many long standing open problems in extremal combinatorics, e.g. [5–8, 45, 48, 49, 59, 61, 82, 83, 86].

Another application of graph limits that we would like to mention here belongs to computer science. A property testing algorithm is an algorithm that determines with high probability a property or approximates a parameter of a large input based on a constant size sample; such algorithms started to be systematically studied in the 1990's [41–43, 88], also see, e.g., the surveys in [40]. The theory of graph limits led to an analytic characterization of properties and parameters that can be computed in this way [67, 70]. In particular, it is possible to define a notion of distance, which is called cut distance, between large graphs of not necessarily the same order; this notion extends to the setting of graphons representing graph limits. Inputs that are close in the cut distance cannot be distinguished using property testing algorithms. The converse, which is also true, can be exploited to provide a characterization of properties and parameters amenable to such algorithms [67,70].

The area of limits of sparse graphs, such as graphs of bounded degree, is less developed than the area of limits of dense graphs. Several notions of convergence of such graphs were proposed and the sparse graph convergence is considered to be significantly less understood than the convergence of dense graphs. Still, the area of sparse limits offers one of the most fundamental open problems on graph limits: the conjecture of Aldous and Lyons [1]. This conjecture gives a necessary and sufficient condition on a local neighbourhood distribution to correspond to a sequence of graphs, and is essentially equivalent to Gromov's question whether all countable discrete groups are sofic. We will cover basic notions concerning the sparse graph convergence, including the conjecture of Aldous and Lyons, in Section 5.

2 Preliminaries

In this section, we introduce basic notation used throughout the paper. The set of all positive integers is denoted by $\mathbb{N}$, the set of all non-negative integers by $\mathbb{N}_0$, and the set of integers between 1 and k (inclusive) by $[k]$. All measures considered in this paper are Borel measures on $\mathbb{R}^d$, $d \in \mathbb{N}$. If a set $X \subseteq \mathbb{R}^d$ is measurable, then we write $|X|$ for its measure, and if X and Y are two measurable sets, then we write $X \sqsubseteq Y$ if $|X \setminus Y| = 0$.

All graphs considered in this paper are simple graphs without loops. If G is a graph, we write $|G|$ for its order, i.e., the number of its vertices, and $||G||$ for its size, i.e., the number of its edges.

For completeness, we next give a brief overview of results from the probability theory that we particularly need in our exposition; we refer the reader to, e.g., [3] for further details. We start with the Borel-Cantelli lemma.

Lemma 2.1 (Borel-Cantelli lemma). *Let $(E_n)_{n\in\mathbb{N}}$ be a sequence of probability events. If the sum of probabilities of E_n, $n \in \mathbb{N}$, is finite, i.e.,*

$$\sum_{n\in\mathbb{N}} \mathbb{P}(E_n) < \infty \,,$$

then the probability that infinitely many of the events E_n occur is zero.

We next define a notion of a martingale. Fix a probability space Ω and let $(X_n)_{n\in\mathbb{N}}$ be a sequence of random real variables on Ω. The sequence $(X_n)_{n\in\mathbb{N}}$ forms a *martingale* if the expected value of each X_n is equal to a real number X_0 and

$$\mathbb{E}(X_{n+1}|X_1,\ldots,X_n) = X_n \text{ for every } n \in \mathbb{N},$$

i.e., the expected value of X_{n+1} conditioned on the values of $X_1,\ldots,X_n$ is the value of X_n. With a slight abuse of notation, X_0 can be understood to be the random variable on Ω equal to X_0 everywhere. For a martingale $(X_n)_{n\in\mathbb{N}}$, we can bound the probability of a large deviation of X_n from its expected value.

Theorem 2.2 (Azuma-Hoeffding inequality). *Let $(X_n)_{n\in\mathbb{N}}$ be a martingale with $\mathbb{E}X_n = X_0$ for all $n \in \mathbb{N}$, and let $(c_n)_{n\in\mathbb{N}}$ be a sequence of reals. If it holds for every $n \in \mathbb{N}$ that $|X_n - X_{n-1}| \le c_n$ with probability one, then*

$$\mathbb{P}\left(|X_n - X_0| \ge t\right) \le 2e^{\frac{-t^2}{2\sum_{k=1}^n c_k^2}}$$

for every $n \in \mathbb{N}$ and every $t \in \mathbb{R}$.

Finally, we will need the following corollary of Doob's Martingale Convergence Theorem.

Corollary. *Let $(X_n)_{n\in\mathbb{N}}$ be a martingale on a probability space Ω with probability μ. If there exists $K \in \mathbb{R}$ such that $\mathbb{E}|X_n| < K$ for every $n \in \mathbb{N}$, then there exists a random variable X on Ω such that*

$$\lim_{n\to\infty} X_n(\omega) = X(\omega)$$

for μ-almost all $\omega \in \Omega$.

3 Dense graph limits

In this section, we are primarily concerned with limits of dense graphs, i.e., graphs where the number of edges is quadratic in the number of vertices. If G and H are two graphs, the *density* of H in G, denoted by $d(H,G)$, is the probability that a randomly chosen subset of $|H|$ vertices of G induces a subgraph isomorphic to H. A sequence $(G_n)_{n\in\mathbb{N}}$ of graphs is *convergent* if the sequence of densities $d(H,G_n)$ converges for every graph H. In what follows, we will only consider convergent sequences $(G_n)_{n\in\mathbb{N}}$ of graphs such that the number of vertices of G_n tends to infinity.

Simple examples of convergent sequences of graphs include the sequence of complete graphs K_n, the sequence of complete bipartite graphs $K_{n,n}$ with parts of equal size and the sequence of complete bipartite graphs $K_{\lfloor\alpha n\rfloor,n}$ for $\alpha \in (0,1)$. A less trivial example of a convergent sequence of graphs is the sequence of Erdős-Rényi random graphs $G_{n,p}$. Recall that the *Erdős-Rényi random graph* $G(n,p)$, $n \in \mathbb{N}$ and $p \in [0,1]$, is the graph with n vertices such that any two of its vertices are joined by an edge with probability p independently of all the other pairs of vertices. The convergence of this sequence of graphs can be shown using the Borel-Cantelli lemma (Lemma 2.1) and the Azuma-Hoeffding inequality (Theorem 2.2).

Assume now that $(G_n)_{n\in\mathbb{N}}$ is a sequence of sparse graphs, which means that the number of edges of G_n is $o(|G_n|^2)$, i.e.,

$$\lim_{n\to\infty} \frac{||G_n||}{|G_n|^2} = 0.$$

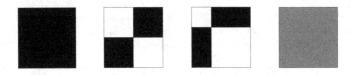

Figure 1: Graphons that are limits of the sequences $(K_n)_{n\in\mathbb{N}}$, $(K_{n,n})_{n\in\mathbb{N}}$, $(K_{n,2n})_{n\in\mathbb{N}}$ and $(G(n,1/2))_{n\in\mathbb{N}}$.

Consequently, the density $d(H, G_n)$ of any non-edgeless graph H converges to zero and the density $d(H, G_n)$ of any edgeless graph H converges to one. Hence, the sequence $(G_n)_{n\in\mathbb{N}}$ is convergent in the sense that we have defined earlier. Consequently, it holds that any sequence of sparse graphs is convergent regardless of its structure. This is the reason why the notion of convergence that we have just defined is of interest for dense graphs. Notions of convergence appropriate for sparse graphs will be described in Section 5.

Another way of defining convergent sequences of graphs is to consider homomorphic densities of graphs. If G and H are two graphs, the *homomorphic density* of H in G, which is denoted by $t(H, G)$, is the probability that a random map from the vertex set of H to the vertex set of G is a homomorphism from H to G. A simple application of the Principle of Inclusion and Exclusion shows that $d(H, G)$ is determined by the values of $t(H', G)$ for all spanning subgraphs H' of H, and $t(H, G)$ is determined by the values of $d(H', G)$ for all supergraphs H' of H with the same number of vertices. Hence, a sequence $(G_n)_{n\in\mathbb{N}}$ of graphs is convergent if and only if if the sequence of homomorphic densities $t(H, G_n)$ converges for every graph H.

We next introduce an analytic object that is used to represent a convergent sequence of graphs. This object is called a graphon. A *graphon* is a symmetric measurable function $W : [0,1]^2 \to [0,1]$, where symmetric stands for the property that $W(x,y) = W(y,x)$ for all $x, y \in [0,1]$. One can think of a graphon as a continuous analogue of the adjacency matrix of a graph; this analogy provides a good first intuition when working with graphons, however, the matter is more complex as we will see in the following. The analogy with adjacency matrices also motivates some of the definitions that follow.

A graphon can be viewed as a recipe for creating a random graph as we now present. If W is a graphon, then a W-*random graph of order n* is the random graph obtained by sampling n points $x_1, \ldots, x_n$ independently and uniformly in the unit interval $[0,1]$ and joining the i-th vertex and the j-th vertex of the graph by an edge with probability $W(x_i, x_j)$. Note that if W is the graphon equal to $p \in [0,1]$ for all $x, y \in [0,1]$, then the W-random graph of order n is the Erdős-Rényi random graph $G(n, p)$. Graphons are usually depicted in the unit square with values being different shades of gray, where white represents zero and black represents one. The origin of the coordinate system is usually in the top left corner to follow the analogy with adjacency matrices. An example of such visualization can be found in Figure 1.

We now relate graphons to convergent sequences of graphs. Let the *density* of a graph H in a graphon W be the probability that the W-random graph of order $|H|$ is isomorphic to H; this probability is denoted by $d(H, W)$. It can be shown that the following holds:

$$d(H, W) = \frac{|H|!}{|\mathrm{Aut}(H)|} \int_{[0,1]^{|H|}} \prod_{v_i v_j \in E(H)} W(x_i, x_j) \prod_{v_i v_j \notin E(H)} (1 - W(x_i, x_j)) \, \mathrm{d}x_1 \cdots x_{|H|}$$

Figure 2: Examples of step graphons.

where $V(H) = \{v_1, \ldots, v_{|H|}\}$ and $\mathrm{Aut}(H)$ is the automorphism group of H. We say that a graphon W is the *limit* of a convergent sequence $(G_n)_{n \in \mathbb{N}}$ of graphs if

$$d(H, W) = \lim_{n \to \infty} d(H, G_n)$$

for every graph H. Examples of graphons that are limits of some simple convergent sequences of graphs are given in Figure 1. In what follows, we will also consider a special type of graphons called step graphons: a graphon W is a *step graphon* if there exist an integer k and a partition of $[0, 1]$ into k measurable sets $A_1, \ldots, A_k$ such that the graphon W is constant on $A_i \times A_j$ for all $i, j \in [k]$. Examples of step graphons can be found in Figure 2.

It is natural to ask whether every convergent sequence of graphs has a limit, whether this limit is unique (if it exists), and whether every graphon is a limit of a convergent sequence of graphs. We start with the latter of these questions, which is simpler to answer, and we discuss the former of the questions later in this section.

Theorem 3.1. *Let W be a graphon and let G_n be a W-random graph of order n, $n \in \mathbb{N}$. The sequence $(G_n)_{n \in \mathbb{N}}$ is convergent and the graphon W is its limit with probability one.*

Proof. Fix a graph H and an integer n such that $n \geq |H|$. The probability that a particular $|H|$-tuple of vertices of G_n induces a copy of H is $d(H, W)$. The linearity of expectation implies that the expected number of copies of H in G_n is equal to $d(H, W)\binom{n}{|H|}$. We next estimate the probability of a large deviation from this expected value. Let X_i, $i = 0, \ldots, n$, be the random variable equal to the expected number of copies of H after the first i choices of the vertices of G_n are made in the interval $[0, 1]$ and the edges between the first i vertices are fixed when constructing the W-random graph of order n. Observe that X_n is just the number of copies of H in G_n and X_0 is equal to $d(H, W)\binom{n}{|H|}$.

Since the random variables $X_0, \ldots, X_n$ form a martingale, we can apply the Azuma-Hoeffding inequality (Theorem 2.2) with $c_i \leq n^{|H|-1}$ and get that

$$\mathbb{P}(|X_n - X_0| \geq t) \leq 2e^{\frac{-t^2}{2n^{2|H|-1}}}$$

for every $t \in \mathbb{R}$. Substituting $t = \varepsilon n^{|H|}$, we get that

$$\mathbb{P}\left(|X_n - X_0| \geq \varepsilon n^{|H|}\right) \leq 2e^{-\varepsilon^2 n/2},$$

which yields that

$$\mathbb{P}\left(|d(H, G_n) - d(H, W)| \geq |H|!2^{|H|}\varepsilon\right) \leq 2e^{-\varepsilon^2 n/2}$$

if $n \geq 2|H|$. The Borel-Cantelli lemma implies that the sequence $(d(H, G_n))_{n \in \mathbb{N}}$ is convergent with probability one and its limit is $d(H, W)$. In particular, the sequence $(G_n)_{n \in \mathbb{N}}$ is convergent and the graphon W is its limit with probability one. $\square$

Proving that there exists a limit graphon for every convergent sequence of graphs is harder. We will present here the proof by Lovász and Szegedy from [69]. The proof uses weak regularity of graphs introduced by Frieze and Kannan in [35]; this notion is weaker than the more well-known notion of Szemerédi regularity. However, it is simpler and sufficient for our purposes. If G is a graph and S and T are two subsets of its vertices, then $e(S,T)$ denotes the number of pairs of vertices $s \in S$ and $t \in T$ joined by an edge and $d(S,T)$ denotes the corresponding density, i.e., $d(S,T) = \frac{e(S,T)}{|S| \cdot |T|}$. A partition $V_1, \ldots, V_k$ of a vertex set of a graph G is an *equipartition* if $|\,|V_i| - |V_j|\,| \leq 1$ for every $i, j \in [k]$, and it is *weak ε-regular* if it is an equipartition and it holds that

$$\left| e(S,T) - \sum_{i,j=1}^{k} d(V_i, V_j)\, |S \cap V_i|\, |T \cap V_j| \right| \leq \varepsilon |G|^2$$

for any two subsets S and T of the vertex set of G. Frieze and Kannan [35] proved the following theorem.

Theorem 3.2. *For every $\varepsilon \in (0,1)$, there exists $K = 2^{O(\varepsilon^{-2})}$ such that every graph G has a weak ε-regular partition with at most K parts.*

We will need a strengthening of Theorem 3.2, whose proof follows the same lines as the proof of Theorem 3.2. We say that a partition $V_1', \ldots, V_{k'}'$ of a vertex set of a graph G is a *refinement* of a partition $V_1, \ldots, V_k$ if for every $j \in [k']$, there exists $i \in [k]$ such that $V_j' \subseteq V_i$.

Theorem 3.3. *For every $\varepsilon \in (0,1)$, there exists $K = 2^{O(\varepsilon^{-2})}$ such that every equipartition of the vertex set of a graph G into k parts can be refined to a weak ε-regular partition with at most $K \cdot k$ parts.*

Finally, weak regular partitions are related to subgraph densities as follows [35].

Theorem 3.4. *For every graph H and every $\delta \in (0,1)$, there exists $\varepsilon \in (0,1)$ such that if G is a graph with at least ε^{-1} vertices and $V_1, \ldots, V_k$ is a weak ε-regular partition of its vertex set, then*

$$\left| d(H,G) - \frac{|H|!}{|\mathrm{Aut}(H)|k^{|H|}} \sum_{i_1,\ldots,i_{|H|}=1}^{k} \prod_{v_j v_{j'} \in E(H)} d(V_{i_j}, V_{i_{j'}}) \prod_{v_j v_{j'} \notin E(H)} (1 - d(V_{i_j}, V_{i_{j'}})) \right| \leq \delta$$

where $V(H) = \{v_1, \ldots, v_{|H|}\}$.

We are now ready to prove that every convergent sequence of graphs can be represented by a graphon.

Theorem 3.5 (Lovász and Szegedy [69]). *Let $(G_n)_{n \in \mathbb{N}}$ be a convergent sequence of graphs. There exists a graphon W that is a limit of the sequence $(G_n)_{n \in \mathbb{N}}$.*

Proof. Fix a convergent sequence $(G_n)_{n \in \mathbb{N}}$, and set $\varepsilon_\ell = 2^{-\ell}$ for $\ell \in \mathbb{N}$. For every graph G_n in the sequence, fix a weak ε_1-regular partition $V_1^{n,1}, \ldots, V_{k_{n,1}}^{n,1}$ of its vertex set; such a partition exists by Theorem 3.2. Suppose that we have already fixed a weak ε_ℓ-regular partition $V_1^{n,\ell}, \ldots, V_{k_{n,\ell}}^{n,\ell}$ of G_n for some $n \in \mathbb{N}$ and $\ell \in \mathbb{N}$. By Theorem 3.3, there exists a weak $\varepsilon_{\ell+1}$-regular partition $V_1^{n,\ell+1}, \ldots, V_{k_{n,\ell+1}}^{n,\ell+1}$ of G_n that is a refinement of the partition

$V_1^{n,\ell}, \ldots, V_{k_{n,\ell}}^{n,\ell}$. By reordering the sets in the partition, we can assume that if $V_i^{n,\ell+1} \subseteq V_j^{n,\ell}$, $V_{i'}^{n,\ell+1} \subseteq V_{j'}^{n,\ell}$ and $i < i'$, then it holds that $j \leq j'$. We will refer to this property as the *ordering property*. Note that Theorems 3.2 and 3.3 yield the existence of a constant K_ℓ, $\ell \in \mathbb{N}$, such that $k_{n,\ell} \leq K_\ell$ for every $n \in \mathbb{N}$ and every $\ell \in \mathbb{N}$.

For every $n \in \mathbb{N}$ and $\ell \in \mathbb{N}$, associate the graph G_n with a $(k_{n,\ell} \times k_{n,\ell})$-matrix $A^{n,\ell}$ such that the entry $A_{ij}^{n,\ell}$ is equal to $d(V_i^{n,\ell}, V_j^{n,\ell})$. Next choose a subsequence $(G_n')_{n \in \mathbb{N}}$ of the sequence $(G_n)_{n \in \mathbb{N}}$ such that the following holds for every $\ell \in \mathbb{N}$:

- all but finitely values of $k_{n,\ell}$ are the same, and

- the matrices $A^{n,\ell}$ coordinate-wise converge.

Note that $k_{n,\ell}$ can have only values between 1 and K_ℓ, which implies that it is possible to choose a subsequence satisfying the first of the two properties. For such a subsequence, all but finitely many matrices $A^{n,\ell}$ have the same size and since their coordinates are reals between 0 and 1, it is possible to choose a subsequence of the former subsequence that also satisfies the second property. So, the subsequence $(G_n')_{n \in \mathbb{N}}$ indeed exists.

Let k_ℓ be the value that appears infinitely often among the values $k_{n,\ell}$ for the subsequence $(G_n')_{n \in \mathbb{N}}$. Further, let A^ℓ be the $(k_\ell \times k_\ell)$-matrix that is the coordinate-wise limit of the matrices $A^{n,\ell}$ for the subsequence $(G_n')_{n \in \mathbb{N}}$. Theorem 3.4 implies that the following holds for every graph H:

$$\lim_{n \to \infty} d(H, G_n') = \lim_{\ell \to \infty} \frac{|H|!}{|\mathrm{Aut}(H)| k_\ell^{|H|}} \sum_{i_1, \ldots, i_{|H|} = 1}^{k_\ell} \prod_{v_j v_{j'} \in E(H)} A_{i_j, i_{j'}}^\ell \prod_{v_j v_{j'} \notin E(H)} (1 - A_{i_j, i_{j'}}^\ell)$$

where $V(H) = \{v_1, \ldots, v_{|H|}\}$. Since $(G_n')_{n \in \mathbb{N}}$ is a subsequence of the sequence $(G_n)_{n \in \mathbb{N}}$, it follows that

$$\lim_{n \to \infty} d(H, G_n) = \lim_{\ell \to \infty} \frac{|H|!}{|\mathrm{Aut}(H)| k_\ell^{|H|}} \sum_{i_1, \ldots, i_{|H|} = 1}^{k_\ell} \prod_{v_j v_{j'} \in E(H)} A_{i_j, i_{j'}}^\ell \prod_{v_j v_{j'} \notin E(H)} (1 - A_{i_j, i_{j'}}^\ell). \tag{3.1}$$

The matrices A^ℓ yield random variables X_ℓ on $[0, 1)^2$ defined as follows:

$$X_\ell(x, y) = A_{\lfloor x \cdot k_\ell \rfloor + 1, \lfloor y \cdot k_\ell \rfloor + 1}^\ell.$$

By the ordering property, the random variables X_ℓ, $\ell \in \mathbb{N}$, form a martingale. Hence, Corollary 2 implies that there exists a measurable function W from $[0, 1]^2$ to $[0, 1]$ such that

$$W(x, y) = \lim_{\ell \to \infty} X_\ell(x, y)$$

for almost every $(x, y) \in [0, 1)^2$. Observe that the following holds for every m and every $J \subseteq [m]^2$:

$$\int_{[0,1]^m} \prod_{jj' \in J} W(x_j, x_{j'}) \, \mathrm{d}x_1 \cdots x_m = \lim_{\ell \to \infty} \int_{[0,1)^m} \prod_{jj' \in J} X_\ell(x_j, x_{j'}) \, \mathrm{d}x_1 \cdots x_m. \tag{3.2}$$

Since it also holds for every $\ell \in \mathbb{N}$, every $m \in \mathbb{N}$ and every $J \subseteq [m]^2$ that

$$\frac{1}{k_\ell^m} \sum_{i_1, \ldots, i_m = 1}^{k_\ell} \prod_{jj' \in J} A_{i_j, i_{j'}}^\ell = \int_{[0,1)^m} \prod_{jj' \in J} X_\ell(x_j, x_{j'}) \, \mathrm{d}x_1 \cdots x_m,$$

it follows that

$$d(H, W) = \lim_{n \to \infty} d(H, G_n)$$

by (3.1) and (3.2).   □

The proof that we have presented here is not the only proof of the existence of a limit graphon of a convergent sequence of graphs that is known. The existence of the limit graphon can be derived from the representation theorem on symmetrically exchangeable random variables due to Aldous [2] and Hoover [52] and further developed by Kallenberg [56]; see [4, 27] for further details. Another way of proving the existence of a limit graphon is using the arguments concerning a suitable measure space defined using the ultraproduct of graphs in the sequence as presented by Elek and Szegedy in [32]. More recently, another approach was given by Doležal, Grebík, Hladký, Rocha and Rozhoň [28–30]: in a certain sense, they consider all weak* accumulation points of zero-one step graphons associated with the graphs in the sequence and define a certain "structuredness" order on them such that the most structured points are limit graphons.

While graphons were originally developed to represent large graphs, there are various mathematical properties of graphons that are of their own interest to study. Among many such properties, we would like to mention the notion of weakly norming graphs and relate it to one of the most important open problems in extremal graph theory—Sidorenko's Conjecture. This beautiful conjecture of Erdős and Simonovits [90] and of Sidorenko [89] asserts, in the language of graphons, that $t(K_2, W)^{\|H\|} \leq t(H, W)$ for every bipartite graph H and every graphon W and every graphon W, i.e., a quasirandom graph minimizes the density of H among all graphs with the same edge density. Sidorenko [89] confirmed the conjecture for trees, cycles and bipartite graphs with one of the sides having at most three vertices; it is interesting that the case of paths is equivalent to the Blakley-Roy inequality for matrices, which was proven in [10]. Additional graphs were added to the list of graphs satisfying the conjecture by Conlon, Fox and Sudakov [21], by Hatami [47], and by Szegedy [93]. More general results concerning recursively described classes of bipartite graphs were obtained by Conlon, Kim, Lee and Lee [22], by Kim, Lee and Lee [58], by Li and Szegedy [64] and by Szegedy [92]. In particular, Szegedy [92] has described a class of graphs called thick graphs that satisfy the conjecture. More recently, Conlon and Lee [24] showed that the conjecture is satisfied by bipartite graphs such that one of the parts has many vertices of maximum degree. Sidorenko's Conjecture is also known to hold in the local sense [67, Proposition 16.27], i.e., it holds for graphons W close to the constant graphon; a stronger statement with uniform quantitative bounds has recently been proven by Fox and Wei [34].

We say that a graph H is *weakly norming* if the function $\|W\|_H = t(H, W)^{1/\|H\|}$ is a norm on the space of graphons. A stronger notion of norming graphs concerns a generalization of graphons to functions on $[0, 1]^2$ that do not need to be non-negative on $[0, 1]^2$, however, we prefer not deviating from the main topic of our survey and we avoid giving further details here. It is easy to show that every weakly norming graph satisfies Sidorenko's conjecture, and some results on Sidorenko's conjecture actually deal with this stronger property of graphs. Hatami [47] characterized weakly norming graphs as those satisfying a certain Hölder-type inequality involving graphs edge-decorated by graphons; also see [23,62] for additional results on weakly norming graphs. However, it is interesting that the property of being weakly norming is equivalent to a generalization of the property concerned in Sidorenko's conjecture. To state this link precisely, we need several definitions. Let

$\mathcal{P} = \{J_1, \ldots, J_k\}$ be a partition of the interval $[0,1]$ into non-null measurable sets. If W is a graphon, then the graphon $W^{\mathcal{P}}$ is defined as the average on the parts in $\mathcal{P}$, i.e.,

$$W^{\mathcal{P}}(x,y) = \frac{1}{|J_i| \cdot |J_j|} \int_{J_i \times J_j} W(s,t) \mathrm{d}s\mathrm{d}t$$

where J_i and J_j are the unique parts from $\mathbb{P}$ such that $x \in J_i$ and $y \in J_j$. We say that a graph H has the *step Sidorenko property* if $t(H, W^{\mathcal{P}}) \leq t(H, W)$ for every graphon W and every finite partition $\mathcal{P}$. Considering the partition $\mathcal{P}$ with a single part implies that every graph that has the step Sidorenko property satisfies Sidorenko's conjecture. The converse is not true; the graph obtained from C_4 by adding a new vertex adjacent to one of the vertices of the cycle is known to satisfy Sidorenko's conjecture but does not have the step Sidorenko property [62]. However, a graph H is weakly norming if and only if H has the step Sidorenko property. The proof of one of the implications can be found in [67, Proposition 14.13] and the other implication has recently been proven by Doležal et al. in [28].

We conclude this section by describing an analytic object representing k-uniform hypergraphs. While it may be natural to expect this object to be a function from $[0,1]^k$ to $[0,1]$, the situation is more complex for the same reasons why graph regularity does not straightforwardly generalize to the setting of hypergraphs. A *k-hypergraph* is a hypergraph where every edge contains exactly k vertices. In the analogy to graphs, the *density* of an ℓ-vertex hypergraph H in a hypergraph G is the probability that a randomly chosen subset of ℓ vertices of G induces a subhypergraph isomorphic to H. A sequence $(G_n)_{n \in \mathbb{N}}$ of hypergraphs is *convergent* if the density of every hypergraph H in the hypergraphs G_n converges.

We next define an analytic object, which we call k-hypergraphon. A *k-hypergraphon* is a measurable function W from $[0,1]^{2^k-2}$ to $[0,1]$ such that the $2^k - 2$ variables of W are associated with the $2^k - 2$ proper subsets of $[k]$ and satisfy that $W(\vec{x}) = W(\pi(\vec{x}))$ for every $\vec{x} \in [0,1]^{2^k-2}$ and every permutation $\pi \in S_k$, where $\pi(J) = \{\pi(j), j \in J\}$ for $J \subseteq [k]$. Observe that the definition of a 2-hypergraphon coincide with the definition of a graphon. Given a k-hypergraphon W, we may define a W-random k-hypergraph of order n as follows. Fix n vertices and assign to every ℓ-element subset of vertices, $\ell \in [k-1]$, independently and uniformly a number from the unit interval $[0,1]$. The vertices $v_1, \ldots, v_k$ form an edge with probability $W(\vec{x})$ where the coordinate of x associated with a $J \subseteq [k]$, $J \notin \{\emptyset, [k]\}$, is equal to the number assigned to the $|J|$-tuple of vertices $\{v_j, j \in J\}$. Again, we define the density of a k-hypergraph H with n vertices as the probability that a W-random k-hypergraph of order n is isomorphic to H, and say that a k-hypergraphon W is a limit of a convergent sequence of k-hypergraphs if the density of every k-hypergraph H in W is the limit density of H in the sequence. The existence of a limit k-hypergraph for every convergent sequence of k-hypergraphs was established by Elek and Szegedy [32] using the ultraproduct argument that we have mentioned earlier in relation to graphons, however, the existence of a limit hypergraphon can also be proven using arguments similar to those that we have presented in the graph setting earlier as shown by Zhao [96].

4   Finite forcibility

In this section, we will discuss in what sense a limit graphon of a convergent sequence of graphs is unique and when its structure is determined by finitely many densities. We will say that two graphons W_1 and W_2 are *weakly isomorphic* if $d(H, W_1) = d(H, W_2)$ for every graph H, i.e., the graphons W_1 and W_2 are limits of the same sequences of graphs.

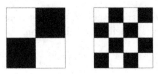

Figure 3: Two weakly isomorphic graphons.

For example, the graphons depicted Figure 3 are weakly isomorphic; they both are a limit of the sequence $(K_{n,n})_{n\in\mathbb{N}}$ of complete bipartite graphs with parts of equal sizes.

The following is a general way of constructing weakly isomorphic graphons. Let φ be a measure preserving map from $[0,1]$ to $[0,1]$, i.e., $|\varphi^{-1}(A)| = |A|$ for every measurable subset A of $[0,1]$. If W is a graphon, we define a graphon W^{φ} by setting $W^{\varphi}(x,y) = W(\varphi(x),\varphi(y))$. A standard measure theory argument yields that $d(H,W) = d(H,W^{\varphi})$, i.e., the graphons W and W^{φ} are weakly isomorphic. For example, consider the following measure preserving map:

$$\varphi(x) = \begin{cases} 2x & \text{if } x \leq 1/2, \\ 2x-1 & \text{otherwise.} \end{cases}$$

If W_1 and W_2 are the two graphons depicted in Figure 3, then $W_2 = W_1^{\varphi}$.

Borgs, Chayes and Lovász [14], also see [67, Chapter 13] for further discussion, have shown that the above way of constructing weakly isomorphic graphons is in a certain sense the only way of obtaining weakly isomorphic graphons (note that the following two theorems are not obviously equivalent since the maps φ_1 and φ_2 need not be bijective).

Theorem 4.1. *If W_1 and W_2 are weakly isomorphic graphons, then there exist a graphon W and measure preserving maps φ_1 and φ_2 such that the graphons W^{φ_1} and W_1 are equal almost everywhere, and W^{φ_2} and W_2 are equal almost everywhere.*

Theorem 4.2. *If W_1 and W_2 are weakly isomorphic graphons, then there exist measure preserving maps φ_1 and φ_2 such that the graphons $W_1^{\varphi_1}$ and $W_2^{\varphi_2}$ are equal almost everywhere.*

In general, it is necessary to know the densities $d(G,W)$ of all graphs G in a graphon W to know the structure of W. For example, see [37] for a more detailed discussion, if W is the graphon depicted in Figure 4, then for every finite set $\mathcal{G}$ of graphs, there exists a graphon W' such that $d(H,W) = d(H,W')$ for every $H \in \mathcal{G}$ but W and W' are not weakly isomorphic, i.e., there exists a graph H' such that $d(H,W) \neq d(H',W)$. On the other hand, the classical results on quasirandom graphs due to Thomasson [94] and Chung, Graham and Wilson [19] yield that if a graphon W satisfies that $t(K_2,W) = p$ and $t(C_4,W) = p^4$ for some $p \in [0,1]$, then W is equal to p almost everywhere. In particular, there are graphons such that their structure is determined by finitely many densities. In the rest of the section, we will be interested in such graphons.

The ideas presented in the previous paragraph leads to the following definition: a graphon W is *finitely forcible* if there exists a finite set $\mathcal{G}$ of graphs such that any graphon W' satisfying $d(H,W) = d(H,W')$ for every graph $H \in \mathcal{G}$ is weakly isomorphic to W; such a set $\mathcal{G}$ is called a *forcing family* of W. In particular, the constant graphon is finitely forcible and its forcing family is $\{K_2, C_4, K_4 \setminus e, K_4\}$ (note that $t(C_4,W)$ is determined by $d(C_4,W)$, $d(K_4 \setminus e, W)$ and $d(K_4,W)$), and the graphon given in Figure 4 is not finitely forcible. In fact, finitely forcible graphons are rather rare in the sense that the set of finitely

Figure 4: A graphon that is not finitely forcible.

forcible graphons is of the first category in the space $L^2\left([0,1]^2\right)$ as shown by Lovász and Szegedy [71].

The study of finitely forcible graphons is motivated by the link to extremal combinatorics captured in the following (folklore) proposition.

Proposition 4.3. *Let W_0 be a finitely forcible graphon. There exists a linear combination of subgraph densities such that W_0 is its unique (up to a weak isomorphism) minimizer, i.e., there exist $\alpha_1, \ldots, \alpha_k \in \mathbb{R}$ and graphs $H_1, \ldots, H_k$ such that the graphon W_0 minimizes the expression*

$$\min_W \sum_{i=1}^k \alpha_i d(H_i, W)$$

and any graphon minimizing this expression is weakly isomorphic to W_0.

Examples of finitely forcible graphons include many graphons that appear as optimal solutions of problems in extremal graph theory. For example, Lovász and Sós [68], also see [91], showed that every step graphon is finitely forcible. A more systematic study of finitely forcible graph limits was initiated by Lovász and Szegedy in [71]. In particular, they showed that if p is a polynomial in x and y such that the function $W : [0,1]^2 \to [0,1]$ defined as

$$W(x,y) = \begin{cases} 1 & \text{if } p(x,y) \geq 0, \text{ and} \\ 0 & \text{otherwise,} \end{cases}$$

is symmetric, then W is a finitely forcible graphon.

Inspired by the known examples of finitely forcible graphons, Lovász and Szegedy [71] conjectured that all finitely forcible graphons posses a simple structure in the sense that we now describe. To state this precisely, we need the following definition. For a graphon W and $x \in [0,1]$, define a function $f_x^W : [0,1] \to [0,1]$ to be

$$f_x^W(y) := W(x,y).$$

Since the function f_x^W belongs to $L^1([0,1])$ for almost every $x \in [0,1]$, the graphon W naturally defines a probability measure μ on $L^1([0,1])$ [71]. The space $T(W)$ is formed by the support of the measure μ equipped with the topology inherited from $L^1([0,1])$, and is referred to as the *space of typical vertices* of W. A vertex x of the graphon W is called *typical* if $f_x^W \in T(W)$. Lovász and Szegedy [71, Conjectures 9 and 10] conjectured the following; we cite both conjectures verbatim.

Conjecture 4.4. *If W is a finitely forcible graphon, then $T(W)$ is a compact space. (We can't even prove that $T(W)$ is locally compact.)*

Conjecture 4.5. *If W is a finitely forcible graphon, then $T(W)$ is finite dimensional. (We intentionally do not specify which notion of dimension is meant here—a result concerning any variant would be interesting.)*

The interest in Conjecture 4.5 comes from the following link to weak regularity partitions of graphons. It is possible to define a different notion of the space of typical vertices of a graphon as follows. If f and g are two functions from $L^1([0,1])$, we define

$$d_W(f,g) := \int_{[0,1]} \left| \int_{[0,1]} W(x,y)(f(y) - g(y)) dy \right| dx,$$

and refer to $d_W(f,g)$ as the *similarity distance* of the functions f and g. Note that the similarity distance d_W depends on the graphon W. The space $\overline{T}(W)$ is formed by the closure (with respect to d_W) of the support of the measure μ, which we have defined earlier, equipped with the topology given by the metric d_W. The structure of the space $\overline{T}(W)$ is related to weak regularity partitions of W as follows [67, Chapter 13]: if the Minkowski dimension of $\overline{T}(W)$ (with respect to the metric d_W) is d, then W has a weak ε-regular partition with $O(\varepsilon^{-d})$ parts for every $\varepsilon > 0$. Note that the number of parts of a weak ε-regular partition may need to be $2^{\Theta(\varepsilon^{-2})}$, and this is the best possible as shown by Conlon and Fox [20].

Conjectures 4.4 and 4.5 were disproved in [39] and [38], respectively. More specifically, a construction of a finitely forcible graphon W such that $T(W)$ fails to be locally compact was given in [39] (the graphon can be found in Figure 5) and a construction of a finitely forcible graphon W such that $T(W)$ contains a space homeomorphic to $[0,1]^{\mathbb{N}}$ in [38] (this graphon is depicted in Figure 6). A stronger counterexample to Conjecture 4.5 was given in [25], where the authors constructed a finitely forcible graphon W such that any weak ε-regular partition must have a number of parts almost exponential in ε^{-2} for infinitely many $\varepsilon > 0$, which is close to the general lower bound. This graphon can be found in Figure 7. This line of research culminated with the following general result of Cooper et al. [26] (the graphon is visualized in Figure 8), which we state as Theorem 4.6. To state the result, we need the following definition: if W_1 and W_2 are two graphons and $X \subseteq [0,1]$ a non-null measurable set, then we say that W_1 is a *subgraphon* of W_2 induced by X if there exist measure-preserving maps $\varphi_1 \colon X \to [0, |X|)$ and $\varphi_2 \colon X \to X$ such that

$$W_1\left(|X|^{-1} \cdot \varphi_1(x), |X|^{-1} \cdot \varphi_1(y)\right) = W_2\left(\varphi_2(x), \varphi_2(y)\right)$$

for almost every $(x,y) \in X \times X$.

Theorem 4.6. *For every graphon W_F, there exists a finitely forcible graphon W_0 such that W_F is a subgraphon of W_0 induced by a $1/14$ fraction of the vertices of W_0.*

Theorem 4.6 provides a universal framework for constructing finitely forcible graphons with very complex structure, including counterexamples to Conjectures 4.4 and 4.5. In view of Proposition 4.3, Theorem 4.6 says that problems on minimizing a linear combination of subgraph densities, which are among the problems of the simplest kind in extremal graph theory, may have unique optimal solutions with highly complex structure. Given the general nature of Theorem 4.6, it is surprising that the forcing family for the graphon W_0 in Theorem 4.6 is the same for all choices of W_F, i.e., the structure of W_0 is controlled by choosing the densities of the graphs in the forcing family only.

It is natural to ask whether the fraction $1/14$ given in Theorem 4.6 can be improved. The techniques presented in [26] would easily yield that the fraction $1/14$ can be replaced by $1/2 - \varepsilon$ for any $\varepsilon > 0$. A recent result given in [60] shows that it is possible to improve this fraction to be arbitrarily close to 1.

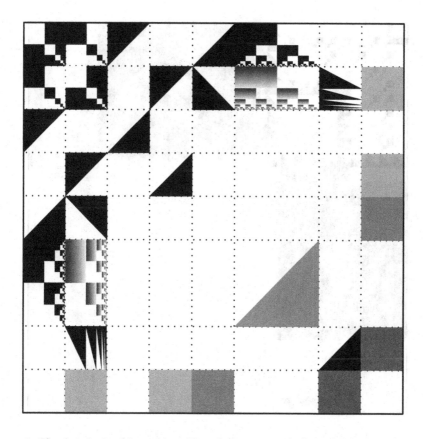

Figure 5: The finitely forcible graphon W with the space $T(W)$ of typical vertices that is not compact constructed in [39].

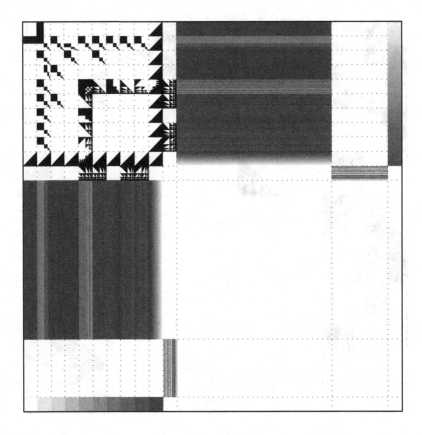

Figure 6: The finitely forcible graphon W with the space $T(W)$ of typical vertices containing a subspace homeomorphic to $[0,1]^{\mathbb{N}}$ that was constructed in [38].

Figure 7: The finitely forcible graphon W constructed in [25]. Any weak ε-regular partition of W must have a number of parts almost exponential in ε^{-2} for infinitely many $\varepsilon > 0$.

Figure 8: Visualization of the universal construction of complex finitely forcible graphon given in [26].

Theorem 4.7. *For every $\varepsilon > 0$ and every graphon W_F, there exists a finitely forcible graphon W_0 such that W_F is a subgraphon of W_0 induced by a $1 - \varepsilon$ fraction of the vertices of W_0.*

Recall that the forcing family in Theorem 4.6 was the same for all choices of W_F. However, the forcing family in Theorem 4.7 depends on ε and this dependance is necessary as shown in [60].

We now briefly outline the ideas used in the proofs that the graphons depicted in Figures 5–8 are finitely forcible. The arguments are based on the method of decorated constraints, which was developed in [39] and formalized in [38], and which builds on the flag algebra method of Razborov. Each of the graphons depicted in Figures 5–8 have the property that the interval $[0,1]$ is split into finitely many sets $X_1, \ldots, X_k$ such that the integral

$$\int_{[0,1]} W(x,y) \, \mathrm{d}y$$

is the same for all x from the same set X_i, i.e., the degrees of the vertices in each of the parts are the same. We will refer to these sets as *parts* and to graphons with this structure as *partitioned* graphons. The flag algebra arguments can be used to show that there is a polynomial combination of densities that is zero if and only if a graphon has a given number of parts with given sizes and vertices of given degrees. In particular, the structure of a partitioned graphon can be forced by finitely many densities. The method of decorated constraints uses the power of the flag algebra method to restrict the structure inside and between the parts of a partitioned graphon by constraints that are simple to analyse even for complex graphons such as those in Figures 5–8.

We would like to conclude this section with a recent result concerning the relation of finitely forcible graphons and optimal solutions of problems in extremal graph theory. As a motivation, let us have a look at several classical results in extremal graph theory. One of the oldest results in extremal graph theory is the theorem of Mantel [73], which says that the maximum number of edges of an n-vertex triangle-free graph is $\lfloor n/2 \rfloor \cdot \lceil n/2 \rceil$ and the maximum is attained only by the balanced complete bipartite graph, i.e., the graph $K_{\lfloor n/2 \rfloor, \lceil n/2 \rceil}$. In the language of graph limits, Mantel's theorem says that the maximum value of $d(K_2, W)$ among all graphons W with $d(K_3, W) = 0$ is $1/2$ and every graphon achieving this maximum is weakly isomorphic to the graphon representing (large) complete bipartite graphs with parts of equal sizes. Mantel's theorem was extended by Turán [95] to graphs avoiding complete graphs of arbitrary sizes and by Erdős and Stone [33] to all graphs. In the language of graph limits, we obtain that, for every graph H, the maximum value of $d(K_2, W)$ among all graphons W with $t(H, W) = 0$ is equal to $\frac{\chi(H)-2}{\chi(H)-1}$ and every graphon achieving this maximum is weakly isomorphic to the graphon representing (large) complete $(\chi(H) - 1)$-partite graphs with parts of equal sizes.

Since the graphon representing (large) complete $(\chi(H) - 1)$-partite graphs with parts of equal sizes is finitely forcible, it may be tempting to think that the converse of Proposition 4.3 could hold, i.e., the optimal configurations for every extremal graph theory problem are asymptotically unique. However, the following shows that this is not true. Let us consider the problem of minimizing the sum $d(K_3, W) + d(\overline{K_3}, W)$, i.e., the sum of the induced densities of K_3 and its complement. A classical result of Goodman [44] implies that this sum is minimized by any graphon such that

$$\int_{[0,1]} W(x,y) \mathrm{d}y = \frac{1}{2}$$

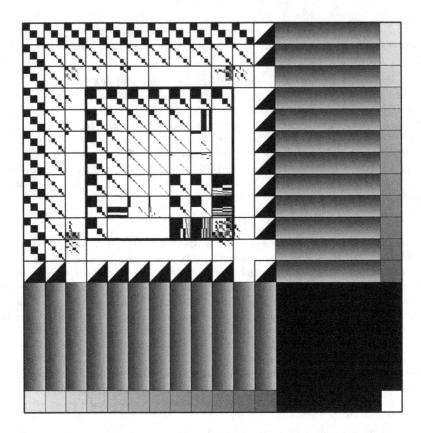

Figure 9: Visualization of the graphons forming the family $\mathcal{W}$ in Theorem 4.9. The family is obtained by varying densities in the sqaure in the third row and the third column in a controlled way.

for almost every $x \in [0,1]$, i.e., by any graphon representing graphs where almost every vertex has degree close to the number of vertices divided by two. However, the structure of an optimal solution can be made unique by adding additional constraints. For example, any graphon W that minimizes the sum and that satisfies $d(K_3, W) = 0$ corresponds to (large) complete bipartite graphs with parts of equal sizes, any graphon W that minimizes the sum and that satisfies $d(\overline{K_3}, W) = 0$ corresponds to (large) graphs that are the union of two complete graphs of equal sizes, or any graphon W that minimizes the sum and that satisfies $t(C_4, W) = 1/16$ is equal to $1/2$ almost everywhere, i.e., it corresponds to Erdős-Rényi random graphs $G_{n,1/2}$.

A conjecture of Lovász asserts that the phenomenon that we have just described is a more general one. The conjecture has been the most frequently quoted conjecture concerning dense graph limits, it also sometimes appeared as a question, and we include only some of the many references to its statement.

Conjecture 4.8 (Lovász [65, Conj 3], [66, Conj 9.12], [67, Conj 16.45], and [71, Conj 7]).

Let $H_1, \ldots, H_\ell$ be graphs and $d_1, \ldots, d_\ell$ reals. If there exists a convergent sequence of graphs with the limit density of H_i equal to d_i, $i = 1, \ldots, \ell$, then there exists such a sequence that its limit graphon is finitely forcible.

Informally speaking, the conjecture says that "every extremal problem has a finitely forcible optimum", see [67, p. 308]. The conjecture has been recently disproved in [46] using the universal construction of complex finitely forcible graph limits from [26]. More precisely, the authors proved the following theorem in [46].

Theorem 4.9. *There exists a family of graphons $\mathcal{W}$, graphs $H_1, \ldots, H_\ell$ and reals $d_1, \ldots, d_\ell$ such that*

- *a graphon W is weakly isomorphic to a graphon contained in $\mathcal{W}$ if and only if $d(H_i, W) = d_i$ for every $i \in [\ell]$, and*

- *no graphon in $\mathcal{W}$ is finitely forcible, i.e., for all graphs $H'_1, \ldots, H'_r$ and reals $d'_1, \ldots, d'_r$, the family $\mathcal{W}$ contains either zero or infinitely many graphons W with $d(H'_i, W) = d'_i$, $i \in [r]$.*

The family $\mathcal{W}$ of graphons from Theorem 4.9 is visualized in Figure 9. Unlike in the results that we have mentioned earlier, the graphons in the family $\mathcal{W}$ have a part that depends on a countable vector $z \in [0, 1]^{\mathbb{N}}$ and analytic tools are applied to understand and to restrict the behaviour of graphons in the family $\mathcal{W}$. We remark that Theorem 4.9 can be further generalized [46] in the way that all graphons in the family $\mathcal{W}$ have the same value of a given graphon parameter that behaves nicely on the space of graphons. An example of such a parameter may be the graphon entropy, i.e., informally speaking, there are problems in extremal graph theory with no single "typical" graphon at the exponential scale.

5 Sparse graph limits

In this section, we give a brief overview of the main notions of convergence for sparse graphs. We restrict our attention to graphs with bounded maximum degree though many of the presented concepts can be extended to more general settings. We will also be less technical than in the previous sections, primarily focusing on presenting the main ideas behind the relevant concepts.

As we said earlier, the theory of limits of sparse graphs is developed in a less satisfactory way than the theory of limits of dense graphs. While this can be caused by the lack of our understanding of the structure of sparse graphs, many believe that there is no perfect notion of convergence because of the nature of sparse graphs. Such a perfect notion of convergence should be able to distinguish graphs with different local and global structures, i.e., sequences of graphs such that their local or global properties differ substantially should not be convergent. The notion should also be robust enough that sublinear modifications of graphs in the sequence do not affect the convergence, i.e., a convergent sequence should stay convergent if a sublinear number of edges is added or removed. Finally, the notion should ideally allow representing convergent sequences of sparse graphs with an analytic object that captures the interplay between local and global properties, similarly to the way that graphons in the dense setting capture the interplay between subgraph densities (a local property) and regularity partitions (a global property). In what follows, we present several notions of convergence for sparse graphs that have been studied and we will discuss their mutual relation and demonstrate their power on examples of particular sequences of graphs that do or do not converge with respect to these notions.

   The most widely used notion of convergence in relation to graphs with bounded degrees is the one defined by Benjamini and Schramm [9], known as *Benjamini-Schramm convergence*, shortly BS-convergence, and also as *left convergence*. Suppose that $(G_n)_{n \in \mathbb{N}}$ is a sequence of graphs with maximum degree at most Δ. For every $d \in \mathbb{N}$, let $\mathcal{G}^v(d, \Delta)$ be the set of all rooted graphs with maximum degree Δ where all vertices have distance at most d from the root. Note that the set $\mathcal{G}^v(d, \Delta)$ is finite for every pair d and Δ. By choosing a root in G_n randomly and restricting the graph G_n to the d-neighbourhood of the root, i.e., the vertices at distance at most d from the root, we get a (finite) probability distribution on graphs from $\mathcal{G}^v(d, \Delta)$. Let $p_{n,d} \in [0,1]^{\mathcal{G}^v(d,\Delta)}$ be the corresponding vector of probabilities. We say that the sequence $(G_n)_{n \in \mathbb{N}}$ is *BS-convergent* if the sequence $(p_{n,d})_{n \in \mathbb{N}}$ converges for every d. Benjamini-Schramm convergent sequences of graphs can be associated with an analytic representation called a graphing [31], however, we omit further details concerning this representation here and explore the view on limits of BS-convergent sequences in terms of distribution on rooted neighbourhoods of vertices.

   Every BS-convergent sequence yields a probability measure on the space $\mathcal{G}^v(\Delta)$ of (not necessarily finite) rooted graphs with maximum degree Δ. The topology on $\mathcal{G}^v(\Delta)$ is generated by clopen sets of rooted graphs with the same d-neighbourhood of the root for some d, and the limit probabilities from the definition of Benjamini-Schramm convergence give a probability measure on the corresponding σ-algebra on $\mathcal{G}^v(\Delta)$ by Carathéodory's Extension Theorem. In what follows, we will just write $\mathcal{G}^v$ instead of $\mathcal{G}^v(\Delta)$ when Δ is clear from the context.

   It is not true that every probability measure μ on $\mathcal{G}^v$ corresponds to a BS-convergent sequence of graphs. Let us fix $\Delta = 3$, i.e., we restrict our attention to graphs with maximum degree three in the following exposition. Let T be the infinite rooted tree where the vertices at even levels (including the root) have degree three and the vertices at odd levels have degree two. If $\mu(\{T\}) = 1$, then there is no BS-convergent sequence of graphs corresponding to μ. Indeed, graphs in such a sequence would have almost all vertices of degree three but almost every vertex of degree three would have neighbours of degree two only—this is clearly impossible.

   We now describe a condition on a probability measure μ that is necessary in order that μ corresponds to a BS-convergent sequence of graphs. We start with defining a different probability measure μ' on $\mathcal{G}^v$ as

$$\mu'(S) = \frac{\int\limits_{S} \delta(G) \mathrm{d}G}{\int\limits_{\mathcal{G}^v} \delta(G) \mathrm{d}G},$$

where the integration is with respect to the measure μ and $\delta(G)$ for $G \in \mathcal{G}^v$ is the degree of the root of G (we may assume that $\delta(G) > 0$ with non-zero probability, i.e., μ' is well-defined, since otherwise μ clearly corresponds to a BS-convergent sequence of graphs).

   We next define a probability measure μ_e on rooted graphs $\mathcal{G}^e$ with one distinguished edge at the root. Choose a rooted graph $G \in \mathcal{G}^v$ according to μ' and make randomly one of the edges incident with the root distinguished. This defines the probability measure μ_e on rooted graphs $\mathcal{G}^e$. Another probability measure μ'_e on $\mathcal{G}^e$ can be obtained from μ_e by choosing a random graph $G \in \mathcal{G}^e$ according to μ_e and making the other end of the distinguished edge to be the root. If μ corresponds to a BS-convergent sequence of graphs, then the probability measures μ_e and μ'_e are the same. The conjecture that is known as the conjecture of Aldous and Lyons [1] asserts that this necessary condition is also sufficient for a probability measure μ on $\mathcal{G}^v$ to correspond to a BS-convergent sequence of graphs. We

Figure 10: The construction of an infinite graph presented in relation to Benjamini-Schramm convergence: a part the original directed infinite tree and the corresponding part of the obtained undirected graph.

remark that this conjecture is closely related to a question of Gromov whether all countable discrete groups are sofic, see [67, Chapter 19].

It is tempting to think that the following condition, which is weaker than the one presented in the previous paragraph, can also be sufficient for a probability measure μ to correspond to a BS-convergent sequence of graphs. Let μ_v be the probability distribution on $\mathcal{G}^v$ obtained as follows: sample a rooted graph with a distinguished edge based on μ_e and keep the root, i.e., forget that any edge of the sampled graph is distinguished. Note that μ_v differs from μ if $\mu(\{T\}) > 0$ where T is the single vertex graph with its only vertex being the root, and μ_v and μ are the same if $\mu(\{T\}) = 0$. We define μ'_v based on μ'_e in the analogous way. Informally speaking, μ'_v is the distribution obtained from μ_v by rerooting to a random neighbour of the root (in an appropriately weighted way). Clearly, if μ corresponds to a BS-convergent sequence of graphs, then the probability measures μ_v and μ'_v are the same. It may be tempting to think that if μ_v and μ'_v are the same for a measure μ on $\mathcal{G}^v$, then μ corresponds to a BS-convergent sequence of graphs. However, this is not true as we explain in the next paragraph.

Consider an infinite tree T_0 where every vertex has degree three and each edge is directed in such a way that each vertex has out-degree exactly one (note that this determines the tree T_0 completely) and let T be the infinite (undirected) graph obtained from T_0 by joining two vertices by an edge if they are joined by a directed path of length one or two; see Figure 10 for an illustration of the construction. Since T_0 is vertex-transitive, T is also vertex-transitive. In particular, if $\mu(\{T\}) = 1$, then the corresponding measures μ_v and μ'_v are the same. However, there is no sequence $(G_n)_{n\in\mathbb{N}}$ of graphs such that μ is the resulting measure on $\mathcal{G}^v$. To see this, we proceed as follows. Assume that a sequence $(G_n)_{n\in\mathbb{N}}$ of graphs with maximum degree eight is BS-convergent and μ is the resulting measure on $\mathcal{G}^v$. A vertex of G_n is *typical* if its 2-neighbourhood is the same as the 2-neighbourhood of vertices in T. For $\varepsilon > 0$, consider $n \in \mathbb{N}$ such that the 2-neighbourhood of all but $\varepsilon|G_n|$ vertices of G_n are typical. We consider each typical vertex v of G_n and orient some of the edges incident with v as follows. The vertex v is incident with exactly three edges e_1, e_2 and e_3 contained in three triangles and all but a single pair of these three edges are contained in a common triangle, i.e., we can assume by symmetry that e_1 and e_2 are contained in a common triangle and e_1 and e_3 are contained in a common triangle. We now orient the edge

e_1 from the vertex v and the edges e_2 and e_3 towards v. Because the 2-neighbourhood of v is the same as the 2-neighbourhood of the vertices in T, no edge is oriented in two different ways. Observe that the sum of in-degrees of the vertices of G_n is at least $2(1-\varepsilon)|G_n|$ (each typical vertex has two incoming edges) but the sum of out-degrees is at most $(1+8\varepsilon)|G_n|$ (each typical vertex has a single outgoing edge and each vertex that is not typical can have at most eight such edges). However, this is impossible if $\varepsilon < 1/10$. We conclude that there is no BS-convergent sequence of graphs such that μ is the resulting measure on $\mathcal{G}^v$.

Benjamini-Schramm convergence has the drawback that we next describe. Let us consider a setting of graphs with maximum degree three.

Example 5.1. Let $(G_n)_{n\in\mathbb{N}}$ be a sequence of graphs such that G_n is a random $(2n)$-vertex cubic graph when n is odd, and G_n is a random $(2n)$-vertex cubic bipartite graph when n is even.

We claim that the sequence from Example 5.1 is BS-convergent with probability one. Indeed, the probability that a randomly chosen vertex of a random cubic graph is contained in a cycle of length k tends to 0 for any fixed integer k. The same is true for random cubic bipartite graphs. Hence, the sequence from Example 5.1 is BS-convergent and the corresponding probability measure μ on $\mathcal{G}^v$ satisfies that $\mu(\{T\}) = 1$ for the infinite rooted cubic tree T. However, the independence number of a random n-vertex cubic graph is at most $0.455n$ with probability tending to one [74], i.e., it is bounded away from $n/2$. In other words, Example 5.1 shows that BS-convergence is not robust enough to distinguish bipartite graphs from graphs that are far from being bipartite. We consider one more example.

Example 5.2. Let $(G_n)_{n\in\mathbb{N}}$ be a sequence of graphs such that G_n is H_n when n is odd, and G_n is the union of two copies of H_n when n is even, where $(H_n)_{n\in\mathbb{N}}$ is a BS-convergent sequence of cubic expanders (to obtain $(H_n)_{n\in\mathbb{N}}$, consider a sequence of cubic expanders and one of its convergent subsequences).

Since $(H_n)_{n\in\mathbb{N}}$ is BS-convergent, the sequence $(G_n)_{n\in\mathbb{N}}$ is also BS-convergent. This example shows that BS-convergence is not robust to distinguish well-connected graphs, which appear in the sequence $(G_n)_{n\in\mathbb{N}}$ on even positions, from disconnected graphs, which appear in the sequence on odd positions.

To overcome the phenomenon demonstrated by Examples 5.1 and 5.2, a finer notion of convergence called local-global convergence was proposed in [11] and further studied in [50]. This notion of convergence takes into account possible partitions of vertex sets of graphs in a sequence. Formally, let $\mathcal{G}^v(d, k, \Delta)$ be the set of all rooted k-vertex-coloured graphs with maximum degree Δ (the vertex colouring need not be proper) such that every vertex is at distance at most d from the root. For a graph G with maximum degree Δ, let $P_{d,k}(G)$ be the set of all vectors from $[0,1]^{\mathcal{G}^v(d,k,\Delta)}$ that corresponds to the probability distribution on d-neighbourhoods for all k-vertex-colourings of G. A sequence $(G_n)_{n\in\mathbb{N}}$ of graphs with maximum degree Δ is *local-global convergent* if the sets $(P_{d,k}(G_n))_{n\in\mathbb{N}}$ converge in the Hausdorff metric for every $d \in \mathbb{N}$ and $k \in \mathbb{N}$, i.e., for every $\varepsilon > 0$, there exists n_0 such that the Hausdorff distance of $P_{d,k}(G_i)$ and $P_{d,k}(G_j)$ is at most ε for every $i, j \geq n_0$. Recall, that the *Hausdorff distance* of two subsets A and B of $\mathbb{R}^D$ is

$$\max\{\sup_{x\in A}\inf_{y\in B} d(x,y), \sup_{x\in B}\inf_{y\in A} d(x,y)\},$$

where $d(x,y)$ is the distance between points x and y (in this definition, it does not matter which of the standard metrics on $\mathbb{R}^D$ we use, so, we can use the L_1-metric for example).

Informally speaking, the definition says that the sequence $(G_n)_{n \in \mathbb{N}}$ is local-global convergent if and only if for every k-colouring of G_i, there exists a k-colouring of G_j with a close statistic of d-neighbourhoods assuming that both i and j are sufficiently large.

Observe that if a sequence of graphs is local-global convergent, it is also BS-convergent (set $k = 1$ in the definition). However, the converse is not necessarily true: neither of the sequences given in Examples 5.1 and 5.2 is local-global convergent. In Example 5.1, an $(2n)$-vertex random cubic bipartite graphs has a vertex-colouring with two colours, say red and blue, such that the number of red vertices is n and there are no red-red edges. However, a 2-vertex-colouring with a neighbourhood statistic close to this 2-vertex-colouring does not exist for $(2n)$-vertex random cubic graphs with high probability since the size of their largest independent set is at most $0.91n$ with high probability as we have mentioned earlier. In Example 5.2, the union of two n-vertex (cubic) expanders has a 2-vertex-colouring such that each colour is used on half of the vertices and all edges are monochromatic but no n-vertex cubic expander has a 2-vertex-colouring with a neighbourhood statistic close to this 2-vertex-colouring.

Another notion of convergence related to BS-convergence is that of right convergence. Let H be a complete graph with a loop at each vertex such that all its vertices and edges are assigned positive weights. We refer to such a graph H as to a *target*. We remark that such graphs are also often called soft cores, while graphs, where non-negative weights are allowed are called hard cores. The definition that we use here is weaker than the original definition, which was using hard cores instead of soft cores, however, every sequence of graphs that is convergent in the definition that we use can be modified by changing a sublinear number of edges to a sequence of graphs convergent in the original (stronger) definition, see [12] for further details.

The number of weighted homomorphisms from a graph G to H, denoted by $\hom(G, H)$, is

$$\sum_{f:V(G) \to V(H)} \prod_{v \in V(G)} w(f(v)) \prod_{vv' \in E(G)} w(f(v)f(v')),$$

where w is the weight function of H. For a homomorphism f, the corresponding summand in the expression above is referred as the *weight* of the homomorphism f. A sequence $(G_n)_{n \in \mathbb{N}}$ of graphs is *right convergent* if the fraction

$$\frac{\log \hom(G_n, H)}{|G_n|}$$

convergences for every target H. It can be shown [13], also see [72], that if a sequence $(G_n)_{n \in \mathbb{N}}$ of graphs with bounded maximum degree is right convergent, then it is also BS-convergent. However, the converse is not true since the sequence given in Example 5.1 is not right convergent, i.e., informally speaking, right convergence can distinguish graphs close to being bipartite and those far from being bipartite. To see that the sequence given in Example 5.1 is not right convergent, consider the target graph H_K, $K \in \mathbb{N}$, with two vertices v and w such that the weight of the vertex v is K, the weight of the loop at v is $1/K$, the weights of w, the loop at w and the edge vw are one. Observe that G_n has a homomorphism to H of weight K^m if and only if the independence number of G_n is m. This can be used to show that

$$\lim_{K \to \infty} \lim_{n \to \infty} \frac{\log \hom(G_n, H)}{|G_n|} = \lim_{n \to \infty} \frac{\alpha(G_n)}{|G_n|}.$$

Hence, the sequence given in Example 5.1 is not right convergent.

We next consider the following modification of Example 5.2.

Example 5.3. Let $(G_n)_{n\in\mathbb{N}}$ be a sequence of graphs such that G_n is H_n when n is odd, and G_n is the union of two copies of H_n when n is even, where $(H_n)_{n\in\mathbb{N}}$ is a right convergent sequence of cubic expanders (to obtain $(H_n)_{n\in\mathbb{N}}$, consider a sequence of cubic expanders and one of its convergent subsequences).

Observe that it holds that

$$\frac{\log \hom(G, H)}{|G|} = \frac{\log \hom(G \cup G, H)}{|G \cup G|}$$

for every graph G and every target H, where $G \cup G$ stands for the union of two disjoint copies of G. Consequently, Example 5.3 shows that right convergence does not imply local-global convergence.

Another notion of convergence of sparse graphs, which is entirely based on possible vertex partitions, was proposed by Bollobás and Riordan in [11]. A k-*partition* of a graph G is a partition of its vertex set into k subsets. The *statistic* of a k-partition $\mathbf{P} = (P_1, \ldots, P_k)$ is a vector $s(\mathbf{P}) \in \mathbb{R}^{k+\binom{k+1}{2}}$ whose first k coordinates are the relative sizes $p_i = \frac{|P_i|}{|G|}$ of the parts and the remaining $\binom{k+1}{2}$ coordinates are the edge densities $e_{ij} = \frac{e(P_i, P_j)}{|G|}$ between the parts (including the cases when $i = j$), where $e(P_i, P_j)$ stands for the number of edges between parts P_i and P_j. Note that the normalization here is different than the one used in Section 3 when dealing with dense graphs. Let $P_k(G) \subseteq \mathbb{R}^{k+\binom{k+1}{2}}$ be the set of statistics $s(\mathbf{P})$ of all k-partitions $\mathbf{P}$ of a graph G. A sequence $(G_n)_{n\in\mathbb{N}}$ of graphs with bounded maximum degree is *partition convergent* if the sequence $(P_k(G_n))_{n\in\mathbb{N}}$ converges in the Hausdorff metric for every $k \in \mathbb{N}$. Observe that local-global convergence of a sequence of graphs trivially implies partition convergence but Example 5.2 and its modification considered in the previous paragraph yield that neither BS-convergence nor right convergence implies partition convergence.

We will now show that there exists a sequence $(G_n)_{n\in\mathbb{N}}$ of graphs that is partition convergent but is not BS-convergent (and so is neither right convergent nor local-global convergent). Consider the following sequence of 2-regular graphs.

Example 5.4. Let $(G_n)_{n\in\mathbb{N}}$ be a sequence of graphs such that G_n is the union of n cycles of length four, i.e., the graph $n\, C_4$, when n is odd, and it is the union of n cycles of length six, i.e., the graph $n\, C_6$, when n is even.

The sequence from Example 5.4 is clearly not BS-convergent, however, the sequence $(P_k(G_n))_{n\in\mathbb{N}}$ converges in the the Hausdorff metric for every $k \in \mathbb{N}$, i.e., the sequence from Example 5.4 is partition convergent. We sketch the argument for $k = 2$. Let $U_2 \subseteq \mathbb{R}^5$ be the set of all non-negative real vectors $(p_1, p_2, e_{11}, e_{12}, e_{22})$ such that $p_1 + p_2 = 1$, $p_1 = e_{11} + e_{12}/2$ and $p_2 = e_{22} + e_{12}/2$. Observe that $P_2(G) \subseteq U_2$ for every 2-regular graph G. It can be shown that both the sequence $(P_2(n\, C_4))_{n\in\mathbb{N}}$ and the sequence $(P_2(n\, C_6))_{n\in\mathbb{N}}$ converge to U_2 in the Hausdorff metric, i.e., the statistics of the partitions into two parts of the vertices of the graphs in these sequences converge to the set of all possible statistics of the partitions into two parts of the vertices of 2-regular graphs.

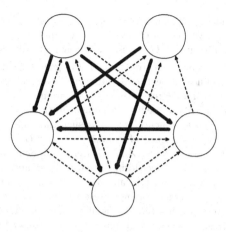

Figure 11: The relation between the presented notions of convergence of bounded degree graphs. The bold arrows represent that the notion at the tail of an arrow implies the other and the dashed arrows that this is not the case in general. When an arrow is missing, the relation between the notions is not known.

We refer the reader to Figure 11 for the relation between the notions of convergence of sparse graphs that we have already discussed and the notion of large deviation convergence that we introduce next. The recent notion of large deviation convergence, which was introduced in [12], is a common refinement of right convergence and partition convergence. A sequence $(G_n)_{n\in\mathbb{N}}$ of graphs with bounded maximum degree is *LD-convergent* if the following limit exists (while possibly being infinite)

$$I_k(x) = \lim_{\varepsilon\to 0}\lim_{n\to\infty} -\frac{\log\frac{|\{\mathbf{P}\text{ such that }\|s(\mathbf{P})-x\|_1\leq\varepsilon\}|}{k^{|G_n|}}}{|G_n|}$$

for every k and $x \in \mathbb{R}^{k+\binom{k+1}{2}}$, where $s(\mathbf{P})$ is the statistic of a k-partition $\mathbf{P}$ as defined earlier. Note that $I_k(x) \in [0, \log k] \cup \{\infty\}$. On the intuitive level, one can think that the number of k-partitions of G_n, if n is large, with statistic close to x is approximately $k^{|G_n|} \cdot e^{-I_k(x)|G_n|}$. If a sequence $(G_n)_{n\in\mathbb{N}}$ is LD-convergent, then it is also partition convergent. In fact, the sequence $(P_k(G_n))_{n\in\mathbb{N}}$ converges to the set $\{x \mid I_k(x) < \infty\}$ in the Hausdorff metric. A more involved argument shows that every LD-convergent sequence of graphs is right convergent [12], which implies that it is also BS-convergent. We would like to emphasize here that it is important here that we consider targets with positive weights only (soft cores). If the definition of right convergence uses targets with non-negative weights (hard cores), when LD-convergence does not imply right convergence. An example showing this is the sequence $(C_n)_{n\in\mathbb{N}}$ of cycles with alternating parities that can be shown to be LD-convergent but it is not right convergent when targets are allowed to have elements with zero weight (a cycle can be homomorphically mapped to K_2 if and only if its length is even).

The final notion of convergence of graphs that we would like to mention is the notion of first order convergence introduced in [76, 77, 80] and further studied in [18, 36, 57, 78, 79]. This notion is an attempt to provide a universal notion of graph convergence that can be applied both in the sparse and in the dense settings. If ψ is a first order formula with k free variables and G is a (finite) graph, then the *Stone pairing* $\langle \psi, G \rangle$ is the probability that a uniformly chosen k-tuple of vertices of G satisfies ψ. A sequence $(G_n)_{n \in \mathbb{N}}$ of graphs is *first order convergent* if the limit $\lim_{n \to \infty} \langle \psi, G_n \rangle$ exists for every first order formula ψ.

It is not hard to show that every first order convergent sequence of dense graphs is convergent in the sense defined in Section 3 and every first order convergent sequence of graphs with bounded maximum degree is Benjamini-Schramm convergent. Neither of the opposite implications is true. We present an argument in the case of Benjamini-Schramm convergence. Let $(G_n)_{n \in \mathbb{N}}$ be a sequence of graphs such that G_n is the union of n copies of K_2, and let $(G'_n)_{n \in \mathbb{N}}$ be a sequence of graphs such that $G'_n = G_n$ if n is even and $G'_n = G_n \cup K_1$ if n is odd. The sequence $(G'_n)_{n \in \mathbb{N}}$ is Benjamini-Schramm convergent but not first order convergent: if ψ is a first order formula that is true if and only if a graph contains an isolated vertex, then the values $\langle \psi, G'_n \rangle$ alternate between zero and one. This example also shows that first order convergence is not preserved by constant size modifications of graphs in the sequence: the sequence $(G_n)_{n \in \mathbb{N}}$ is first order convergent unlike $(G'_n)_{n \in \mathbb{N}}$.

Some first order convergent sequence graphs can be represented by an analytic object called a *modeling* but not every first order convergent sequence of graphs has such a representation [76, 77]; an interesting example of a sequence of a first order convergent sequence of graphs with no modeling is the sequence of Erdős-Rényi random graphs $G_{n,1/2}$ that has no modeling with probability one. In general, a sequence of dense graphs converging to a graphon W has a modeling if and only if the graphon W is random-free [76, 77], i.e., $W(x, y) \in \{0, 1\}$ for almost every $(x, y) \in [0, 1]^2$. A nice conjecture of Nešetřil and Ossona de Mendez [76, 77] asserted the following: if $\mathcal{G}$ is a nowhere-dense class of graphs (see [75] for the definition and further exposition), then any first order convergent sequence of graphs from $\mathcal{G}$ can be represented by a modeling. Another conjecture of Nešetřil and Ossona de Mendez [79] asserted that every residual first order convergent sequence of graphs has a limit modeling; a sequence $(G_n)_{n \in \mathbb{N}}$ of graphs is *residual* if for every $d \in \mathbb{N}$ and $\varepsilon > 0$, there exists n_0 such that the number of vertices at distance at most d from any vertex in G_n, $n \geq n_0$, is at most $\varepsilon |G_n|$. Both conjectures were proven in [78], however, their stronger forms asserting that every first order convergent sequence of graphs from a nowhere-dense class of graphs $\mathcal{G}$ has a limit modeling obeying a property called the finitary mass transport principle (see [77] for the definition of this property) and that every residual first order convergent sequence of graphs has a limit modeling obeying the finitary mass transport principle remain open and present very interesting problems.

Acknowledgement

The authors would like to thank the anonymous reviewer for the many detailed comments, which helped to improve the survey significantly.

References

[1] David Aldous and Russell Lyons, *Processes on unimodular random networks*, Electronic Journal of Probability **12** (2007), 1454–1508.

[2] David J. Aldous, *Representations for partially exchangeable arrays of random variables*, Journal of Multivariate Analysis **11** (1981), 581–598.

[3] Noga Alon and Joel H. Spencer, *The probabilistic method*, John Wiley & Sons, New York, 2004.

[4] Tim Austin, *On exchangeable random variables and the statistics of large graphs and hypergraphs*, Probability Surveys **5** (2008), 80–145.

[5] Rahil Baber, *Turán densities of hypercubes*, preprint, arXiv:1201.3587 (2012).

[6] Rahil Baber and John Talbot, *Hypergraphs do jump*, Combinatorics, Probability and Computing **20** (2011), 161–171.

[7] Rahil Baber and John Talbot, *A solution to the 2/3 conjecture*, SIAM Journal on Discrete Mathematics **28** (2014), 756–766.

[8] József Balogh, Ping Hu, Bernard Lidický, and Hong Liu, *Upper bounds on the size of 4-and 6-cycle-free subgraphs of the hypercube*, European Journal of Combinatorics **35** (2014), 75–85.

[9] Itai Benjamini and Oded Schramm, *Recurrence of distributional limits of finite planar graphs*, Electronic Journal of Probability **6** (2001), paper 23.

[10] G.R. Blakley and Prabir Roy, *A Hölder type inequality for symmetric matrices with nonnegative entries*, Proceedings of the American Mathematical Society **16** (1965), 1244–1245.

[11] Béla Bollobás and Oliver Riordan, *Sparse graphs: metrics and random models*, Random Structures & Algorithms **39** (2011), 1–38.

[12] Christian Borgs, Jennifer Chayes, and David Gamarnik, *Convergent sequences of sparse graphs: A large deviations approach*, Random Structures & Algorithms **51** (2017), 52–89.

[13] Christian Borgs, Jennifer Chayes, Jeff Kahn, and László Lovász, *Left and right convergence of graphs with bounded degree*, Random Structures & Algorithms **42** (2013), 1–28.

[14] Christian Borgs, Jennifer Chayes, and László Lovász, *Moments of two-variable functions and the uniqueness of graph limits*, Geometric and Functional Analysis **19** (2010), 1597–1619.

[15] Christian Borgs, Jennifer Chayes, László Lovász, Vera T Sós, Balázs Szegedy, and Katalin Vesztergombi, *Graph limits and parameter testing*, Proceedings of the 38th Annual ACM Symposium on Theory of Computing (STOC'06), ACM, 2006, pp. 261–270.

[16] Christian Borgs, Jennifer T Chayes, László Lovász, Vera T Sós, and Katalin Vesztergombi, *Convergent sequences of dense graphs I: Subgraph frequencies, metric properties and testing*, Advances in Mathematics **219** (2008), 1801–1851.

[17] Christian Borgs, Jennifer T Chayes, László Lovász, Vera T Sós, and Katalin Vesztergombi, *Convergent sequences of dense graphs II. Multiway cuts and statistical physics*, Annals of Mathematics (2012), 151–219.

[18] Demetres Christofides and Daniel Král', *First-order convergence and roots*, Combinatorics, Probability and Computing **25** (2016), 213–221.

[19] F.R.K. Chung, R.L. Graham, and R.M. Wilson, *Quasi-random graphs*, Combinatorica **9** (1989), 345–362.

[20] David Conlon and Jacob Fox, *Bounds for graph regularity and removal lemmas*, Geometric and Functional Analysis **22** (2012), 1191–1256.

[21] David Conlon, Jacob Fox, and Benny Sudakov, *An approximate version of Sidorenko's conjecture*, Geometric and Functional Analysis **20** (2010), 1354–1366.

[22] David Conlon, Jeong Han Kim, Choongbum. Lee, and Joonkyung Lee, *Some advances on Sidorenko's conjecture*, Journal of the London Mathematical Society **98** (2018), 593–608.

[23] David Conlon and Joonkyung Lee, *Finite reflection groups and graph norms*, Advances in Mathematics **315** (2017), 130–165.

[24] David Conlon and Joonkyung Lee, *Sidorenko's conjecture for blow-ups*, preprint, arXiv:1809.01259 (2018).

[25] Jacob W. Cooper, Tomáš Kaiser, Daniel Král', and Jonathan A. Noel, *Weak regularity and finitely forcible graph limits*, Transactions of the American Mathematical Society **370** (2018), 3833–3864.

[26] Jacob W. Cooper, Daniel Král', and Taísa Martins, *Finitely forcible graph limits are universal*, Advances in Mathematics **340** (2018), 819–854.

[27] Persi Diaconis and Svante Janson, *Graph limits and exchangeable random graphs*, Rendiconti di Matematica e delle sue Applicazioni **28** (2008), 33–61.

[28] Martin Doležal, Jan Grebík, Jan Hladký, Israel Rocha, and Václav Rozhoň, *Cut distance identifying graphon parameters over weak* limits*, preprint, arXiv:1809.03797 (2018).

[29] Martin Doležal, Jan Grebík, and Jan Hladký, Israel Rocha, and Václav Rozhoň, *Relating the cut distance and the weak* topology for graphons*, preprint, arXiv:1806.07368 (2018).

[30] Martin Doležal and Jan Hladký, *Cut-norm and entropy minimization over weak* limits*, preprint, arXiv:1701.09160 (2017).

[31] Gábor Elek, *Note on limits of finite graphs*, Combinatorica **27** (2007), 503–507.

[32] Gábor Elek and Balazs Szegedy, *Limits of hypergraphs, removal and regularity lemmas. A non-standard approach*, preprint, arXiv:0705.2179 (2007).

[33] P. Erdős and A.H. Stone, *On the structure of linear graphs*, Bulletin of the American Mathematical Society **52** (1946), 1087–1091.

[34] Jacob Fox and Fan Wei, *On the local approach to Sidorenko's conjecture*, Electronic Notes in Discrete Mathematics **61** (2017), 459–465.

[35] Alan Frieze and Ravi Kannan, *Quick approximation to matrices and applications*, Combinatorica **19** (1999), 175–200.

[36] Jakub Gajarský, Petr Hliněný, Tomáš Kaiser, Daniel Král', Martin Kupec, Jan Obdržálek, Sebastian Ordyniak, and Vojtěch Tůma, *First order limits of sparse graphs: Plane trees and path-width*, Random Structures & Algorithms **50** (2017), 612–635.

[37] Roman Glebov, Andrzej Grzesik, Tereza Klimošová, and Daniel Král', *Finitely forcible graphons and permutons*, Journal of Combinatorial Theory, Series B **110** (2015), 112–135.

[38] Roman Glebov, Tereza Klimošová, and Daniel Král', *Infinite dimensional finitely forcible graphon*, Proceedings of the London Mathematical Society, to appear.

[39] Roman Glebov, D. Král', and Jan Volec, *Compactness and finite forcibility of graphons*, preprint, arXiv:1309.6695 (2013).

[40] Oded Goldreich, *Property testing: current research and surveys*, Lecture Notes in Computer Science 6390, Springer, 2010.

[41] Oded Goldreich, Shari Goldwasser, and Dana Ron, *Property testing and its connection to learning and approximation*, Proceedings of the 37th Annual Symposium on Foundations of Computer Science (FOCS'96), IEEE, 1996, pp. 339–348.

[42] Oded Goldreich, Shari Goldwasser, and Dana Ron, *Property testing and its connection to learning and approximation*, Journal of the ACM (JACM) **45** (1998), 653–750.

[43] Oded Goldreich and Luca Trevisan, *Three theorems regarding testing graph properties*, Random Structures & Algorithms **23** (2003), 23–57.

[44] A. W. Goodman, *On sets of acquaintances and strangers at any party*, American Mathematical Monthly **66** (1959), 778–783.

[45] Andrzej Grzesik, *On the maximum number of five-cycles in a triangle-free graph*, Journal of Combinatorial Theory, Series B **102** (2012), 1061–1066.

[46] Andrzej Grzesik, Daniel Král', and László Miklós Lovász, *Elusive extremal graphs*, preprint, arXiv:1807.01141 (2018).

[47] Hamed Hatami, *Graph norms and Sidorenko's conjecture*, Israel Journal of Mathematics **175** (2010), 125–150.

[48] Hamed Hatami, Jan Hladký, Daniel Král', Serguei Norine, and Alexander Razborov, *Non-three-colourable common graphs exist*, Combinatorics, Probability and Computing **21** (2012), 734–742.

[49] Hamed Hatami, Jan Hladký, Daniel Král', Serguei Norine, and Alexander Razborov, *On the number of pentagons in triangle-free graphs*, Journal of Combinatorial Theory, Series A **120** (2013), 722–732.

[50] Hamed Hatami, László Lovász, and Balázs Szegedy, *Limits of locally–globally convergent graph sequences*, Geometric and Functional Analysis **24** (2014), 269–296.

[51] Jan Hladký, András Máthé, Viresh Patel, and Oleg Pikhurko, *Poset limits can be totally ordered*, Transactions of the American Mathematical Society **367** (2015), 4319–4337.

[52] Douglas N. Hoover, *Relations on probability spaces and arrays of random variables*, Institute for Advanced Study, Princeton, 1979..

[53] Carlos Hoppen, Yoshiharu Kohayakawa, Carlos Gustavo T. de A. Moreira, Balázs Ráth, and Rudini Menezes Sampaio, *Limits of permutation sequences*, Journal of Combinatorial Theory, Series B **103** (2013), 93–113.

[54] Carlos Hoppen, Yoshiharu Kohayakawa, Carlos Gustavo T. de A. Moreira, and Rudini Menezes Sampaio, *Testing permutation properties through subpermutations*, Theoretical Computer Science **412** (2011), 3555–3567.

[55] Svante Janson, *Poset limits and exchangeable random posets*, Combinatorica **31** (2011), 529–563.

[56] Olav Kallenberg, *Symmetries on random arrays and set-indexed processes*, Journal of Theoretical Probability **5** (1992), 727–765.

[57] František Kardoš, Daniel Král', Anita Liebenau, and Lukáš Mach, *First order convergence of matroids*, European Journal of Combinatorics **59** (2017), 150–168.

[58] Jeong Han Kim, Choongbum Lee, and Joonkyung Lee, *Two approaches to Sidorenko's conjecture*, Transactions of the American Mathematical Society **368** (2016), 5057–5074.

[59] Daniel Král', Chun-Hung Liu, Jean-Sébastien Sereni, Peter Whalen, and Zelealem B. Yilma, *A new bound for the 2/3 conjecture*, Combinatorics, Probability and Computing **22** (2013), 384–393.

[60] Daniel Král', László Miklós Lovász, Jonathan A. Noel, and Jakub Sosnovec, *Finitely forcible graphons with an almost arbitrary structure*, preprint, arXiv:1809.05973 (2018).

[61] Daniel Král', Lukáš Mach, and Jean-Sébastien Sereni, *A new lower bound based on Gromov's method of selecting heavily covered points*, Discrete & Compututational Geometry **48** (2012), 487–498.

[62] Daniel Král', Taísa Martins, Pétér Pál Pach, and Marcin Wrochna, *The step Sidorenko property and non-norming edge-transitive graphs*, Journal of Combinatorial Theory, Series A **162** (2019), 34–54.

[63] Daniel Král' and Oleg Pikhurko, *Quasirandom permutations are characterized by 4-point densities*, Geometric and Functional Analysis **23** (2013), 570–579.

[64] J.L. Xiang Li and Balazs Szegedy, *On the logarithimic calculus and Sidorenko's conjecture*, Combinatorica, to appear.

[65] László Lovász, *Graph homomorphisms: Open problems*, manuscript (2008).[2]

[66] László Lovász, *Very large graphs*, Current Developments in Mathematics **2008** (2009), 67–128.

[67] László Lovász, *Large networks and graph limits*, Colloquium Publications, Volume 60, American Mathematical Society, 2012.

---

[2]See http://www.cs.elte.hu/~lovasz/problems.pdf

[68] László Lovász and Vera Sós, *Generalized quasirandom graphs*, Journal of Combinatorial Theory, Series B **98** (2008), 146–163.

[69] László Lovász and Balázs Szegedy, *Limits of dense graph sequences*, Journal of Combinatorial Theory, Series B **96** (2006), 933–957.

[70] László Lovász and Balázs Szegedy, *Testing properties of graphs and functions*, Israel Journal of Mathematics **178** (2010), 113–156.

[71] László Lovász and Balázs Szegedy, *Finitely forcible graphons*, Journal of Combinatorial Theory, Series B **101** (2011), 269–301.

[72] László Miklós Lovász, *A short proof of the equivalence of left and right convergence for sparse graphs*, European Journal of Combinatorics **53** (2016), 1–7.

[73] W. Mantel, *Problem 28*, Wiskundige Opgaven **10** (1907), 60–61.

[74] D.B. McKay, *Independent sets in regular graphs of high girth*, Ars Combininatoria **23A** (1987), 179–185.

[75] Jaroslav Nešetřil and Patrice Ossona de Mendez, *Sparsity: Graphs, structures, and algorithms*, Algorithms and Combinatorics, Volume 28, Springer, 2012.

[76] Jaroslav Nešetřil and Patrice Ossona de Mendez, *A unified approach to structural limits (with applications to the study of graphs limits with bounded tree-depth)*, Memoirs of the American Mathematical Society, to appear.

[77] Jaroslav Nešetřil and Patrice Ossona de Mendez, *A unified approach to structural limits, and limits of graphs with bounded tree-depth*, preprint, arXiv:1303.6471 (2013).

[78] Jaroslav Nešetřil and Patrice Ossona de Mendez, *Existence of modeling limits for sequences of sparse structures*, preprint, arXiv:1608.00146 (2016).

[79] Jaroslav Nešetřil and Patrice Ossona de Mendez, *Modeling limits in hereditary classes: Reduction and application to trees*, Electronic Journal of Combinatorics **23** (2016), 2–52.

[80] Jaroslav Nešetřil and Patrice Ossona de Mendez, *A model theory approach to structural limits*, Commentationes Mathematicae Universitatis Carolinae **53** (2012), 581–603.

[81] V.S. Nikiforov, *The number of cliques in graphs of given order and size*, Transactions of the American Mathematical Society **363** (2011), 1599–1618.

[82] Oleg Pikhurko and Alexander Razborov, *Asymptotic structure of graphs with the minimum number of triangles*, Combinatorics, Probability and Computing **26** (2017), 138–160.

[83] Oleg Pikhurko and Emil R. Vaughan, *Minimum number of k-cliques in graphs with bounded independence number*, Combinatorics, Probability and Computing **22** (2013), 910–934.

[84] Alexander Razborov, *Flag algebras*, Journal of Symbolic Logic **72** (2007), 1239–1282.

[85] Alexander Razborov, *On the minimal density of triangles in graphs*, Combinatorics, Probability and Computing **17** (2008), 603–618.

[86] Alexander Razborov, *On 3-hypergraphs with forbidden 4-vertex configurations*, SIAM Journal on Discrete Mathematics **24** (2010), 946–963.

[87] Christian Reiher, *The clique density theorem*, Annals of Mathematics (2016), 683–707.

[88] Ronitt Rubinfeld and Madhu Sudan, *Robust characterizations of polynomials with applications to program testing*, SIAM Journal on Computing **25** (1996), 252–271.

[89] Alexander Sidorenko, *A correlation inequality for bipartite graphs*, Graphs and Combinatorics **9** (1993), 201–204.

[90] Miklós Simonovits, *Extremal graph problems, degenerate extremal problems, and supersaturated graphs*, Progress in graph theory, Academic Press, Toronto, 1984, pp. 419–437.

[91] Joel Spencer, *Quasirandom multitype graphs*, An Irregular Mind, Bolyai Society Mathematical Studies, Volume 21, Springer, 2010, pp. 607–617.

[92] Balazs Szegedy, *An information theoretic approach to Sidorenko's conjecture*, preprint, arXiv:1406.6738 (2014).

[93] Balazs Szegedy, *On Sidorenko's conjecture for determinants and Gaussian Markov random fields*, preprint, arXiv:1701.03632 (2017).

[94] Andrew Thomason, *Pseudorandom graphs*, Random graphs '85, North-Holland Mathematics Studies, Volume 144, North-Holland, Amsterdam, 1987, pp. 307–331.

[95] Paul Turán, *On an extremal problem in graph theory*, Matematikai és Fizikai Lapok **48** (1941), 436–452.

[96] Yufei Zhao, *Hypergraph limits: A regularity approach*, Random Structures & Algorithms **47** (2015), 205–226.

Faculty of Mathematics and Computer Science,
Jagiellonian University,
Łojasiewicza 6, 30-348 Kraków,
Poland.
Andrzej.Grzesik@uj.edu.pl

Faculty of Informatics,
Masaryk University,
Botanická 68A, 602 00 Brno,
Czech Republic,
and
Mathematics Institute,
DIMAP and Department of Computer Science,
University of Warwick,
Coventry CV4 7AL, UK.
dkral@fi.muni.cz

Topological connectedness and independent sets in graphs

Penny Haxell

Abstract

An abstract simplicial complex $\mathcal{C}$ is said to be *k-connected* if for each $-1 \leq d \leq k$ and each continuous map f from the sphere S^d to $||\mathcal{C}||$ (the body of the geometric realization of $\mathcal{C}$), the map f can be extended to a continuous map from the ball B^{d+1} to $||\mathcal{C}||$. In 2000 a link was discovered between the topological connectedness of the independence complex of a graph and various other important graph parameters to do with colouring and partitioning. When the graph represents some other combinatorial structure, for example when it is the line graph of a hypergraph $\mathcal{H}$, this link can be exploited to obtain information such as lower bounds on the matching number of $\mathcal{H}$. Since its discovery there have been many other applications of this phenomenon to combinatorial problems. The aim of this article is to outline this general method and to describe some of its applications.

1 Introduction

For many combinatorial problems, one needs to understand the independent sets in a graph. This is a very difficult problem in general, indeed even basic parameters such as the size of a largest independent set are hard to determine or even approximate (unless P=NP). In this article we will describe a particular method that can be used in certain circumstances to obtain important information on the independent sets in a graph, and describe some of its applications.

The typical problem we will investigate concerns the following.

Definition 1.1. Let G be a graph, and let $\mathcal{P} = \{V_1, \ldots, V_m\}$ be a partition of the vertex set $V(G)$. An *independent transversal* (IT) of G with respect to $\mathcal{P}$ is an independent set $\{v_1, \ldots, v_m\}$ in G such that $v_i \in V_i$ for each i.

This is a very general notion, and many combinatorial problems can be formulated by asking whether a given graph with a given vertex partition has an IT. Indeed the SAT problem can be formulated in these terms (see e.g. [47]), and so we cannot expect to find an efficient characterisation of those G for which an IT exists. However, results giving sufficient conditions for the existence of an IT can still be very useful in many applications.

Here we will explore a perhaps surprising connection between the problem of finding an IT in a vertex-partitioned graph, and a parameter of graphs that has a topological definition, namely the *topological connectedness of the independence complex*. Our discussion here will be based on what is essentially the *homotopic* definition of connectedness given in the abstract. It is also possible to derive all this material in other ways, either by using a *homological* approach (as was done in [60,61]), or at a more basic level using only combinatorial language involving sets and parity [45]. All of these approaches are roughly equivalent, but the one we choose to use here is perhaps the most intuitive, as explained in Section 1.1.

We remark that the term "connectivity" instead of "connectedness" is frequently used in the literature. In this article we choose to use the latter (following [59]), since we will be discussing graphs, and "connectivity" has a different standard meaning in graph theory.

The main aims for this article are as follows. We start by giving some intuition about why topological considerations might be relevant to combinatorial problems such as proving the existence of an IT (Section 1.1). In Section 2 we outline why the specific notion

Figure 1: Two 2-simplices whose intersection fails to satisfy condition (2) in the definition of geometric simplicial complex.

of topological connectedness can be useful for this purpose. The main tools for estimating connectedness, and the principal graph parameters that provide lower bounds on this parameter, are described in Section 3. Section 4 is devoted to some of the many applications of this method, including a few of the original results that motivated its development, and some that are much more recent. Our exposition is by no means comprehensive, and there remain many applications that we will not even be able to touch upon in this article. We end with concluding remarks and open problems in Section 5.

1.1   A tale of two complexes

Unless stated otherwise, graphs in this article will be simple, i.e. without loops or multiple edges. The set $\mathcal{I}(G)$ of all independent sets in a graph G is called the *independence complex* of G. This is an example of an *abstract simplicial complex* on $V(G)$, i.e. a set $\mathcal{C}$ of subsets of $V(G)$ with the property that whenever $\sigma \in \mathcal{C}$ and $\tau \subset \sigma$ then $\tau \in \mathcal{C}$. (Soon we will drop the term "abstract" and just say *simplicial complex*.) The elements of an abstract simplicial complex $\mathcal{C}$ are called its *simplices*, and the *dimension* of a simplex σ is $|\sigma| - 1$. In general $\mathcal{I}(G)$ is a "nasty" simplicial complex: as noted in the introduction, even very basic questions about it are hard to answer, indeed even determining its *dimension* (the largest dimension of any simplex in $\mathcal{I}(G)$) is computationally infeasible (unless P=NP).

On the other hand, another important class of abstract simplicial complexes comes from the notion of a geometric simplicial complex, and indeed this is the source for terminology such as "simplex" and "dimension". A *simplex* is the convex hull σ of a set of affinely independent points in real space (called the *vertices* of σ). A *face* of σ is the convex hull of some subset of its vertices. A *geometric simplicial complex* $\mathcal{K}$ is a family of simplices in real space such that (1) if τ is a face of a simplex $\sigma \in \mathcal{K}$ then $\tau \in \mathcal{K}$ and (2) if $\sigma, \sigma' \in \mathcal{K}$ then $\sigma \cap \sigma'$ is a face of both σ and σ'. (See Figures 1 and 2.) The set of all vertex sets of simplices of a geometric simplicial complex $\mathcal{K}$ is called the *vertex scheme* of $\mathcal{K}$. Note then that this is an abstract simplicial complex on the *point set* $V(\mathcal{K})$ of $\mathcal{K}$ (the union of all vertex sets of its simplices). The union of all simplices in $\mathcal{K}$ is called the *body* of $\mathcal{K}$, denoted by $\|\mathcal{K}\|$. We will normally regard geometric simplicial complexes as abstract ones, via their vertex schemes. In particular any simplex with $n + 1$ vertices has dimension n. We often use the definite article and say "the n-simplex Δ_n" when referring to a generic n-dimensional (geometric) simplex.

A *triangulation* of Δ_n is a geometric simplicial complex $\mathcal{K}$ such that $\|\mathcal{K}\| = \Delta_n$. Informally we can think of a triangulation as a partition of Δ_n into smaller simplices that fit together in a nice way (see Figure 3). In contrast to the independence complex of a graph, a triangulation of the simplex Δ_n is a "nice" simplicial complex for our purposes, as we will see in a moment.

If we regard the partition $\mathcal{P}$ in Definition 1.1 as an assignment of a *colour i* to the vertices in V_i for each i then, in the language of simplicial complexes, an IT in G is a *multicoloured*

Figure 2: A two-dimensional geometric simplicial complex. Of its maximal simplices, two are 2-simplices (shaded) and six are 1-simplices.

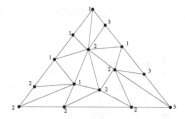

Figure 3: A triangulation of the 2-simplex Δ_2. (To be consistent with Figures 1 and 2 each small triangle should be shaded, but here and in later figures shading of triangulations is omitted for clarity.) The given colouring satisfies the conditions of Sperner's Lemma, so for example the vertices on the bottom one-dimensional face must be coloured 2 or 3, while those in the interior can receive any of $\{1, 2, 3\}$.

simplex (or sometimes called a *rainbow simplex*) in the independence complex $\mathcal{I}(G)$, meaning that it contains one vertex of each of the m colours assigned by $\mathcal{P}$. Thus we are looking for natural sufficient conditions that will guarantee the existence of a multicoloured simplex in the "nasty" simplicial complex $\mathcal{I}(G)$.

If instead we were trying to show the existence of a multicoloured simplex in a colouring of the points in the "nice" simplicial complex given by a triangulation $\mathcal{T}$ of the $(m-1)$-simplex Δ_{m-1}, then we would have a powerful tool available to us in the form of Sperner's Lemma [67]. This well-known classical result from combinatorial topology gives a simple condition on the colouring that guarantees the existence of a multicoloured simplex in $\mathcal{T}$ (see Figure 3).

Lemma 1.2 (Sperner's Lemma). *Let $\mathcal{T}$ be a triangulation of the n-simplex Δ_n. Suppose the points of $\mathcal{T}$ are coloured, such that*

- *each vertex of Δ_n receives a different colour, and*

- *each point of $\mathcal{T}$ on a face σ of Δ_n receives one of the colours that appear on the vertices of σ.*

Then there exists an n-dimensional multicoloured simplex in $\mathcal{T}$.

Thus if we could somehow transfer the problem of finding a multicoloured simplex in $\mathcal{I}(G)$ to finding one in some suitable triangulation $\mathcal{T}$ of Δ_{m-1}, then we might be able to exploit this valuable tool. The main theme of this article is to show that this strategy is indeed possible, and the means by which the "nasty" situation can be transformed into the "nice" one will be the notion of topological connectedness.

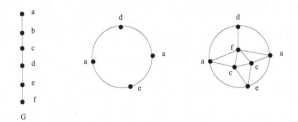

Figure 4: (Left) a graph G, (centre) an $\mathcal{I}(G)$-labelled triangulation of S^1, and (right) a filling of it.

1.2 Simplicial complexes and connectedness

There are several natural simplicial complexes one can associate with a graph G. For example, (if G is simple) one can regard G itself as a simplicial complex, with one-dimensional simplices $E(G)$ and zero-dimensional simplices $V(G)$. The *neighbourhood complex*, whose simplices are the sets of vertices in G having a common neighbour, was shown by Lovász [58] to be important for proving lower bounds on the chromatic number. However, in this article our main focus will be on the independence complex $\mathcal{I}(G)$ defined in the introduction.

We will write B^k for the k-dimensional ball, and S^{k-1} for its boundary, the $(k-1)$-dimensional sphere. The boundary S^{-1} of B^0 is just the empty set. Extending the notion of triangulation described in Section 1.1, a *triangulation* of a topological space X is a geometric simplicial complex whose body is homeomorphic to X. Here we will consider only PL-triangulations of the sphere S^{k-1} or the ball B^k, where PL stands for *piecewise linear*. This technical property is necessary to guarantee certain key properties of triangulations (see e.g. [28]). In fact the main properties of a triangulation of B^k that are actually essential for our purposes (see e.g. [45]) are the following.

- it is *pure*, meaning that all maximal simplices are of the same dimension (k), and

- every $(k-1)$-simplex is contained in two k-simplices if it is not in the boundary, and in one k-simplex if it is in the boundary.

For example, the geometric simplicial complex shown in Figure 2 is not pure, while the one in Figure 3 is a triangulation of B^2 since Δ_2 is homeomorphic to B^2.

A function $\ell : V(\mathcal{C}_1) \to V(\mathcal{C}_2)$ between two simplicial complexes $\mathcal{C}_1$ and $\mathcal{C}_2$ is called a *simplicial map* if for every $\sigma \in \mathcal{C}_1$, the set $\ell(\sigma) = \{\ell(v) : v \text{ is a vertex of } \sigma\}$ is a simplex of $\mathcal{C}_2$. The pair $(\mathcal{C}_1, \ell)$ is called a $\mathcal{C}_2$-*labelled simplicial complex* and the map ℓ is a *labelling* of $\mathcal{C}_1$ by $\mathcal{C}_2$. We will make use of this notion specifically when $\mathcal{C}_2$ is the independence complex of a graph G and $\mathcal{C}_1$ is a triangulation of S^{k-1} or B^k. In this setting, to check whether a map ℓ is a simplicial map, it suffices to verify that for each 1-simplex in the 1-*skeleton* of $\mathcal{C}_1$ (the subcomplex of $\mathcal{C}_1$ consisting of all simplices of dimension at most one), its vertices x and y are labelled with non-adjacent vertices in G, i.e. $\ell(x)\ell(y) \notin E(G)$.

Given a $\mathcal{C}$-labelled triangulation $(\mathcal{T}, \ell)$ of S^{k-1}, a *filling* of $(\mathcal{T}, \ell)$ is a $\mathcal{C}$-labelled triangulation $(\mathcal{T}', \ell')$ of B^k whose boundary is $\mathcal{T}$ and $\ell' : V(\mathcal{T}') \to V(\mathcal{C})$ is a simplicial map with $\ell'|_{V(\mathcal{T})} = \ell$ (see Figure 4).

Our main parameter was introduced by Aharoni and Berger in [8], and is defined as follows.

Definition 1.3. Let $\mathcal{C}$ be a simplicial complex. Then $\eta(\mathcal{C})$ is the maximum d such that, for every $0 \leq k \leq d - 1$, every $\mathcal{C}$-labelled triangulation of S^{k-1} has a filling.

Remark 1. Since every abstract simplicial complex $\mathcal{C}$ can be regarded as a topological space via its *geometric realization* (the simplicial complex whose vertex scheme is $\mathcal{C}$, which is unique up to homeomorphism), it turns out that $\eta(\mathcal{C}) - 2$ is essentially the topological connectedness of $\mathcal{C}$ as it is defined in the abstract (see e.g. Proposition 2.8 in [68]). In other words $\mathcal{C}$ as a topological space is $(\eta(\mathcal{C}) - 2)$-connected but not $(\eta(\mathcal{C}) - 1)$-connected. We will not need to make use of this equivalence, as everything that follows will be in terms of Definition 1.3, but it does justify the use of the term "connectedness" when referring to the parameter η.

For the independence complex $\mathcal{I}(H)$ of a graph H, the following basic properties of $\eta(\mathcal{I}(H))$ are easily derived from the definitions.

Lemma 1.4.

1. *A graph H is nonempty if and only if $\eta(\mathcal{I}(H)) > 0$.*

2. *If the graph H contains an isolated vertex, then $\eta(\mathcal{I}(H)) = \infty$.*

Proof. For Part 1, suppose H is nonempty. Since S^{-1} is the empty set and B^0 is a single point v, every $\mathcal{I}(H)$-labelled triangulation of S^{-1} (which is empty) can be filled to B^0 with the labelling $f(v) = x$, where x is an arbitrary vertex of H. Conversely, if $\eta(\mathcal{I}(H)) > 0$ then such a filling exists, so such a vertex exists as well.

For the second part, suppose x is an isolated vertex in H. For any k and any $\mathcal{I}(H)$-labelled triangulation $(\mathcal{T}, \ell)$ of S^{k-1}, a filling of $(\mathcal{T}, \ell)$ is given by placing one point v inside B^k and joining it to every simplex of $\mathcal{T}$, and extending ℓ to ℓ' by setting $\ell'(v) = x$. Since x is not adjacent to any vertex of H, this gives a labelling as required. $\square$

For a slightly less trivial example, the independence complex of the complete bipartite graph $K_{s,t}$ is the disjoint union of an $(s-1)$-dimensional simplex and a $(t-1)$-dimensional simplex (together with all their faces). Then $\eta(\mathcal{I}(K_{s,t})) = 1$, since the labelling of the two points of S^0 with vertices from opposite sides of $K_{s,t}$ does not have a filling.

2 Finding an IT using topological connectedness

The main focus of this article is on the following theorem and its applications. The idea of Theorem 2.1 was implicit in [18], where the topological approach to IT-type problems was first introduced to give an extension of Hall's Theorem to hypergraphs (see Section 4.1.1). The formulation in terms of topological connectedness was given explicitly in [8].

Theorem 2.1. *Let G be a graph and let $\mathcal{P} = \{V_1, \ldots, V_m\}$ be a partition of $V(G)$. Suppose that for each subset S of $\mathcal{P}$, the subgraph $G_S = G[\bigcup_{V_i \in S} V_i]$ of G induced by $\bigcup_{V_i \in S} V_i$ satisfies $\eta(\mathcal{I}(G_S)) \geq |S|$. Then G has an IT with respect to $\mathcal{P}$.*

This is a special case of a more general statement for the existence of a multicoloured simplex in an arbitrary simplicial complex whose vertices are coloured (see e.g. [8,45,60]), in other words there is no need to restrict to the independence complex of a graph. However, as this case is the only one we will need in this article, we will not discuss the more general version.

Here we give an informal outline of how to prove Theorem 2.1 using the strategy of Section 1.1. Given G and $\mathcal{P}$ with m classes, construct (somehow) a triangulation $\mathcal{T}$ of the $(m-1)$-dimensional simplex Δ_{m-1}. Associate the m vertices x_i of Δ_{m-1} with the m vertex classes V_i of $\mathcal{P}$. Suppose we can also construct a simplicial map f from $\mathcal{T}$ to $\mathcal{I}(G)$ such that, for each point x of $\mathcal{T}$, the image $f(x)$ of x is a vertex of G in some V_i such that x_i is a vertex of the *support* of x (the face of Δ_{m-1} of smallest dimension that contains x). We say that f is *consistent* if it has this property. Note that f induces a natural *colouring* of the points of $\mathcal{T}$, namely, a point x of $\mathcal{T}$ gets the colour i where $f(x) \in V_i$. If f is consistent then this colouring together with $\mathcal{T}$ satisfy the conditions of Sperner's Lemma.

If we succeed in executing the above, then Sperner's Lemma tells us that $\mathcal{T}$ contains a multicoloured elementary $(m-1)$-simplex τ. Thus the m vertices w_i of τ are coloured with all m colours, implying that the corresponding vertices $v_i = f(w_i) \in V_i$ in G for each i. Because τ is a simplex and f is a simplicial map, we know that $\{f(w_i) : 1 \le i \le m\}$ forms a simplex in $\mathcal{I}(G)$, in other words, $\{v_1, \ldots, v_m\}$ is independent in G. Hence this is the required IT.

We are therefore left with the problem of how to construct the triangulation $\mathcal{T}$ and the simplicial map f, given the graph G and the partition $\mathcal{P}$. The overall approach is to begin with the vertices (the zero-dimensional faces) of the $(m-1)$-simplex Δ_{m-1}, and construct $\mathcal{T}$ by defining its restriction to each face of Δ_{m-1} one by one, in order of dimension. At the same time we will define the simplicial map f on the points of $\mathcal{T}$ in the face currently under construction. This strategy is illustrated in Figure 5, for the graph G shown in the first frame.

To begin, fix the vertices $x_1, \ldots, x_m$ of Δ_{m-1}. Define each $f(x_i)$ as follows. By assumption, for the singleton subset $S = \{V_i\}$ of $\{V_1, \ldots, V_m\}$ we have that $\eta(G_{\{V_i\}}) \ge 1$, in other words (see Lemma 1.4(1)), the vertex class V_i of G is nonempty. We may therefore define the simplicial map f on the vertex x_i of Δ_{m-1} by $f(x_i) = v$ for an arbitrary $v \in V_i$ (Frame 2 of Figure 5). Note that this ensures that f is consistent on the face $\{x_i\}$ for each i. This completes the definition of $\mathcal{T}$ and f on the zero-dimensional faces of Δ_{m-1}.

Now let σ be a j-dimensional face of Δ_{m-1} (for example the one-dimensional face $\sigma_{1,2}$ with vertex set $\{x_1, x_2\}$ in Figure 5). Suppose $\mathcal{T}$ and f have been defined on every face of Δ_{m-1} of dimension at most $j-1$, such that f is consistent on every such face. In particular then, they have been defined on every proper face of σ. Therefore $\mathcal{T}' = \mathcal{T}|_{\partial\sigma}$ is a triangulation of the boundary $\partial\sigma$ of the j-simplex σ (which is homeomorphic to S^{j-1}), and $f' = f|_{\partial\sigma}$ is a simplicial map from $\mathcal{T}'$ to $\mathcal{I}(G)$. Moreover, since f' is consistent on every proper face of σ, in fact f' is a simplicial map from $\partial\sigma$ to $\mathcal{I}(G_S)$ where $S = \{V_i : x_i \in V(\sigma)\}$. By the assumption that $\eta(\mathcal{I}(G_S)) \ge |S| = j+1$, there exists a filling of $(c\mathcal{T}', f')$, i.e. a triangulation $\mathcal{T}''$ of σ whose restriction to $\partial\sigma$ is $\mathcal{T}'$, and a simplicial map f'' from $\mathcal{T}''$ to $\mathcal{I}(G_S)$ whose restriction to $\partial\sigma$ is f'. (An example of a filling for $\sigma_{1,2}$ is shown in the third frame of Figure 5, and for the other two one-dimensional faces of Δ_2 in the next two frames.) We may therefore extend our definition of $\mathcal{T}$ and f by setting $\mathcal{T}|_\sigma = \mathcal{T}''$ and $f|_\sigma = f''$. Then f is consistent on σ by definition. This completes the definition of $\mathcal{T}$ and f and therefore fills in the last piece in the proof strategy. For its completion in the example in Figure 5, the sixth frame shows a filling of the whole Δ_2, whose induced Sperner colouring is shown the seventh frame, in which a multicoloured simplex guaranteed by Sperner's Lemma is marked with an asterisk. The IT $\{b, c, e\}$ in G corresponding to this multicoloured simplex is marked in the last frame.

Theorem 2.1 also has the following defect version (see e.g. [4, 45, 49]), which is needed for some applications.

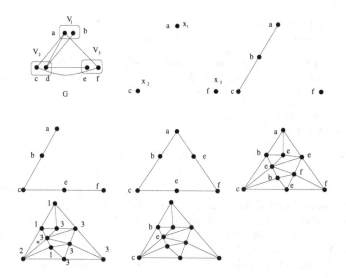

Figure 5: The strategy of the proof for Theorem 2.1, for the graph G shown. Fillings for the three one-dimensional faces of Δ_2 are shown in Frames 3, 4 and 5, and for the two-dimensional face in Frame 6. Frame 7 shows the induced Sperner colouring, in which a multicoloured simplex is marked. The corresponding independent transversal of G is indicated in the last frame.

Theorem 2.2. *Let $d \geq 0$ be given. Let G be a graph and let $\mathcal{P} = \{V_1, \ldots, V_m\}$ be a partition of $V(G)$. Suppose that for each subset S of $\mathcal{P}$, the subgraph G_S satisfies $\eta(\mathcal{I}(G_S)) \geq |S| - d$. Then G has a (partial) IT of at least $m - d$ classes of $\mathcal{P}$.*

3 Lower bounds on connectedness

3.1 Tools for proving lower bounds

In order to apply Theorem 2.1, one needs ways to prove that $\eta(\mathcal{I}(G_S))$ is large. This is not in general an easy matter, but there are a few tools at our disposal. In this section we first describe two such tools, that can be used to prove lower bounds on $\eta(\mathcal{I}(H))$ for general graphs H. We will see in Section 3.2 how they can be used to relate certain graph parameters to $\eta(\mathcal{I}(H))$ (that are, in certain circumstances, easier to determine than $\eta(\mathcal{I}(H))$ itself).

3.1.1 Special triangulations In [18] the strategy outlined in Section 2 was executed directly, i.e. suitable triangulations of the simplex were explicitly constructed. This approach of direct construction was also used for example in [17, 49, 68].

The statement below (which is from [49]) combines the main idea of the proof of Theorem 4.3 implicit in [18] with a construction from [17]. A different construction with similar properties was given in [18]. For a graph H and subset $I \subseteq V(H)$, we say that a simplex σ with vertex labels from $V(H)$ is I-*blocking* if each element of I is adjacent (in H) to some label of σ.

Lemma 3.1. *Let I be an independent set in a graph H. Then every $\mathcal{I}(H)$-labelled triangulation of S^{k-1} with no I-blocking simplices has a filling.*

Proof. Let $(\mathcal{T}, \ell)$ be an $\mathcal{I}(H)$-labeled triangulation of S^{k-1} with no I-blocking simplices. By Lemma 1.1 of [17] (see also Lemma 2.2 of [18]) there exists a triangulation $\mathcal{T}'$ of B^k with boundary $\mathcal{T}$ with the following properties.

(1) For every $x \in V(\mathcal{T}') \setminus V(\mathcal{T})$, the neighbours of x in the 1-skeleton of $\mathcal{T}'$ that lie on S^{k-1} form a (possibly empty) simplex σ_x of $\mathcal{T}$.

(2) Any 1-simplex of $\mathcal{T}'$ with endpoints in $V(\mathcal{T})$ is also a 1-simplex in $\mathcal{T}$.

We give an $\mathcal{I}(H)$-labelling ℓ' of $\mathcal{T}'$ that extends ℓ, by defining it on each interior point $x \in V(\mathcal{T}') \setminus V(\mathcal{T})$. Given x, since σ_x is not I-blocking, we may define $\ell'(x)$ to be an arbitrary vertex of I that is not adjacent to any label of σ_x. To check that this gives an $\mathcal{I}(H)$-labelling, consider an arbitrary 1-simplex τ of $\mathcal{T}'$. If the vertices x and y of τ are both in $V(\mathcal{T})$ then τ is also a 1-simplex of $\mathcal{T}$ by (2) and hence $\ell'(x)\ell'(y) = \ell(x)\ell(y) \notin E(H)$. If $x \in V(\mathcal{T}') \setminus V(\mathcal{T})$ and $y \in V(\mathcal{T})$, then $\ell'(x)\ell'(y) \notin E(H)$ by our choice of $\ell'(x)$. If both vertices x and y are in the interior $V(\mathcal{T}') \setminus V(\mathcal{T})$, then $\ell'(x)$ and $\ell'(y)$ are both in I, which is an independent set. □

3.1.2 Meshulam's Theorem

Our second tool is Meshulam's Theorem [61], which relates $\eta(\mathcal{I}(H))$ to that of two subgraphs of H, one obtained by deleting an edge, the other by what is called *exploding* an edge. For a graph H and an edge $e \in E(H)$, we write $H - e$ for the subgraph of H obtained by deleting e. We denote the edge *explosion* of e by $H * e$, which is the subgraph of H that remains after deleting from H both endpoints of e and all of their neighbours.

Theorem 3.2 (Meshulam's Theorem). *Let H be a graph and let e be an edge of H. Then*

$$\eta(\mathcal{I}(H)) \geq \min(\eta(H - e), \eta(H * e) + 1).$$

This result (in a different formulation) is proved in [61]. For more on Meshulam's Theorem see e.g. [2, 13, 15], and [62], Section 5.3.

3.2 Graph parameters

We may now identify some of the main parameters of a graph G that relate to the connectedness parameter $\eta(\mathcal{I}(G))$.

3.2.1 General bounds

The first two parameters are defined in terms of domination. We say that a set U of vertices of G *(strongly) dominates* G if every vertex of G is adjacent to a vertex in U. (The more usual definition of domination just requires this property for each vertex of $V(G) \setminus U$, but here we mean the stronger notion.) For a graph G we define the *independence domination* number $i(G)$ to be the smallest ℓ for which every independent set in G can be dominated by a set of at most ℓ vertices in G. The following forms part of the main theorem in [18] (see Lemma 4.3). To illustrate the use of Lemma 3.1 and Theorem 3.2, we will show how to derive it from each of them (using the proof ideas from [18] and [60] respectively).

Theorem 3.3. *For every nonempty graph G we have $\eta(\mathcal{I}(G)) \geq i(G)$.*

Proof. (using Lemma 3.1:) Let k be such that $0 \leq k \leq i(G) - 1$. Let $(\mathcal{T}, \ell)$ be an arbitrary $\mathcal{I}(G)$-labelled triangulation of S^{k-1}. Each simplex σ of $\mathcal{T}$ has at most k vertices, so the label set on σ is an independent set U in G of size at most $i(G) - 1$.

By definition of $i(G)$ there exists an independent set I in G that is not dominated by any set U of vertices of G with $|U| \leq i(G) - 1$, so in particular no simplex of $(\mathcal{T}, \ell)$ is I-blocking. Thus by Lemma 3.1 we obtain that $(\mathcal{T}, \ell)$ has a filling.

Since every $(\mathcal{T}, \ell)$ has a filling, by definition we find that $\eta(\mathcal{I}(G)) \geq i(G)$ as required. $\square$

Proof. (using Theorem 3.2:) We prove the statement by induction on $|E(G)|$. If $|E(G)| = 0$ then since G is nonempty, it contains an isolated vertex. Hence by Lemma 1.4(2) we have $\eta(\mathcal{I}(G)) = \infty$.

Assume $|E(G)| \geq 1$ and that the theorem is true for graphs with fewer edges. We may also assume as above that G has no isolated vertices. Then since every independent set W in G is dominated by a set of size at most $|W|$ (by taking one neighbour for each vertex in W) we know that $i(G)$ is finite. Moreover $i(G) \geq 1$ since no nonempty set can be dominated by the empty set.

By definition of $i(G)$, there exists an independent set W in G that is not dominated by any set of size at most $i(G) - 1$. Then $W \neq \emptyset$. Since no vertex of W is isolated, we may choose an edge $e = vw$ incident to some $w \in W$.

In the graph $G - e$, the set W is still independent, and it still cannot be dominated in $G - e$ by any set U with $|U| \leq i(G) - 1$. Therefore W witnesses the fact that $i(G-e) \geq i(G)$, and so by induction $\eta(\mathcal{I}(G - e)) \geq i(G - e) \geq i(G)$.

In the graph $G * e$, the set $W \setminus N(v)$ is independent (where $N(v)$ denotes the neighbourhood of v). Suppose $W \setminus N(v)$ is dominated in $G * e$ by a set U'. Then W is dominated in G by $U' \cup \{v\}$, and hence $|U'| \geq i(G) - 1$. Therefore $W \setminus N(v)$ witnesses the fact that $i(G * e) \geq i(G) - 1$. Certainly $|E(G * e)| < |E(G)|$ since in particular e is not an edge of $G * e$, so by induction $\eta(\mathcal{I}(G * e)) \geq i(G * e) \geq i(G) - 1$.

Therefore by Theorem 3.2 we find that $\eta(\mathcal{I}(G) \geq i(G)$, as required. $\square$

In fact in [18] the stated result is somewhat stronger, in that the dominating sets U in the definition of $i(G)$ can also be assumed to be independent. This fact can be read out of either of the proofs given.

The *domination* number $\gamma(G)$ is the smallest size of a (strongly) dominating set U in G. Then Theorem 3.2 implies the following (proved here as in [60]).

Theorem 3.4. *For every graph G we have $\eta(\mathcal{I}(G)) \geq \gamma(G)/2$.*

Proof. If $V(G) = \emptyset$ the statement clearly holds, so we may assume $V(G) \neq \emptyset$. Again we proceed by induction on $|E(G)|$. If $|E(G)| = 0$ then (since G has an isolated vertex) we have $\eta(\mathcal{I}(G)) = \infty$.

Assume $|E(G)| \geq 1$ and that the theorem is true for graphs with fewer edges. Let $e = xy$ be an edge of G. Suppose on the contrary that $\eta(\mathcal{I}(G)) < \gamma(G)/2$. Then by Theorem 3.2 either (i) $\eta(\mathcal{I}(G - e)) < \gamma(G)/2$, or (ii) $\eta(\mathcal{I}(G * e)) < \gamma(G)/2 - 1$.

By induction we know that $\gamma(G-e)/2 \leq \eta(\mathcal{I}(G-e))$. Thus if (i) holds then $\gamma(G-e)/2 < \gamma(G)/2$, which is clearly false since any set that dominates $G - e$ also dominates G.

Also by induction we have that $\gamma(G * e)/2 \leq \eta(\mathcal{I}(G * e))$. So if (ii) holds then $\gamma(G * e)/2 < \gamma(G)/2 - 1$. Note however that for any set U that dominates $G * e$, the set $U \cup \{x, y\}$ dominates G. Therefore $\gamma(G * e) \geq \gamma(G) - 2$, so (ii) cannot hold either.

We conclude that $\eta(\mathcal{I}(G)) \geq \gamma(G)/2$ as required. $\square$

Another parameter that leads to a lower bound on $\eta(\mathcal{I}(G))$ comes from the largest eigenvalue $\lambda(G)$ of the Laplacian of G. Recall that the *Laplacian* of a graph G is the matrix $D(G) - A(G)$ where $D(G)$ is the diagonal matrix of degrees of $V(G)$ and $A(G)$ is the adjacency matrix (indexed in the same order as $D(G)$). It was proved in [12] that $\eta(\mathcal{I}(G))$ is bounded below by the quantity $|V(G)|/\lambda(G)$ for every graph G. (To be precise, this was proved for the *homological* definition of the connectedness parameter, but this coincides with $\eta(\mathcal{I}(G))$ whenever the latter is at least three, see e.g. [6].)

Finally we mention that Theorem 3.2 immediately implies that the following (somewhat artificial) graph parameter is a lower bound for $\eta(\mathcal{I}(G))$ for every graph G (see e.g. [15]). We define $\psi(G)$ by setting $\psi(G) = 0$ if $V(G) = \emptyset$ and $\psi(G) = \infty$ if $V(G) \neq \emptyset$ but $E(G) = \emptyset$. Otherwise

$$\psi(G) = \max_{e \in E(G)} \{\min\{\psi(G - e), \psi(G \ast e) + 1\}\}.$$

A description of the parameter $\psi(G)$ using game terminology was used to obtain lower bounds for $\eta(\mathcal{I}(G))$ in several different settings, for example [13, 15, 19].

It was conjectured in [15] that $\eta(\mathcal{I}(G)) = \psi(G)$ for every G, but this was shown to be false in [2], where constructions are given to show that in fact these parameters can be arbitrarily far apart.

3.2.2 Special graph classes

Stronger lower bounds on $\eta(\mathcal{I}(G))$ are known when G belongs to certain special classes of graphs. Here we mention a few of these briefly, without going into the details.

Recall that a graph is *chordal* if it contains no induced cycle of length greater than three. A classical theorem of Gavril [40] states that every chordal graph has a *tree representation*, meaning that it is the intersection graph of a family of subtrees of a tree. This fact is used in [14] to show in particular that every chordal graph G of maximum degree d satisfies $\eta(\mathcal{I}(G)) \geq |V(G)|/(d + 1)$.

A graph is $K_{1,k}$-*free* if it does not contain $K_{1,k}$ (the star with k leaves) as an induced subgraph. When $k = 3$ the standard term is *claw-free*. The parameter $\eta(\mathcal{I}(G))$ for claw-free graphs is studied in [36], and for more general $K_{1,k}$-free graphs in [6]. By proving an upper bound on the maximum eigenvalue $\lambda(G)$ of the Laplacian of G, in [6] a lower bound for $\eta(\mathcal{I}(G))$ is obtained using the quantity $|V(G)|/\lambda(G)$, as mentioned in Section 3.2.1. However, the authors go on to prove a stronger lower bound for $\eta(\mathcal{I}(G))$ using an argument similar to Theorem 3.2 as follows (again here the homological definition of η is used, but the authors also discuss how the same results can be obtained for the homotopic version).

Theorem 3.5. *Let G be a $K_{1,k}$-free graph of maximum degree d, where $k \geq 3$. Then*

$$\eta(\mathcal{I}(G)) \geq \frac{(k - 1)|V(G)|}{d(2k - 3) + k - 1}.$$

This gives the same result as that of [36] for claw-free graphs (i.e. when $k = 3$).

Many more detailed results are known on the independence complexes of special classes of graphs. For example, cycles and paths are studied in [56], forests in [35], and various classes including claw-free graphs and triangle-free graphs in [25].

4 Applications

When combined with Theorem 2.1, each of the lower bounds for connectedness described in Section 3.2 gives a sufficient condition for a vertex-partitioned graph to have an independent transversal. For example Theorems 3.3 and 3.4 lead to the following.

Corollary 4.1. *Let G be a graph and let $\mathcal{P} = \{V_1, \ldots, V_m\}$ be a partition of $V(G)$.*

1. *Suppose that for each subset S of $\mathcal{P}$ we have $i(G_S) \geq |S|$. Then G has an IT with respect to $\mathcal{P}$.*

2. *Suppose that for each subset S of $\mathcal{P}$ we have $\gamma(G_S) \geq 2|S| - 1$. Then G has an IT with respect to $\mathcal{P}$.*

Remark 2. All known proofs of Corollary 4.1(1) use the notion of topological connectedness in some form. However, this is not the case for Corollary 4.1(2), which was proved earlier by a purely combinatorial argument ([42], see also e.g. [27, 47, 50]). In fact the combinatorial proof gives more information, since it can be used to derive certain structural information about the small dominating set that must exist if G has no IT (see e.g. [15]), and suitable minor weakenings of the assumptions even allow for algorithmic proofs ([24, 41]).

Since the number of vertices dominated by a set U is at most the sum of the degrees of the vertices in U, Corollary 4.1(2) has the following immediate consequence.

Corollary 4.2. *Let G be a graph with maximum degree Δ and let $\mathcal{P} = \{V_1, \ldots, V_m\}$ be a partition of $V(G)$. If $|\bigcup_{V_i \in S} V_i| > (2|S| - 2)\Delta$ for each $S \subseteq \mathcal{P}$, then G has an IT with respect to $\mathcal{P}$.*

In particular if $|V_i| \geq 2\Delta$ for each i then G has an IT. This answered a question of Bollobás, Erdős and Szemerédi from 1975 [30]. The bound given in Corollary 4.2 is best possible, since for every Δ there exist graphs for which $|V_i| = 2\Delta - 1$ for each i that do not have independent transversals [68].

4.1 Hypergraph matching

Throughout this subsection, the use of the term hypergraph allows the possibility of multiple edges. Thus here a *hypergraph* $\mathcal{H}$ is a pair (V, E), where $V = V(\mathcal{H})$ is the set of *vertices*, and $E = E(\mathcal{H})$ is a multiset of subsets of vertices, called the *edges* of $\mathcal{H}$. The number of times a subset $e \subseteq V$ appears in E is called the *multiplicity* of e. If every edge of $\mathcal{H}$ has size r, we say that $\mathcal{H}$ is an *r-graph*. Thus a 2-graph is simply a (multi)graph, possibly with multiple edges (but not loops). An edge $e \in E$ is said to be *parallel* to an edge $f \in E$ if their underlying vertex subsets are the same. In particular, every edge is parallel to itself.

A *matching* in a hypergraph $\mathcal{J}$ is a set of disjoint edges. We write $\nu(\mathcal{J})$ for the size of a largest matching in $\mathcal{J}$. The *matching complex* $\mathcal{M}(\mathcal{J})$ of $\mathcal{J}$ is the independence complex of its *line graph* $L(\mathcal{J})$, the graph with vertex set $E(\mathcal{J})$ and edge set $\{ef : e \cap f \neq \emptyset\}$. Thus the simplices of $\mathcal{M}(\mathcal{J})$ are precisely the matchings in $\mathcal{J}$. Then the following result (implicit in [18]) is an easy consequence of Theorem 3.3.

Lemma 4.3. *Let $\mathcal{J}$ be an r-graph. Then $\eta(\mathcal{M}(\mathcal{J})) \geq \lceil \nu(\mathcal{J})/r \rceil$.*

Proof. Let M be a matching in $\mathcal{J}$ of size $\nu(\mathcal{J})$. Then M is an independent set in $L(\mathcal{J})$. We claim that M certifies that $i(\mathcal{M}(\mathcal{J})) \geq \lceil \nu(\mathcal{J})/r \rceil$, which implies by Theorem 3.3 that $\eta(\mathcal{M}(\mathcal{J})) = \eta(\mathcal{I}(L(\mathcal{J}))) \geq \lceil \nu(\mathcal{J})/r \rceil$ as required. To verify our claim, we need to show that M is not dominated by any set U of vertices of $L(\mathcal{J})$ (i.e. edges of $\mathcal{J}$) with $|U| \leq \lceil \nu(\mathcal{J})/r \rceil - 1$. For any such set U, the total number of vertices of $\mathcal{J}$ contained in the edges of U is at most $r|U| \leq |M| - 1$. Therefore there exists an edge of M that is disjoint from every edge of U, and hence U does not dominate M in $L(\mathcal{J})$, thus completing the proof. $\square$

4.1.1 Hall's Theorem for hypergraphs Hall's fundamental matching theorem tells us that a bipartite graph G with vertex classes A and X has a matching of size $|A|$ if and only if *Hall's condition* holds: that for every subset $S \subseteq A$, the *neighbourhood* $N(S) = \{x \in X : sx \in E(G)$ for some $s \in S\}$ is *big enough*, i.e. $|N(S)| \geq |S|$.

The original work [18] that introduced the methods of this article gave a hypergraph analogue of Hall's Theorem. To formulate this, we will say that a hypergraph $\mathcal{H}$ is *bipartite* if it has a vertex partition into classes A and X such that $|e \cap A| = 1$ for every edge e of $\mathcal{H}$. Corresponding to the notion of neighbourhood, the *link* $\mathrm{lk}_{\mathcal{H}}(S)$ of a subset $S \subseteq A$ is the hypergraph with vertex set X and edge set $\{e \setminus S : e \in E(\mathcal{H}), e \cap A \subseteq S\}$. (Note that $\mathrm{lk}_{\mathcal{H}}(S)$ may have parallel edges, even when $\mathcal{H}$ does not.) As we will see, the requirement that the neighbourhood is big enough in Hall's condition translates in the hypergraph setting to the condition that the connectedness of the matching complex of $\mathrm{lk}_{\mathcal{H}}(S)$ should be big enough.

To see this, we describe the notion of matching in a bipartite hypergraph in terms of independent transversals. For a bipartite hypergraph $\mathcal{H}$, let $G^{\mathcal{H}}$ be the line graph of $\mathrm{lk}_{\mathcal{H}}(A)$, i.e. $G^{\mathcal{H}}$ has a vertex for each edge of $\mathcal{H}$, and two are adjacent if and only if they intersect in X. The vertex partition $\mathcal{P}$ of $G^{\mathcal{H}}$ is given by the elements of A: we put edges e and f of $\mathcal{H}$ into the same part of $\mathcal{P}$ if and only if they contain the same vertex of A. Then a set M of edges of $\mathcal{H}$ is a matching in $\mathcal{H}$ if and only if it forms an IT in $G^{\mathcal{H}}$ of $|M|$ of the parts of $\mathcal{P}$. In this setting, the defect version Theorem 2.2 of Theorem 2.1 translates to the following statement.

Theorem 4.4. *Let $d \geq 0$ be given, and let $\mathcal{H}$ be a bipartite hypergraph with vertex classes A and X. Suppose that for each subset S of A, the link $\mathrm{lk}_{\mathcal{H}}(S)$ of S satisfies $\eta(\mathcal{M}(\mathrm{lk}_{\mathcal{H}}(S)) \geq |S| - d$. Then $\mathcal{H}$ has a matching of size at least $|A| - d$.*

For bipartite r-graphs, each $\mathrm{lk}_{\mathcal{H}}(S)$ is an $(r-1)$-graph. Thus Theorem 4.4 with $d = 0$ together with Lemma 4.3 immediately imply the following natural analogue of Hall's Theorem for r-graphs from [18].

Theorem 4.5. *Let $\mathcal{H}$ be a bipartite r-graph with vertex classes A and X. Suppose that $\nu(\mathrm{lk}_{\mathcal{H}}(S)) \geq (r-1)(|S| - 1) + 1$ for every $S \subseteq A$. Then $\mathcal{H}$ has a matching of size $|A|$.*

In fact Theorem 4.5 is best possible, since there exist bipartite r-graphs $\mathcal{H}$ as above for which each $\mathrm{lk}_{\mathcal{H}}(S)$ has a matching of size at least $(r-1)(|S| - 1)$, but $\mathcal{H}$ has no matching of size $|A|$ (see [3]).

One significant difference between Hall's Theorem and Theorem 4.5 is that the condition in Theorem 4.5 is not necessary. Using the above approach, it is not difficult to verify the following version of the theorem (see [18]) that is both necessary and sufficient.

Theorem 4.6. *Let $\mathcal{H}$ be a bipartite r-graph with vertex classes A and X. Then $\mathcal{H}$ has a matching of size $|A|$ if and only if for every $S \subseteq A$ there exists a matching $\mathcal{M}_S$ in $\mathrm{lk}_{\mathcal{H}}(S)$, such that $\mathcal{M}_S$ is not dominated in $G^{\mathcal{H}}$ by any set of $|S| - 1$ edges of $\bigcup_{T \subset S} \mathcal{M}_T$.*

4.1.2 Ryser's Conjecture One of the most striking consequences of Theorem 2.1 has been Aharoni's proof of the tripartite case of an old and notoriously difficult open problem due to Ryser [65]. This conjecture seeks to generalise the classical result of König that every bipartite graph G has a cover of size $\nu(G)$. Here a *cover* of a graph or hypergraph is a set of vertices that intersects every edge. Denoting by $\tau(\mathcal{H})$ the size of a smallest cover of $\mathcal{H}$, we see from the definitions that $\tau(\mathcal{H}) \geq \nu(\mathcal{H})$ for every $\mathcal{H}$, and that if $\mathcal{H}$ is an r-graph then $\tau(\mathcal{H}) \leq r\nu(\mathcal{H})$. Ryser conjectured in 1967 that this trivial upper bound can be improved

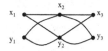

Figure 6: The truncated Fano plane.

to $(r-1)\nu(\mathcal{H})$ whenever $\mathcal{H}$ is r-partite. The r-graph $\mathcal{H}$ is said to be r-*partite* if its vertices can be partitioned into r classes such that every edge has exactly one vertex in each class.

The conjectured statement would be best possible, if true, whenever $r-1$ is a prime power. The standard example showing this is the *intersecting* r-graph (i.e. with matching number one) formed by removing one point from the projective plane of order $r-1$. In particular, for $r=3$ it is easy to construct a 3-partite 3-graph with matching size one and cover number two. In this case the standard example is the *truncated Fano Plane* F (also called the *Pasch configuration*), the 3-partite 3-graph with six vertices x_1 x_2, x_3, y_1, y_2, y_3 and four edges $x_1x_2x_3$, $x_1y_2y_3$, $y_1x_2y_3$, $y_1y_2x_3$, where the sets $\{x_i, y_i\}$ are the vertex classes (see Figure 6). Removing any single edge from F gives a second example, called the *loose 3-cycle*.

Ryser's Conjecture remains wide open in general, despite the large amount of attention it has received (see e.g. [49] for a history of the problem). Until 2001, only partial results [43, 69, 70] for $r=3$ were known, and no nontrivial bound for any $r \geq 4$. Aharoni's proof of the $r=3$ case represented a major breakthrough on this question. Here we are able to give the full details of this proof, as a simple application of Theorem 4.4.

We begin with the following observation. Note that a 3-partite 3-graph with parts A, B, and C can be viewed as a bipartite 3-graph (in the sense of Section 4.1.1) in which $X = B \cup C$. The link $\mathrm{lk}_{\mathcal{H}}(S)$ of each $S \subseteq A$ is then a bipartite (multi)graph with vertex classes B and C.

Lemma 4.7. *Let $\mathcal{H}$ be a 3-partite 3-graph with parts A, B, and C. Then for each $S \subseteq A$, the bipartite graph $\mathrm{lk}_{\mathcal{H}}(S)$ satisfies $\nu(\mathrm{lk}_{\mathcal{H}}(S)) \geq |S| - |A| + \tau(\mathcal{H})$.*

Proof. For each $S \subseteq A$, a cover of $\mathcal{H}$ is given by $(A \setminus S) \cup T$ where T is a minimum cover of $\mathrm{lk}_{\mathcal{H}}(S)$. Since $|T| = \nu(\mathrm{lk}_{\mathcal{H}}(S))$ by König's Theorem, we obtain

$$\tau(\mathcal{H}) \leq |A \setminus S| + |T| = |A| - |S| + \nu(\mathrm{lk}_{\mathcal{H}}(S)).$$

□

Now Ryser's Conjecture for $r=3$ follows easily from Lemma 4.3 and Theorem 4.4.

Theorem 4.8 (Aharoni's Theorem). *Let $\mathcal{H}$ be a 3-partite 3-graph. Then $\tau(\mathcal{H}) \leq 2\nu(\mathcal{H})$.*

Proof. Let A, B and C denote the vertex classes of $\mathcal{H}$. For each $S \subseteq A$ we know by Lemma 4.3 that $\eta(\mathcal{M}(\mathrm{lk}_{\mathcal{H}}(S))) \geq \nu(\mathrm{lk}_{\mathcal{H}}(S))/2$. Lemma 4.7 tells us that $\nu(\mathrm{lk}_{\mathcal{H}}(S)) \geq |S| - |A| + \tau(\mathcal{H})$, and hence $\nu(\mathrm{lk}_{\mathcal{H}}(S)) \geq 2|S| - 2|A| + \tau(\mathcal{H})$ since $|S| \leq |A|$. Thus we conclude that

$$\eta(\mathcal{M}(\mathrm{lk}_{\mathcal{H}}(S))) \geq (2|S| - 2|A| + \tau(\mathcal{H}))/2 = |S| - (|A| - \tau(\mathcal{H})/2).$$

Therefore $\mathcal{H}$ satisfies the conditions of Theorem 4.4 with $d = |A| - \tau(\mathcal{H})/2$, which implies that $\mathcal{H}$ has a matching of size at least $|A| - d \geq \tau(\mathcal{H})/2$, as required. □

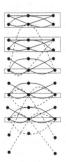

Figure 7: A home-base hypergraph.

4.1.3 Extremal hypergraphs for Ryser's Conjecture

While Aharoni's Theorem is a beautiful simple proof of the Ryser conjecture for the $r = 3$ case, it does not give much information on which hypergraphs $\mathcal{H}$ actually attain the bound $\tau(\mathcal{H}) = 2\nu(\mathcal{H})$. Let us call an r-partite r-graph $\mathcal{H}$ *Ryser-extremal* if $\tau(\mathcal{H}) = (r-1)\nu(\mathcal{H})$. The set of r for which Ryser-extremal r-graphs exist at all is not well understood, see e.g. [1,7].

For the $r = 3$ case however, a complete characterisation of Ryser-extremal 3-graphs was proved in [49]. It is the family of *home-base* hypergraphs, which can be described as follows (see Figure 7). As noted in Section 4.1.2, the 3-uniform loose 3-cycle C is Ryser-extremal. To construct a home-base hypergraph with matching number k, start with k disjoint copies of C. Next choose some arbitrary m, $0 \le m \le k$, and add to m copies of C an edge consisting of the three degree-1 vertices of the copy, thus completing it to a copy of the truncated Fano plane F. For each of the remaining $k - m$ copies of C, choose an arbitrary number of additional edges e, each of which intersects C in at least two of its degree-2 vertices. The third vertex of e can be an arbitrary element of the remaining vertex class. The set of all hypergraphs obtained in this fashion (after possibly adding parallel edges in some of the copies of F) is the set of home-base hypergraphs.

Theorem 4.9. *Let $\mathcal{H}$ be a 3-partite 3-graph. Then $\tau(\mathcal{H}) = 2\nu(\mathcal{H})$ if and only if $\mathcal{H}$ is a home-base hypergraph.*

The "if" direction is easy to verify. For the other direction, one of the main components of the proof is to characterise the links of the three vertex classes of $\mathcal{H}$. For this purpose the following result about bipartite multigraphs is needed. Here C_4 and P_4 denote the cycle with four vertices and (respectively) the path with four vertices. An *interior* vertex of P_4 is a vertex of degree two in P_4.

Theorem 4.10. *Let G be a bipartite multigraph. Then $\eta\left(\mathcal{M}\left(G\right)\right) = \frac{\nu(G)}{2}$ if and only if G contains a collection of $\nu(G)/2$ pairwise vertex-disjoint subgraphs, each of them a C_4 or a P_4, such that every edge of G is parallel to an edge of one of the C_4's or is incident to an interior vertex of one of the P_4's.*

The main tool used in the proof of Theorem 4.10 is Lemma 3.1.

4.1.4 Stability for matchings in regular 3-partite hypergraphs

One simple and well-known consequence of Hall's Theorem for bipartite graphs is that every *regular* bipartite graph of positive degree has a perfect matching. What should the hypergraph analogue of this statement be? The standard example mentioned in Section 4.1.2 (the projective plane of order $r - 1$ with one point removed) is an $(r - 1)$-regular r-partite r-graph with $r - 1$ vertices in each class, that has matching number one. Taking n disjoint copies of this example shows that (at least when $r - 1$ is a prime power) an $(r - 1)$-regular r-partite r-graph with n vertices in each class could have matching number as low as $n/(r - 1)$. For $r = 3$ it is an immediate consequence of Theorem 4.8 that this is the right answer.

Theorem 4.11. *Let $\mathcal{H}$ be a regular 3-partite 3-graph of positive degree, with n vertices in each class. Then $\mathcal{H}$ has a matching of size at least $\frac{n}{2}$.*

Proof. Suppose $\mathcal{H}$ is r-regular. Observe that $\mathcal{H}$ has rn edges, but each vertex is contained in only r edges. Therefore any vertex cover of $\mathcal{H}$ must have at least $\frac{rn}{r} = n$ vertices, in other words $\tau(\mathcal{H}) \geq n$. By Theorem 4.8 it follows that $\nu(\mathcal{H}) \geq \frac{\tau(\mathcal{H})}{2} \geq \frac{n}{2}$, which completes the proof. $\qquad\square$

Let $s \cdot F$ denote the 3-graph with the same vertices as the truncated Fano plane F, and with each edge e replaced by s parallel copies of e. If $\mathcal{H}$ consists of $\frac{n}{2}$ disjoint copies of $\frac{t}{2} \cdot F$, then $\nu(\mathcal{H}) = \frac{n}{2}$, showing that Theorem 4.11 is best possible for every even t and every even n. In fact it follows from Theorem 4.9 that this is the unique extremal configuration. One may then ask what regular 3-partite 3-graphs $\mathcal{H}$ with class sizes n have $\nu(\mathcal{H})$ close to $\frac{n}{2}$? Are they correspondingly close in structure to the extremal configuration? This is the basic *stability* question associated with Theorem 4.11, which was shown in [48] to have the following positive answer.

Theorem 4.12. *Let $t \geq 2$. Let $\mathcal{H}$ be a t-regular 3-partite 3-graph with n vertices in each class, and let $\varepsilon \geq 0$. If $\nu(\mathcal{H}) \leq (1 + \varepsilon)\frac{n}{2}$, then $\mathcal{H}$ has at least $\left(1 - \left(22t - \frac{77}{3}\right)\varepsilon\right)\frac{n}{2}$ components that are copies of $\frac{t}{2} \cdot F$.*

In particular if t is odd then $\frac{t}{2} \cdot F$ does not exist, implying that in this case

$$\nu(\mathcal{H}) \geq \left(1 + \frac{1}{22t - \frac{77}{3}}\right)\frac{n}{2}.$$

The main lemma used to prove Theorem 4.12 is a result similar to Lemma 4.3 for bipartite multigraphs, which is proved using Theorem 3.2. A *t-regular C_4* is a bipartite multigraph consisting of a cycle of length 4 and edges parallel to the edges of the cycle so that every vertex has degree t. For a graph G and a subgraph J of the line graph $L(G)$ we denote by G^J the subgraph of G consisting of those edges that are in $V(J)$.

Theorem 4.13. *Let G be a bipartite multigraph with maximum degree $t \geq 2$ that does not contain a t-regular C_4 component, and let $J \subseteq L(G)$. Then*

$$\eta(\mathcal{I}(J)) \geq \frac{(2t - 3)\nu(G^J) + |V(J)|}{6t - 7}.$$

In particular, when $J = L(G)$ this is an improvement over the bound on $\eta(\mathcal{M}(G))$ given in Lemma 4.3 whenever $|E(G)| > \frac{2t-1}{2}\nu(G)$.

4.2   Hamilton cycle problems

4.2.1   Cycle-plus-triangles   A *cycle-plus-triangles* graph is a graph formed by adding to a Hamilton cycle on $3k$ vertices an arbitrary set of k disjoint triangles on the same vertex set. Motivated by a specific practical application, Du, Hsu and Hwang [34] asked whether every cycle-plus-triangles graph has an independent set of size k. In other words, does every vertex-partition of C_k into classes of size three have an independent transversal? This problem was later generalised and popularised by Erdős in the 1980's, who conjectured that every cycle-plus-triangles graph is 3-colourable. This conjecture was finally proved by Fleischner and Stiebitz [38] in 1992, and with a different proof by Sachs [66] in 1993. The original question of Du, Hsu and Hwang remained unsolved until the proofs of Fleischner and Stiebitz, and Sachs, of the stronger statement. Both of these proofs are ingenious but quite difficult. Here we can give a solution to the question of Du, Hsu and Hwang that is almost immediate from Theorems 2.1 and 3.3, and in fact is more general (see e.g. [15, 47]).

Theorem 4.14. *Let G be a graph of maximum degree two, in which each cycle is of length divisible by three. Then G has an IT with respect to any vertex-partition into classes of size at least three.*

Proof. Fix a vertex-partition $\mathcal{P}$, and let S be any subset of classes in $\mathcal{P}$. Then G_S has at least $3|S|$ vertices, and its components are paths, and cycles of length divisible by three. Let W_S be a maximum independent set in G_S that is spaced three apart, i.e. for each path component take an endvertex and then every third vertex starting from that end. For each cycle component C_{3k} take an independent set of size k by taking every third vertex. Then $|W_S| \geq |S|$. We claim that W_S witnesses the fact that $i(G_S) \geq |S|$ for each S, and hence by Theorem 3.3 it follows that $\eta(\mathcal{I}(G_S)) \geq |S|$. To verify this claim, let U be an arbitrary set of vertices in G_S of size at most $|S| - 1$. Since W_S is spaced three apart, each vertex of U is adjacent to at most one vertex of W_S. Thus since $|W_S| \geq |S|$, there is a vertex $w \in W_S$ that is not a neighbour of any vertex of U. Therefore G_S is not dominated by U, which shows $i(G_S) \geq |S|$ as required. Therefore by Theorem 2.1 we find that G has an independent transversal. $\square$

4.2.2   Toughness and Hamiltonicity   A graph G is said to be *t-tough* if, for every subset $S \subset V(G)$, the subgraph $G - S$ of G induced by $V(G) \setminus S$ has at most $|S|/t$ components. For example, every Hamiltonian graph is 1-tough. A famous open problem due to Chvátal [33] asks whether there exists an absolute constant t_0 such that every t_0-tough graph is Hamiltonian. (To avoid trivial counterexamples, here we consider only graphs with at least three vertices.)

This conjecture has been proved for various special classes of graphs, including chordal graphs. In [32] it was shown that every 18-tough chordal graph is Hamiltonian, and in [54] the constant 18 was improved to 10. The proof in [54] also makes use of the tree representations of chordal graphs given by Gavril's theorem [40] (see Section 3.2.2). Given a chordal graph G, a bipartite hypergraph $\mathcal{H}$ is constructed (see Section 4.1.1), where the vertex class A is the set of edges of a certain tree T derived from the tree representation of G, and the vertex class X is $V(G)$. The link in $\mathcal{H}$ of each element e of A is a carefully constructed graph called the *overspan* graph of e. It is then proved that if $\mathcal{H}$ has a matching of size $|A|$ then G has a Hamilton cycle. To complete the proof, Theorem 4.5 is applied to show that $\mathcal{H}$ does have such a matching whenever G is 10-tough.

4.3 Colouring

Many graph colouring problems can be formulated as problems about independent transversals. For example, in the *list colouring* problem, each vertex v of a graph G is assigned a *list* $L(v)$ of "permissible colours". An *L-list colouring* is a proper colouring c of the vertices of G such that $c(v) \in L(v)$ for each vertex v. One well-studied parameter is the *list chromatic number* (also called the *choosability*), which for the graph G is the smallest k such that every list assignment L with $|L(v)| \geq k$ for each $v \in V(G)$ admits an L-list colouring.

To formulate list colouring as an IT problem, given a graph G and a list assignment L, construct an auxiliary graph G^L with one vertex class V_v for each $v \in V(G)$. The vertices of V_v are $\{v^\alpha : \alpha \in L(v)\}$. The edges of G^L are $\{v^\alpha w^\beta : vw \in E(G), \alpha = \beta\}$. Then an L-list colouring c corresponds precisely to an IT in G^L given by $\{v^{c(v)} : v \in V(G)\}$.

Given a graph G and a list assignment L, the *maximum colour degree* Δ^L is a parameter (see [63]) that coincides with the maximum degree of G^L. In [63], Reed conjectured that G has an L-list colouring whenever $|L(v)| \geq \Delta^L + 1$ for each $v \in V(G)$. Indeed this is true by Brooks' Theorem when all lists $L(v)$ are the same (so $\Delta^L = \Delta(G)$), and at the other extreme it is true when $\Delta^L = 1$. In fact this conjecture turned out to be false, as was shown in [29], but it could be true with $\Delta^L + 1$ replaced by $\Delta^L + 2$. It is an open problem to determine the best function of Δ^L that would guarantee L-list colourability here.

It follows immediately from Corollary 4.2 that if $|L(v)| \geq 2\Delta^L$ for each $v \in V(G)$ then G is L-list-colourable, and this is the currently the best known general bound. The conjecture has been proven asymptotically in [64], and for chordal graphs it follows from a result in [14] (see e.g. [15]).

The following generalisation of list colouring was introduced in [19]. Let $G_1, \ldots, G_k$ be graphs and set $V = \bigcup_{i=1}^k V(G_i)$. A *cooperative list colouring* is a collection of independent sets A_i in G_i for each i such that $\bigcup_{i=1}^k A_i = V$. If such a collection exists then clearly the A_i may be assumed to be disjoint, so this gives a colouring of V in the sense of being a partition into colour classes, but here each colour class A_i is contributed by the graph G_i. Several results on this notion were established in [19]. In particular, if each G_i has maximum degree at most d, and each vertex of V appears in at least $2d$ of the graphs G_i, then Corollary 4.2 implies that there exists a cooperative list colouring.

Other variations on the notion of graph colouring for which the methods and results described in this article have been useful include the following. In *strong colouring*, one seeks to colour the vertices of a graph G with k colours, while simultaneously colouring an arbitrary union of disjoint copies of the clique K_k on the same vertex set $V(G)$. The smallest k for which this is possible is the *strong chromatic number* $s\chi(G)$ of G. The best known bound on $s\chi(G)$ in terms of $\Delta(G)$ is $3\Delta(G) - 1$ in general [44], and asymptotically $(1 + o(1))2.75\Delta(G)$ [46]. A short proof that $s\chi(G) \leq 3\Delta(G)$ for general graphs, and $s\chi(G) \leq 2\Delta + 1$ for chordal graphs, is given in [15]. The main tool for the proof is Corollary 4.2.

In *circular colouring*, the aim is to minimise the ratio p/q such that a graph can be coloured with colours $\{0, \ldots, p-1\}$, where adjacent elements receive colours that are at least q apart (mod p). Motivated by the *girth conjecture* for snarks due to Jaeger and Swart [52], the following theorem was proved in [55].

Theorem 4.15. *Let $p \in \mathbb{N}$ with $p \geq 2$ and G be a cubic bridgeless graph with girth*

$$g = \begin{cases} 2(2p)^{2p-2} & \text{if } p \geq 2 \text{ is even} \\ 2(2p)^{2p} & \text{if } p \geq 3 \text{ is odd.} \end{cases}$$

Then G admits a proper circular $(3p+1)/p$-edge-colouring.

In the proof, Corollary 4.2 is used to show the existence of an IT in a certain auxiliary graph constructed from G and a perfect matching in G, which is then used to give a colouring with the desired properties.

Clustered colouring is a relaxation of the standard notion of proper graph colouring. A colouring (i.e. a partition of the vertex set) is said to have *clustering c* if each monochromatic component (component of the subgraph induced by a partition class) has at most c vertices. Corollary 4.1(2) and other independent transversal theorems have been used in various results on clustered colouring, for example [22, 50]. See [71] for these and much more on clustered colouring.

Finally we mention that many of these variations on colouring also have *fractional* versions, some of which are studied in e.g. [8, 9, 15].

Most of the colouring applications mentioned in this subsection use Corollary 4.1(2), usually in the form of Corollary 4.2. As mentioned in Section 4, these statements also have combinatorial proofs, and so topological considerations are not essential here. We will therefore not give any further details of these.

5   Remarks and open problems

5.1   Beyond independent transversals

An IT in a vertex-partitioned graph G can be viewed as a set that is simultaneously

- a simplex in the abstract simplicial complex $\mathcal{I}(G)$, and

- a base of the partition matroid induced by the vertex partition of G.

In the *partition matroid*, a set is *independent* if and only if no two vertices are in the same partition class. A *base* is a maximal independent set (all of which have the same size in a matroid). Therefore an IT in a graph can be interpreted as a member of the intersection of a matroid and a simplicial complex on the same vertex set. Study of such intersections was initiated in [8], and many of the results and methods described in this article were shown to have analogues in this much more general setting, including in particular a matroid version of Theorem 2.1. To give one explicit example from [8] (see also [13]), the following is a generalised version of Corollary 4.2.

Theorem 5.1. *Let G be a graph with maximum degree Δ and let $\mathcal{N}$ be a matroid on the vertex set $V(G)$. Suppose $\mathcal{N}$ has 2Δ disjoint bases. Then $\mathcal{N}$ has a base that is a simplex in $\mathcal{I}(G)$.*

Further work in the matroid setting includes for example [13], where the matroidal version of Theorem 2.1 from [8] together with a version of Theorem 3.2 are used to prove the following strengthening of Theorem 5.1.

Theorem 5.2. *Let G be a graph with maximum degree $\Delta \geq 3$ and let $\mathcal{N}$ be a matroid on the vertex set $V(G)$. Suppose $\mathcal{N}$ has 2Δ disjoint bases. Then $\mathcal{N}$ has two disjoint bases that are both simplices in $\mathcal{I}(G)$.*

Instead of an independent transversal, one may ask whether a transversal exists in a vertex-partitioned graph that induces a subgraph with different properties. For example, existence theorems for *forest transversals* and *H-free transversals* for fixed graphs H were proved in [68]. Transversals inducing only small components were studied in [50] and [68]. *Acyclic* transversals in directed graphs were investigated in [9], as well as an acyclic version of strong chromatic number, a fractional version of which was proved. Many of the results in these papers used the topological methods described in this article.

5.2 Open problems

There are many open problems related to the material discussed in the previous sections. For just one example that addresses the connectedness parameter specifically, the following is a question from [6] (see Section 3.2.2).

Conjecture 5.3. *Let G be a $K_{1,k}$-free graph with maximum degree d. Then*

$$\eta(\mathcal{I}(G)) \geq |V|/(d + k - 1).$$

In [6] this is shown to be sharp when $k - 1$ divides d, and is proved for line graphs of simple graphs, using the Laplacian bound described in Section 3.2.1.

The rest of this section loosely organises a selection of open problems into subsections that correspond to some of the topics in Section 4.

5.2.1 Ryser's Conjecture Ryser's conjecture that $\tau(\mathcal{H}) \leq (r-1)\nu(\mathcal{H})$ for every r-partite r-graph $\mathcal{H}$ (see Section 4.1.2) remains unsolved for all $r \geq 4$. The following even stronger conjecture due to Lovász [57] is open even for $r = 3$.

Conjecture 5.4. *Every r-partite r-graph $\mathcal{H}$ contains a set S of at most $r - 1$ vertices with the property that $\nu(\mathcal{H} - S) < \nu(\mathcal{H})$.*

Very little is known about this conjecture, apart from the fact that it holds when $r = 3$ and $\mathcal{H}$ is Ryser-extremal (i.e. $\tau(\mathcal{H}) = 2\nu(\mathcal{H})$). This is a simple consequence of the main result in [49].

Aharoni's proof of Ryser's Conjecture for $r = 3$ (Theorem 4.8) uses two key facts about bipartite graphs G: Lemma 4.3 that $\eta(\mathcal{M}(G)) \geq \nu(G)/2$, and König's Theorem that $\tau(G) = \nu(G)$. If one attempts to prove Ryser's Conjecture for $r = 4$ by following the steps of the proof of Theorem 4.8 for a 4-partite 4-graph $\mathcal{J}$, the analogous facts about 3-partite 3-graphs (applied to the link hypergraphs in $\mathcal{J}$) are (Lemma 4.3) that $\eta(\mathcal{M}(\mathcal{H})) \geq \nu(\mathcal{H})/3$, and (Theorem 4.8) that $\tau(\mathcal{H}) \leq 2\nu(\mathcal{H})$. These combine to give a bound that is even weaker than the trivial bound $\tau(\mathcal{J}) \leq 4\nu(\mathcal{J})$. However, if Lemma 4.3 could be replaced by $\eta(\mathcal{M}(\mathcal{H})) \geq \tau(\mathcal{H})/3$, then this argument would succeed in proving Ryser's Conjecture for $r = 4$. In fact, when $\mathcal{H}$ is itself a Ryser-extremal 3-graph, then $\eta(\mathcal{M}(\mathcal{H})) \geq \tau(\mathcal{H})/3$ does indeed hold. However, unfortunately this is not the case in general. As shown in [49], there exists a 3-partite 3-graph $\mathcal{H}$ with $\tau(\mathcal{H}) = 4$, $\nu(\mathcal{H}) = 3$, and $\eta(\mathcal{M}(\mathcal{H})) = 1$. In [49] it is conjectured that $\eta(\mathcal{M}(\mathcal{H})) \geq \tau(\mathcal{H})/4$ for every 3-partite 3-graph $\mathcal{H}$.

In [8], Aharoni and Berger formulated a version of Ryser's Conjecture for matroids. A special case of this was addressed in [26].

For further partial results, constructions and references for Ryser's Conjecture see e.g. [1, 7, 39, 49].

5.2.2 Matchings in hypergraphs Theorem 4.12 implies that if the t-regular 3-partite 3-graph $\mathcal{H}$ with n vertices in each class contains no copy of $\frac{t}{2} \cdot F$ then

$$\nu(\mathcal{H}) \geq \left(1 + \frac{1}{22t - \frac{77}{3}}\right) \frac{n}{2}.$$

This is not so far from being best possible, as examples of such 3-graphs $\mathcal{H}$ were given in [48] where $\nu(\mathcal{H}) \leq \left(1 + \frac{1}{t+1}\right) \frac{n}{2}$ for even t and $\nu(\mathcal{H}) \leq \left(1 + \frac{1}{t}\right) \frac{n}{2}$ for odd t. It is natural to conjecture that these are the correct expressions for this case.

The constructions in [48] giving these bounds have high edge multiplicity, and it is believed that much stronger bounds should hold for simple hypergraphs (i.e. those without multiple edges). Indeed it was conjectured in [16] that if $\mathcal{H}$ is a simple t-regular 3-partite 3-graph with n vertices in each class then $\nu(\mathcal{H}) \geq \lceil \frac{(t-1)n}{t} \rceil$. However, this conjecture turned out to be too strong, as shown in [5] where such 3-graphs $\mathcal{H}$ are constructed where t is arbitrarily large but $\nu(\mathcal{H}) \leq \frac{2n}{3}$. Therefore the following special case of a conjecture of Alon and Kim [23] would be essentially best possible (noting that, if true, the statement would imply that every simple t-regular 3-partite 3-graph with n vertices in each class has a matching of size at least $(2/3 - o(1))n$).

Conjecture 5.5. *The edges of any 3-uniform simple hypergraph with maximum degree t can be partitioned into $(3/2 + o(1))t$ matchings.*

The best known partial result for the 3-regular case was proved in [31], where it was shown that every simple 3-regular 3-partite 3-graph with n vertices in each class has a matching of size at least $3n/5$.

A large class of open problems about hypergraph matching (some notoriously old and difficult) is the collection of so-called *rainbow matching* problems. Perhaps the most famous is the following conjecture (posed independently and in several slightly different forms) by Ryser, Brualdi and Stein.

Conjecture 5.6. *Every Latin square of order n has a partial transversal of size $n - 1$.*

This can be viewed as a matching problem in a special type of 3-partite 3-graph $\mathcal{H}$ as follows. The vertex classes of $\mathcal{H}$ are three copies R, C and S of $\{1, \ldots, n\}$, corresponding to the rows, columns and symbols respectively of a given Latin square. Each entry a in the Latin square corresponds to exactly one edge of $\mathcal{H}$, which consists of the row, column and symbol of a. Then the properties of Latin squares imply that every pair of vertices in different vertex classes of $\mathcal{H}$ appears in exactly one edge of $\mathcal{H}$. Then the link of each vertex in $\mathcal{H}$ is a matching of size n in a complete bipartite graph. A partial transversal of the Latin square corresponds to a matching in the hypergraph $\mathcal{H}$.

More generally, *rainbow matching* means a set of disjoint edges, each of which comes from a distinct member of a given family of hypergraphs (or graphs). For example the following more general conjecture is posed in [16].

Conjecture 5.7. *Any n matchings of size $n + 1$ in a bipartite graph contain a rainbow matching of size n.*

The methods described in this article have been applied to obtain partial results on rainbow matching problems, for example [11]. For a comprehensive account of many of the better-known rainbow matching conjectures and how they relate to each other see [16].

For a finite set $X \subset \mathbf{R}^d$, the maximal size of a subset of X in general position is denoted by $\varphi(X)$. The following geometric version of Hall's Theorem for hypergraphs was proved in [51].

Theorem 5.8. *For each $d \geq 1$ there exists a function f_d satisfying the following. For every family $\{X_1, \ldots, X_m\}$ of finite sets in $\mathbf{R}^d$ such that*

$$\varphi(\bigcup_{i \in S} X_i) \geq f_d(|S|)$$

for every nonempty subset $S \subseteq \{1, \ldots, m\}$, there exists a subset $\{x_1, \ldots, x_m\}$ in general position, where $x_i \in X_i$ for $1 \leq i \leq m$.

The proof in [51] uses the type of methods described in this article to give an upper bound of $O(k^d)$ on the function $f_d(k)$. The authors pose the problem of finding optimal bounds on this function.

5.2.3 Colouring It is believed (see e.g. [15]) that the correct bound for the strong chromatic number $s\chi(G)$ in terms of the maximum degree $\Delta(G)$ is as follows.

Conjecture 5.9. *Every graph G satisfies $s\chi(G) \leq 2\Delta(G)$.*

However this is not known even for graphs of maximum degree 2 (see e.g. [53]). A fractional version was proved in [15]. A version of the conjecture for matroids was proposed in [8], which can be stated as follows (see [13]).

Conjecture 5.10. *Let G be a graph with maximum degree $\Delta \geq 2$ and let $\mathcal{N}$ be a matroid on the vertex set $V(G)$. Suppose that the maximal number $\pi(\mathcal{N})$ of disjoint bases of $\mathcal{N}$ satisfies $\pi(\mathcal{N}) \geq 2\Delta$. Then $\mathcal{N}$ has $\pi(\mathcal{N})$ disjoint bases, each of which is a simplex of $\mathcal{I}(G)$.*

Thus Theorem 5.1 is a first step towards proving Conjecture 5.10.

Several questions about cooperative list colouring are posed in [19] (see Section 4.3), including the following. Does there exist an absolute constant c such that graphs $G_1, \ldots, G_{d+c}$ have a cooperative list colouring whenever each G_i has maximum degree at most d and $V(G_i) = V$ for each i? (This last condition makes it a *cooperative colouring* instead of a cooperative list colouring.) In [19] it is shown that $c = 1$ is too small for this to be true in general, but $c = 2$ could even be possible, and is proved in some special cases.

Acknowledgements

The author's research was partially supported by NSERC.

References

[1] A. Abu-Khazneh, J. Barát, A. Pokrovskiy and T. Szabó, A family of extremal hypergraphs for Ryser's Conjecture, *Journal of Combinatorial Theory, Series A* **161** (2019), 164–177.

[2] M. Adamaszek and J.A. Barmak, On a lower bound for the connectivity of the independence complex of a graph, *Discrete Mathematics* **311** (2011), 2566–2569.

[3] R. Aharoni, Matchings in n-partite n-graphs, *Graphs and Combinatorics* **1** (1985), 303–304.

[4] R. Aharoni, Ryser's conjecture for tripartite 3-graphs, *Combinatorica* **21** (2001), 1–4.

[5] R. Aharoni, N. Alon, M. Amir, P. Haxell, D. Hefetz, Z. Jiang, G. Kronenberg and A. Naor, Ramsey-nice families of graphs, *European Journal of Combinatorics* **72** (2018), 29–44.

[6] R. Aharoni, N. Alon and E. Berger, Eigenvalues of $K_{1,k}$-free graphs and the connectivity of their independence complexes, *Journal of Graph Theory* **83** (2016), 384–391.

[7] R. Aharoni, J. Barát and I. Wanless, Multipartite hypergraphs achieving equality in Ryser's Conjecture, *Graphs and Combinatorics* **32** (2016), 1–15.

[8] R. Aharoni and E. Berger, The intersection of a matroid and a simplicial complex, *Transactions of the AMS* **358** (2006), 4895–4917.

[9] R. Aharoni, E. Berger and O. Kfir, Acyclic systems of representatives and acyclic colorings of digraphs, *Journal of Graph Theory* **59** (2008), 177–189.

[10] R. Aharoni, E. Berger, D. Kotlar and R. Ziv, Degree conditions for matchability in 3-partite hypergraphs, *Journal of Graph Theory* **87** (2017), 61–71.

[11] R. Aharoni, E. Berger, D. Kotlar and R. Ziv, On a conjecture of Stein, *Abhandlungen aus dem Mathematischen Seminar der Universität Hamburg* **87** (2017), 203–211.

[12] R. Aharoni, E. Berger and R. Meshulam, Eigenvalues and homology of flag complexes and vector representations of graphs, *Geometric and Functional Analysis* **15** (2005), 555–566.

[13] R. Aharoni, E. Berger and P. Sprüssel, Two disjoint independent bases in matroid-graph pairs, *Graphs and Combinatorics* **31** (2015), 1107–1116.

[14] R. Aharoni, E. Berger and R. Ziv, A tree version of König's Theorem, *Combinatorica* **22** (2002), 335–343.

[15] R. Aharoni, E. Berger and R. Ziv, Independent systems of representatives in weighted graphs, *Combinatorica* **27** (2007), 253–267.

[16] R. Aharoni, P. Charbit and D. Howard, On a generalization of the Ryser-Brualdi-Stein Conjecture, *Journal of Graph Theory* **78** (2015), 143–156.

[17] R. Aharoni, M. Chudnovsky and A. Kotlov, Triangulated spheres and colored cliques, *Discrete and Computational Geometry* **28** (2002), 223–229.

[18] R. Aharoni and P. Haxell, Hall's theorem for hypergraphs, *Journal of Graph Theory* **35** (2000), 83–88.

[19] R. Aharoni, R. Holzman, D. Howard and P. Sprüssel, Cooperative colorings and independent systems of representatives, *Electronic Journal of Combinatorics* **22** (2015), P2.27.

[20] N. Alon, The linear arboricity of graphs, *Israel Journal of Mathematics* **62** (1988), 311–325.

[21] N. Alon, The strong chromatic number of a graph, *Random Structures & Algorithms* **3** (1992), 1–7.

[22] N. Alon, G. Ding, B. Oporowski and D. Vertigan, Partitioning into graphs with only small components, *Journal of Combinatorial Theory, Series B* **87** (2003), 231–243.

[23] N. Alon and J.H. Kim, On the degree, size, and chromatic index of a uniform hypergraph, *Journal of Combinatorial Theory, Series A* **77** (1997), 165–170.

[24] C. Annamalai, Finding perfect matchings in bipartite hypergraphs, *Combinatorica* **38** (2018), 1285–1307.

[25] J. Barmak, Star clusters in independence complexes of graphs, *Advances in Mathematics* **241** (2013), 33–57.

[26] E. Berger and R. Ziv, A note on the cover number and independence number in hypergraphs, *Discrete Mathematics* **308** (2008), 2649–2654.

[27] R. Berke, P. Haxell and T. Szabó, Bounded transversals in multipartite graphs, *Journal of Graph Theory* **70** (2012), 318–331.

[28] A. Björner, Topological methods, in *Handbook of Combinatorics* (ed. R. Graham, M. Grötschel, L. Lovász), Elsevier and the MIT Press, Boston (1995).

[29] T. Bohman and R. Holzman, On a list coloring conjecture of Reed, *Journal of Graph Theory* **41** (2002), 106–109.

[30] B. Bollobás, P. Erdős and E. Szemerédi, On complete subgraphs of r-chromatic graphs, *Discrete Mathematics* **13** (1975), 97–107.

[31] N.J. Cavenagh, J. Kuhl and I.M. Wanless, Longest partial transversals in plexes, *Annals of Combinatorics* **18** (2014), 419–428.

[32] G. Chen, H. Jacobson, A. Kézdy and J. Lehel, Tough enough chordal graphs are Hamiltonian, *Networks* **31** (1998), 29–38.

[33] V. Chvátal, Tough graphs and hamiltonian circuits, *Discrete Mathematics* **5** (1973), 215–228.

[34] D.Z. Du, D.F. Hsu and F.K. Hwang, The Hamiltonian property of consecutive-d digraphs, *Mathematical and Computer Modelling* **17** (1993), 61–63.

[35] R. Ehrenborg and G. Hetyei, The topology of the independence complex, *European Journal of Combinatorics* **27** (2006), 906–923.

[36] A. Engström, Independence complexes of claw-free graphs, *European Journal of Combinatorics* **29** (2008), 234–241.

[37] M. Fellows, Transversals of vertex partitions in graphs, *SIAM Journal of Discrete Mathematics* **3** (1990), 206–215.

[38] H. Fleischner and M. Stiebitz, A solution to a colouring problem of P. Erdős, *Discrete Mathematics* **101** (1992), 39–48.

[39] N. Francetić, S. Herke, B.D. McKay and I. M. Wanless, On Ryser's conjecture for linear intersecting multipartite hypergraphs, *European Journal of Combinatorics* **61** (2017), 91–105.

[40] F. Gavril, The intersection graph of subtrees in a tree are exactly the chordal graphs, *Journal of Combinatorial Theory, Series B* **16** (1974), 47–56.

[41] A. Graf and P. Haxell, Finding independent transversals efficiently, preprint.

[42] P. Haxell, A condition for matchability in hypergraphs, *Graphs and Combinatorics* **11** (1995), 245–248.

[43] P. Haxell, A note on a conjecture of Ryser, *Periodica Mathematica Hungarica* **30** (1995), 73–79.

[44] P. Haxell, On the strong chromatic number, *Combinatorics Probability and Computing* **13** (2004), 857–865.

[45] P. Haxell, Independent transversals and hypergraph matchings - an elementary approach, in *Recent Trends in Combinatorics* (ed. A. Beveridge, J. Griggs, L. Hogben, G. Musiker, P. Tetali), *IMA Volume in Mathematics and its Applications*, Springer, New York (2016), pp. 215–233.

[46] P. Haxell, A new bound on the strong chromatic number, *Journal of Graph Theory* **58** (2008), 148–158.

[47] P. Haxell, On forming committees, *American Mathematical Monthly* **118** (2011), 777–788.

[48] P. Haxell and L. Narins, Stability for matchings in tripartite 3-graphs, *Combinatorics Probability and Computing* **27** (2018), 774–793.

[49] P. Haxell, L. Narins and T. Szabó, Extremal Hypergraphs for Ryser's Conjecture, *Journal of Combinatorial Theory, Series A* **158** (2018), 492–547.

[50] P. Haxell, T. Szabó and G. Tardos, Bounded size components - partitions and transversals, *Journal of Combinatorial Theory, Series B* **88** (2003), 281–297.

[51] A. Holmsen, L. Martinez-Sandoval and L. Montejano, A geometric Hall-type theorem, *Proceedings of the AMS* **144** (2016), 503–511.

[52] F. Jaeger and T. Swart, Conjecture 1, in *Combinatorics 79* (ed. M. Deza, I. G. Rosenberg), *Annals of Discrete Mathematics*, **9** North-Holland, Amsterdam (1980), pp. 305.

[53] T.R. Jensen and B. Toft, *Graph Coloring Problems*, Wiley, New York (1995).

[54] A. Kabela and T. Kaiser, 10-tough chordal graphs are Hamiltonian, *Journal of Combinatorial Theory, Series B* **122** (2017), 417–427.

[55] T. Kaiser, D. Král and R. Skrekovski, A revival of the girth conjecture, *Journal of Combinatorial Theory, Series B* **92** (2004), 41–53.

[56] D. Kozlov, Complexes of directed trees, *Journal of Combinatorial Theory, Series A* **88** (1999), 112–122.

[57] L. Lovász, On minimax theorems of combinatorics (in Hungarian), *Matematikai Lapok* **26** (1975), 209–264.

[58] L. Lovász, Kneser's Conjecture, chromatic number, and homotopy, *Journal of Combinatorial Theory, Series A* **25** (1978), 319–324.

[59] J. Matoušek, *Using the Borsuk-Ulam Theorem*, Lectures on Topological Methods in Combinatorics and Geometry, Springer, Berlin (2003).

[60] R. Meshulam, The clique complex and hypergraph matchings, *Combinatorica* **21** (2001), 89–94.

[61] R. Meshulam, Domination numbers and homology, *Journal of Combinatorial Theory, Series A* **102** (2003), 321–330.

[62] L. Narins, Extremal Hypergraphs for Ryser's Conjecture (PhD thesis), Freie Universität Berlin, (2014).

[63] B. Reed, The list colouring constants, *Journal of Graph Theory* **31** (1999), 149–153.

[64] B. Reed and B. Sudakov, Asymptotically the list colouring constants are 1, *Journal of Combinatorial Theory, Series B* **86** (2002), 27–37.

[65] H. Ryser, Neuere Probleme der Kombinatorik, in *Voträge über Kombinatorik Oberwolfach, Mathematisches Forschungsinstitut Oberwolfach, Colloquia Mathematica Societatis János Bolyai*, (1967), pp. 6–91.

[66] H. Sachs, Elementary proof of the cycle-plus-triangles theorem, in *Combinatorics, Paul Erdős is Eighty I* (ed. D. Miklós, V.T. Sós, and T. Szönyi), *Bolyai Society Mathematical Studies, Budapest*, (1993), pp. 347–359.

[67] E. Sperner, Neuer Beweis für die Invarianz der Dimensionszahl und des Gebietes, *Abhandlungen aus dem Mathematischen Seminar der Universität Hamburg* **6** (1928), 265–272.

[68] T. Szabó and G. Tardos, Extremal problems for transversals in graphs with bounded degree, *Combinatorica* **26** (2006), 333–351.

[69] E. Szemerédi and Zs. Tuza, Upper bound for transversals of tripartite hypergraphs, *Periodica Mathematica Hungarica* **13** (1982), 321–323.

[70] Zs. Tuza, On the order of vertex sets meeting all edges of a 3-partite hypergraph, *Ars Combinatoria* **24(A)** (1987), 59–63.

[71] D. Wood, Defective and clustered graph colouring, *Electronic Journal of Combinatorics* (2018), DS23.

Department of Combinatorics and Optimization
University of Waterloo
Waterloo ON Canada N2L 3G1
pehaxell@uwaterloo.ca

Expanders – how to find them, and what to find in them

Michael Krivelevich[1]

Abstract

A graph $G = (V, E)$ is called an expander if every vertex subset U of size up to $|V|/2$ has an external neighborhood whose size is comparable to $|U|$. Expanders have been a subject of intensive research for more than three decades and have become one of the central notions of modern graph theory.

We first discuss the above definition of an expander and its alternatives. Then we present examples of families of expanding graphs and state basic properties of expanders. Next, we introduce a way to argue that a given graph contains a large expanding subgraph. Finally we research properties of expanding graphs, such as existence of small separators, of cycles (including cycle lengths), and embedding of large minors.

1 Introduction

Putting it somewhat informally, a graph G is an expander if every vertex subset U expands outside substantially, i.e., has an external neighborhood whose size is comparable to $|U|$. Since their introduction in the seventies, expanders have become one of the most central, and also one of the most applicable, notions of modern combinatorics. Their uses are numerous and wide-ranging, from extremal problems in graph theory and explicit constructions of graphs with desired properties to design and implementation of lean yet reliable communication networks. The reader is encouraged to consult the comprehensive and very readable survey of Hoory, Linial and Wigderson [25], devoted entirely to expanding graphs and covering many aspects of this subject; the very concise presentation of Sarnak [50] would be a (somewhat...) shorter introduction.

Ever since the seminal paper of Pinsker [47], randomness has been used to argue about expanders and their existence. In many cases, random graphs, sometimes after minor alterations, serve as typical examples of expanders with desired parameters. In fact, expanders are frequently viewed as synonymous to pseudo-random graphs, the latter being graphs whose edge distribution, and other properties, resemble closely those of truly random graphs with the same density. Pseudo-random graphs are covered in survey [38]. The notions of expanders and of pseudo-random graphs, while being close, are quite distinct — for example, a bipartite graph can be an excellent expander, but cannot quite model a random graph. Ideology and tools can be rather similar; in particular, spectral methods are used frequently to argue about both expansion and pseudo-randomness.

In this survey we undertake a (more) qualitative study of expansion and expanding graphs. Instead of looking for best possible quantitative results on expansion, strongest expanders possible, largest eigenvalue ratio etc., we aim here to study inherent properties of expanders, their derivation and consequences. We will rely less on the strength of expansion to derive our results; in a sense our assumption about expansion is fairly weak (we will sometimes even call our expanders weak explicitly). As it turns out, even the rather relaxed definition of expanders we embrace is deep enough to allow for meaningful study and quite interesting consequences.

[1]The author was supported by the Clay Mathematics Institute. In particular, this survey was written as part of the author's role as the Clay Lecturer at the 27th British Combinatorial Conference. The research was also supported in part by USA–Israel BSF grant 2014361 and by ISF grant 1261/17.

The text is organized as follows. In the next section, we introduce a formal definition of expanders we use in this survey, and compare it with alternative but similar definitions. In Section 3 we discuss basic properties of expanding graphs. Examples of families of expanding graphs are presented in Section 4; we use the examples presented also to adjust our expectations, and to see which graph structures can and cannot be hoped to be found in weakly expanding graphs. The connection between expanders and separators is presented in Section 5. In Section 6 we introduce a general tool for proving the existence of a large expanding subgraph in a given graph, this tool is based on the notion of local sparseness; we discuss this notion in the context of sparse random graphs. Amongst the structures one can hope to find in expanders are long paths and cycles, and in Section 7 we show that indeed weak expanders are rich in cycles, in several well defined quantitative senses. Finally, in Section 8 we present results about embedding large minors in expanding graphs.

2   Definition(s) of an expander

Let us start with introducing (the rather standard) notation used throughout this survey. For a graph $G = (V, E)$ and a vertex set $U \subset V$ we denote by $N_G(U)$ the external neighborhood of U in G, i.e.,

$$N_G(U) = \{v \in V \setminus U : v \text{ has a neighbor in } U\}.$$

We also write, for two disjoint subsets $U, W \subset V$,

$$N_G(U, W) = \{v \in W : v \text{ has a neighbor in } U\},$$

thus $N_G(U) = N_G(U, V \setminus U)$. Also, $e_G(U, W)$ stands for the number of edges of G between U and W. Sometimes when the relevant graph G is clear from the context, we will omit the subscript in the above notation.

We can now give the most basic definition of this survey, that of an α-expander.

Definition 2.1. Let $G = (V, E)$ be a graph on n vertices, and let $\alpha > 0$. The graph G is an *α-expander* if $|N_G(U)| \geq \alpha|U|$ for every vertex subset $U \subset V$ with $|U| \leq n/2$.

It is important to note that although the above definition appears to cover only expansion of sets W up to half the order of G, it extends further to sets whose size exceeds $n/2$. Indeed, if $|U| > n/2$, then the set $W = V - (U \cup N_G(U))$ has size $|W| < n/2$, and satisfies $N_G(W) \subseteq N_G(U)$. This allows to bound $|N_G(U)|$ from below as a function of $|U|$, when assuming that G is an α-expander.

Our primary concern in this survey is neither getting strongest or best possible expanders, nor pushing the quantitative notion of expansion to the limit. Our goal is rather different — we mostly assume that we have a decent (one could even say weak) expander G, with α in the definition of an α-expander being a small constant, say, $\alpha = 0.01$, and aim to derive some nice properties of G. In principle, the expansion factor α in the above definition can even be allowed to be a vanishing function of the order n of G, however here we mostly stick to the assumption $\alpha = \Theta(1)$. We should add here that throughout this survey we allow ourselves to use expressions such as "weak expander", "strong expander", etc. rather informally, aiming to indicate the extent of expansion of the corresponding graph; of course formal definitions such as α-expansion should be used to measure the expansion formally and quantitatively.

The choice to stop the expansion assumption at $n/2$ in Definition 2.1 may appear somewhat arbitrary. As we will see shortly, this is certainly a convenient compromise, as it allows us to derive easily some basic properties of G, for example its connectedness. It is quite natural though to give also more general definitions, like the ones below.

Definition 2.2. Let $G = (V, E)$ be a graph, let I be a set of positive integers, and let $\alpha > 0$. The graph G is an (I, α)-*expander* if $|N_G(U)| \geq \alpha|U|$ for every vertex subset $U \subset V$ satisfying $|U| \in I$.

Definition 2.3. Let $G = (V, E)$ be a graph, let k be a positive integer, and let $\alpha > 0$. The graph G is a (k, α)-*expander* if $|N_G(U)| \geq \alpha|U|$ for every vertex subset $U \subset V$ satisfying $|U| \leq k$.

Notice, mostly to exercise the above defined notions, that an α-expander G on n vertices is (exactly) an (I, α)-expander for $I = \{1, \ldots, n/2\}$, and a (k, α)-expander for $k = n/2^2$.

Are the above three notions of expansion radically different? Does an (I, α)-expander on n vertices with I being just a single value, say, $I = \{n/100\}$ have properties much different from an $(n/300, \alpha)$-expander, or from an α-expander? Well, yes in some senses (for example, the former two are easily seen not to guarantee connectedness, while the latter one does). However, since in this survey we will mostly research qualitative properties and will frequently settle for finding substructures in G of size linear in n, without trying too much to optimize the constants involved, in many other senses the above three definitions are morally equivalent. We will provide a formal statement supporting this paradigm shortly.

Definition 2.1 is a fairly commonly used notion of an expander, see, e.g., [2], or [6, Chapter 9]. For the case of d-regular graphs with constant d, or more generally, for graphs with bounded maximum degree, being an α-expander is essentially equivalent (qualitatively) to having edge expansion bounded away from 0. The latter notion can be quantified through the so-called Cheeger constant of a graph G. For a graph $G = (V, E)$ on n vertices and a subset $U \subseteq V$ set $\mathrm{vol}_G(U) = \sum_{v \in U} \deg_G(v)$. Define now

$$h(G) = \min_{\emptyset \neq U \subsetneq V} \frac{e_G(U, V \setminus U)}{\min(\mathrm{vol}_G(U), \mathrm{vol}_G(V \setminus U))},$$

the quantity $h(G)$ is called the *Cheeger constant* of G. (See, e.g., [13] for a general discussion.) Having $h(G)$ bounded away from 0 means the graph is a decent edge expander, and — assuming its maximum degree is bounded — is a decent vertex expander as well.

Let us now address the quantitative comparison between Definitions 2.1, 2.2 and 2.3. As announced, we will see that they are essentially equivalent.

Lemma 2.4. *Let G be an (I, α)-expander on n vertices with $I = \left[\frac{n}{3}, \frac{2n}{3}\right]$. Then there is a vertex subset $Z \subset V(G)$ of size $|Z| < n/3$ such that the graph $G' = G[V \setminus Z]$ is an α-expander.*

Proof. Start with $G' = G$, and as long as there is a vertex subset $A \subset V(G')$ of size $|A| \leq |V(G')|/2$ satisfying $|N_{G'}(A)| < \alpha|A|$, delete A and update $G' := G'[V(G') \setminus A]$. Observe that the disjoint union of the deleted sets Z obviously satisfies $|N_G(Z)| < \alpha|Z|$. Hence by our assumption on the expansion of G we derive that the size of Z is never in I during the execution of the procedure, so either $|Z| < n/3$ or $|Z| > 2n/3$. Suppose that after some iteration the size of Z exceeds $2n/3$ for the first time. Let A be the set deleted

---

[2]Here and later we allow ourselves to treat rounding issues somewhat casually.

at this iteration, and let $Z' = Z \setminus A$. Then $|Z'| < n/3$, and $|A| \le (n - |Z'|)/2$, implying $|Z| = |Z'| + |A| \le |Z'| + (n - |Z'|)/2 = n/2 + |Z'|/2 < n/2 + n/6 = 2n/3$ – a contradiction. This indicates that the deletion procedure halts with $|Z| < n/3$, and the subgraph $G[V \setminus Z]$ has the required expansion property. □

Lemma 2.5. *Let G be a (k, α)-expander. Then G is a $\left(\frac{3k}{2}, \frac{\alpha}{6}\right)$-expander, or there is a subset $V_0 \subset V(G)$ of size $|V_0| \ge \frac{2k}{3}$ such that the induced subgraph $G[V_0]$ is an $(\alpha/2)$-expander.*

Proof. We can assume that G has more than $2k$ vertices, as otherwise the second alternative of the lemma is satisfied with $V_0 = V(G)$. If G does not satisfy the first alternative, then there exists a set X, $k < |X| \le \frac{3k}{2}$ with $|N_G(X)| < \frac{\alpha|X|}{6}$. But then every subset $Y \subset X$ with $\frac{|X|}{3} \le |Y| \le \frac{2|X|}{3}$, satisfies:

$$|N_G(Y, X \setminus Y)| \ge |N_G(Y)| - |N_G(X)| \ge \alpha|Y| - \frac{\alpha|X|}{6} \ge \alpha|Y| - \frac{\alpha|Y|}{2} = \frac{\alpha|Y|}{2}.$$

Feeding the induced subgraph $G[X]$ to Lemma 2.4 produces the required $(\alpha/2)$-expander on at least $2|X|/3 \ge 2k/3$ vertices. □

Lemma 2.6. *Let $G = (V, E)$ be a $(\{k\}, \alpha)$-expander with $0 < \alpha \le 1$. Then there is a vertex subset $Z \subset V$ of size $|Z| < k$ such that the graph $G' = G[V \setminus Z]$ is an $\left(\frac{\alpha k}{3}, \frac{\alpha}{3}\right)$-expander.*

Proof. Start with $G' = G$, and as long as there is a vertex subset $A \subset V(G')$ of size $|A| \le \frac{\alpha k}{3}$ with $|N_{G'}(A)| < \alpha|A|/3$, delete A and update $G' := G'[V(G') \setminus A]$. Observe that the union Z of deleted sets obviously satisfies $|N_G(Z)| < \alpha|Z|/3$. Assume $|Z|$ reaches k after some iteration. Let A be the set deleted at this iteration, and let $Z' = Z - A$. Then $|Z'| < k$, and $|A| \le \alpha k/3$ with $|N_G(A, V - Z')| < \alpha|A|/3$. Choose an arbitrary subset $A' \subset A$ so that $|Z' \cup A'| = k$. Then

$$\alpha k \le |N_G(Z' \cup A')| \le |N_G(Z')| + |N_G(A', V \setminus Z')| \le |N_G(Z')| + |N_G(A, V \setminus Z')| + |A \setminus A'|$$
$$< \frac{\alpha k}{3} + \frac{\alpha|A|}{3} + |A| < \frac{\alpha k}{3} + \frac{\alpha^2 k}{3} + \frac{\alpha k}{3} \le \alpha k$$

– a contradiction. □

We arrive at the following conclusion:

Theorem 2.7. *For every $0 < c, \alpha \le 1$ there are $c', \alpha' > 0$ such that the following holds. Let G be a $(\{k\}, \alpha)$-expander on n vertices for $k = \lfloor cn \rfloor$. Then G contains an induced subgraph G' on at least $c'n$ vertices which is an α'-expander.*

The above theorem indicates clearly that the alternative definitions of expanders, as given by Definitions 2.1–2.3, are basically equivalent on the qualitative level: assuming a one-point expansion of Definition 2.2 with $I = \{k\}$ for $k = \Theta(n)$ delivers an α-expander of Definition 2.1 on linearly many vertices. Hence in the rest of this survey we will mostly work with Definition 2.1.

Observing Theorem 2.7, one may wonder whether assuming expansion of sublinear sets in a graph G, or even more ambitiously of sets of size up to k, where k is much smaller than $n = |V(G)|$, can guarantee the existence of a substantially sized expanding subgraph G' of G. For one, a (k, α)-expander on n vertices can be just a collection of disjoint cliques of size about $(1 + \alpha)k$ each, and so if $k = o(n)$, such G does not contain a linearly sized expanding subgraph. Thus, the assumption $k = \Theta(n)$ in Theorem 2.7 is essential. A much more striking example is provided by Moshkovitz and Shapira [45]. They prove (see Theorem 1 of [45]) that for every n and $k = O(\log \log n)$, there is an n-vertex graph G with all degrees of order k and girth$(G) = \Omega(\log n/k^2)$, in which every induced subgraph H on $t \geq$ girth(G) vertices has a set of edges $E_0 \subseteq E(H)$ of size $|E_0| = o(t)$, whose removal from H creates a graph H' with all connected components having at most $2t/3$ vertices. (Rephrasing it, every induced subgraph H of G with less than girth(G) vertices is a tree, while every induced H with at least girth(G) vertices has a small edge-separator and thus cannot be an expander.) Such a graph G can easily be argued to be a good local expander; for example, every set $U \subset V(G)$ of size $|U| \leq (ck)^{(\text{girth}(G)-1)/2}$ satisfies $|N_G(U)| \geq 2|U|$, for an appropriately chosen constant $c > 0$, see Lemma 2.1 of [52]. Hence local expansion in a graph cannot always be traded for global expansion in a subgraph.

3 Basic properties of α-expanders

This short section discusses the most basic properties of α-expanders. It can serve as a vindication of our choice to stick to Definition 2.1 as the definition of choice for expanders.

For a graph $G = (V, E)$, a vertex $v \in V$ and a natural number t, denote by $B_t(v)$ the ball of radius t around v in G, i.e., $B_t(v)$ is the set of vertices of G at distance at most t from v.

Proposition 3.1. *Let $G = (V, E)$ be an α-expander on n vertices. Then for every $v \in V$ and every natural t, we have: $|B_t(v)| \geq \min\left\{\frac{n}{2}, (1+\alpha)^t\right\}$.*

Proof. By induction on t. Obvious for $t = 0$. For the induction step, observe that $B_t(v) \supseteq B_{t-1}(v)$. If $|B_{t-1}(v)| \geq \frac{n}{2}$, we are done. Otherwise, by the induction hypothesis we have $|B_{t-1}| \geq (1+\alpha)^{t-1}$. Denoting $U = B_{t-1}(v)$, we have: $B_t(v) = U \cup N(U)$, and $|N(U)| \geq \alpha|U|$. It follows that $|B_t(v)| \geq (1+\alpha)|U| \geq (1+\alpha)^t$, as required. $\square$

Corollary 3.2. *Let G be an α-expander on n vertices. Then the diameter $diam(G)$ satisfies:* $\text{diam}(G) \leq \left\lceil \frac{2(\log n - 1)}{\log(1+\alpha)} \right\rceil + 1 = O_\alpha(\log n)$.

Proof. Let $u \neq v \in V(G)$. Growing balls around u, v, we obtain by Proposition 3.1 that for $t = \left\lceil \frac{\log(n/2)}{\log(1+\alpha)} \right\rceil$ the balls $B_t(u), B_t(v)$ have at least $n/2$ vertices each. Then they intersect each other, or are connected by an edge. $\square$

In particular, an α-expander G is connected.

Turning now to the Cheeger constant, observe that an α-expander G of maximum degree $\Delta = \Delta(G)$ satisfies trivially: $h(G) \geq \frac{\alpha}{\Delta}$. Thus, assuming that both $\alpha, \Delta = \Theta(1)$, we derive that in this case the Cheeger constant of G is bounded away from zero.

Going from the Cheeger constant to random walks, we derive, using the classical result of Jerrum and Sinclair [27], that the mixing time T_{mix} of a (lazy) random walk on an α-expander G on n vertices with $\alpha = \Theta(1)$ and bounded maximum degree satisfies $T_{mix} = O(\log n)$. The assumption that the maximum degree of G is bounded is essential here, as for example taking G to be two disjoint cliques of size $n/2$ connected by a nice bounded degree expander, we observe that G is an α-expander with $\alpha = \Theta(1)$, and yet it takes a random walk a linear time in n on average to cross from one clique to the other. Since random walks are essentially outside the scope of this survey, we will not dwell on their aspects anymore, instead referring the interested reader to standard sources on random walks (say, [41]) for background, definitions and discussion.

We can already conclude that the definition of α-expanders we have adopted is capable of guaranteeing easily several basic and desirable properties of the graph. We aim however to set our goals much higher than that, and to derive further properties of α-expanders, such as the existence of linearly long paths and cycles, non-existence of sublinear separators, and embedding of large minors. These properties are to be discussed later in this survey.

4   Examples of α-expanders

In this section we present some archetypal (families of) examples of weak expanders. The aim here is not just to showcase concrete instances of α-expanders, but also to adjust our level of expectations of what can be found in α-expanders upon observing some concrete examples.

1. Bipartite graphs. The complete bipartite graph G with parts of size αn and $(1 - \alpha)n$ for $\alpha \leq 1/2$ is obviously an α-expander. This example can be sparsified by taking G to be a good expander with linearly many edges between parts of size $2\alpha n$ and $(1 - 2\alpha)n$ (for $\alpha \leq 1/4$).

This example shows us that we cannot always expect to find an odd cycle in an expander.

2. Spectral approach. This is a very powerful and appealing approach, allowing to connect between expansion properties of a graph and its eigenvalues. (See, e.g., [12, 13] for a general background on spectral graph theory.) For a graph G with vertex set $V(G) = [n]$, the adjacency matrix $A(G)$ is an n-by-n matrix whose entry a_{ij} is 1 if $(i, j) \in E(G)$ and $a_{ij} = 0$ otherwise. This is a symmetric matrix with n real eigenvalues, usually sorted in the non-increasing order $\lambda_1 \geq \lambda_2 \geq \ldots \geq \lambda_n$; they are usually called the eigenvalues of the graph G itself. For the simplest case of d-regular graphs, the first (or trivial) eigenvalue is $\lambda_1 = d$, and it is the second eigenvalue that governs the graph expansion. Specifically, the following is true (see, e.g., Corollary 9.2.2 of [6]):

Proposition 4.1. *Let G be a d-regular graph on n vertices with the second largest eigenvalue λ. Then G is an α-expander, with $\alpha = \frac{d-\lambda}{2d}$.*

Hence a d-regular graph with eigenvalue gap $d - \lambda_2 = \Theta(1)$ and $d = O(1)$ is a weak expander. This implication is reversible – a regular expander has its two first eigenvalues well separated, see [2].

The spectral approach extends to the general (non-regular) case. (See [13] for an extensive discussion of the subject and all missing definitions.) Let G be a graph on n vertices, and let $0 = \mu_0 \leq \mu_1 \leq \ldots \leq \mu_{n-1}$ be the eigenvalues of the normalized Laplacian $\mathcal{L}(G)$ of G, defined by $\mathcal{L} = I_n - D^{-1/2}AD^{1/2}$, where $D = D(G) = \text{diag}(\deg(v_1), \ldots, \deg(v_n))$ is the degree matrix of G (we use the convention $D_{ii}^{-1} = 0$ when $\deg(v_i) = 0$). We set $\mu(G) = \mu_1$. Let $h(G)$ be the Cheeger constant of G. The famous *Cheeger inequality for graphs* [2, 4, 15] states that

$$\frac{h^2(G)}{2} \leq \mu(G) \leq 2h(G). \tag{4.1}$$

In particular, assuming that G has its degrees bounded, and recalling the prior discussion about expansion and the Cheeger constant, we conclude that having the first non-trivial eigenvalue of the normalized Laplacian $\mu(G)$ bounded away from 0 guarantees expansion. (Here we use the second part of the Cheeger inequality (4.1), the first (and more involved) part will be utilized later, when discussing algorithmic issues.)

3. Stretching edges of a bounded degree expander. This is a systematic way to obtain a (weaker) expander from a given expander G by stretching, or subdividing, the edges of G.

Proposition 4.2. *Let G be an α-expander of maximum degree $\Delta = O(1)$, and let ℓ be a positive integer. Subdividing each edge of G ℓ times produces an $\Omega_\Delta(\alpha/\ell)$-expander G'.*

While being a nice way of producing new expanders, the above proposition should turn all warning lights on. Indeed, even starting with a very strong bounded degree expander G, the new graph G' has no adjacent vertices of degree more than 2. Hence, one *cannot* hope to embed in G' any graph H in which some two vertices of degree at least three are adjacent. Even if H is a little tree with two vertices of degree three adjacent – H is not there in G'. This seems to indicate very gloomy prospects for this line of business, and for this survey in particular – there are seemingly not many structures one can seek to embed in an α-expander... Not all is lost however – weak expanders do have a rich enough structure, and we can argue about the existence of long paths and cycles, and also about embedding minors. This is indeed what we plan to do in the sequel.

4. Random regular graphs. Let $G_{n,d}$ be the probability space of d-regular graphs on n vertices. (See, e.g., [53], or [22, Chapter 10] for background.) For every fixed $d \geq 3$ and growing n a graph G drawn from $G_{n,d}$ is typically an expander. This can be shown directly using the so-called configuration model, as was done by Bollobás in [10]; another possible approach is to invoke Proposition 4.1 and known results about eigenvalues of random regular graphs [20].

5 Expanders and separators

Separators are one of the central concepts in graph theory in general, and in structural graph theory in particular.

Definition 5.1. Let $G = (V, E)$ be a graph on n vertices. A vertex set $S \subset V$ is a *separator* in G if there is a partition $V = A \cup B \cup S$ of the vertex set of G such that $|A|, |B| \leq 2n/3$, and G has no edges between A and B.

Separators serve to measure quantitatively the connectivity of large vertex sets in graphs; the fact that all separators in G are large indicates that it is costly to break G into large pieces not connected by any edge. Observe that if G is a bounded degree graph, then disconnecting G can be done cheaply by removing $O(1)$ neighbors of any vertex $v \in V(G)$; however, if G is well connected, then finding a small sized separator might be impossible.

Separators came into prominence with the celebrated result of Lipton and Tarjan [42], asserting that every planar graph on n vertices has a separator of size $O(\sqrt{n})$. This line of research, advancing the alternative "a small separator or a large minor" to be addressed later in this survey, has been quite fruitful over the years, see, e.g., [5,29,48]. We will return to it later in this survey.

It is easy to argue that expanders do not have small separators.

Proposition 5.2. *Let G be an α-expander on n vertices, and let S be a separator in G. Then $|S| \geq \frac{\alpha n}{3(1+\alpha)}$.*

Proof. Let S be a separator in G of size $|S| = s$, separating A and B, with $|A| = a$, $|B| = b$; we assume $a \leq b \leq 2n/3$. Then $a + s \geq n/3$. Clearly, $N_G(A) \subseteq S$. Since $a \leq n/2$, by the definition of an α-expander we get $s - \alpha a \geq 0$. Multiplying this inequality by $1/\alpha$ and summing with $a + s \geq n/3$, we obtain: $s(1 + 1/\alpha) \geq n/3$, or $s \geq \frac{\alpha n}{3(1+\alpha)}$. $\qquad\square$

Perhaps more surprisingly, it turns out that the opposite implication is true in a well-defined quantitative sense – graphs without small separators contain large induced expanders.

Proposition 5.3. *Let $\beta > 0$ be a constant. If all separators in a graph G on n vertices are of size at least βn, then G contains an induced $(3\beta/2)$-expander on at least $2n/3$ vertices.*

The proof is somewhat similar to Lemmas 2.4 and 2.6.

Proof. Start with $G' = G$, and delete repeatedly subsets A of size at most $|V(G')|/2$ not expanding themselves into $V(G')$ by at least the factor of $3\beta/2$, each time updating $G' := G' - A$. Let Z be the union of the deleted sets, clearly $|N_G(Z)| < 3\beta|Z|/2$. If the size of Z ever reaches $n/3$, focus on the moment it happens for the first time, let A be the last set deleted, and let $Z' = Z \setminus A$. Then $|A| \leq (n - |Z'|)/2$ and $|Z'| \leq n/3$. Combining these two inequalities, we get: $|Z| = |Z'| + |A| \leq 2n/3$. The set $N_G(Z)$ forms a separator in G (separating between Z and $V(G) \setminus (Z \cup N_G(Z))$), hence by the proposition's assumption its size is at least $\beta n \geq 3\beta|Z|/2$ – a contradiction. We conclude that the deletion process stops with $|Z| < n/3$, and the final graph of the process is a $(3\beta/2)$-expander on at least $2n/3$ vertices, as required. $\qquad\square$

The above arguments can easily be extended to sublinear separators and subconstant expansion.

We can learn a very important qualitative lesson here. It turns out that weak expanders and graphs without small separators are essentially the same thing (at least if we do not care that much about multiplicative constants involved). Thus, when aiming to prove results about graphs without sublinear separators, we can choose instead to operate in the realm of weak expanders. This simple yet powerful connection between two central graph theoretic notions (expanders and separators), usually perceived as belonging to quite different worlds (extremal graph theory and structural graph theory, respectively) can be quite fruitful and illuminating.

6 Finding large expanding subgraphs

Given the prominence of expanders and their usability, it is tempting to argue that every graph is an expander, or at least contains a large expander inside. This is obviously too much wishful thinking. Not every graph is an expander, moreover essentially every standard notion of expansion is rather fragile – adding an isolated vertex to a strong expander G produces a non-expanding graph G'. Not every graph, even of average degree bounded away from zero, contains a substantially sized expanding subgraph. Indeed, planar graphs, or more generally graphs of bounded genus have sublinear separators [23, 42]. Hence, applying Proposition 5.2, we conclude that such graphs do not contain expanding subgraphs of super-constant size. Recall also the example of Moshkovitz and Shapira [45], discussed in Section 2.

There is a partial remedy to the problem with this general approach. Shapira and Sudakov [51] and later Montgomery [44] argued that *every* graph G contains a very weak expander G^* of nearly the same average degree. Their notion of expansion is different – the expansion required is gradual in terms of subset sizes; for an m-vertex graph to be a weak expander, vertex sets of size m^c, $0 < c < 1$, should expand by a constant factor, whereas linearly sized vertex sets are required to expand only by about $1/\log m$ factor. (We are rather informal here in our descriptions, see the actual papers [44, 51] for accurate definitions.) Neither of these results guarantees a (weakly) expanding subgraph on linearly many vertices. Much earlier, Komlós and Szemerédi, in their work on topological cliques in graphs [31, 32], presented a fairly general scheme for arguing about the existence of weak expanders in any given graph; their scheme does not provide – naturally – for finding linearly sized expanders, or expanders with constant expansion of subsets.

We now present a fairly simple sufficient condition guaranteeing the existence of a large expanding subgraph in a given graph. This condition is based on the quantitative notion of local sparseness.

Definition 6.1. Let $c_1 > c_2 > 1$, $0 < \beta < 1$. A graph $G = (V, E)$ on n vertices is called a (c_1, c_2, β)-graph if

1. $\frac{|E|}{|V|} \geq c_1$;

2. every vertex subset $W \subset V$ of size $|W| \leq \beta n$ spans fewer than $c_2|W|$ edges.

In words, the above condition says that relatively small sets are sizably sparser than the whole graph. It has been used in recent papers [34, 35] of the author.

How natural or common is this condition? As the (very easy) proposition below shows, most sparse graphs are locally sparse.

Proposition 6.2. Let $c_1 > c_2 > 1$ be reals. Define $\beta = \left(\frac{c_2}{5c_1}\right)^{\frac{c_2}{c_2-1}}$. Let G be a random graph drawn from the probability distribution $G\left(n, \frac{c_1}{n}\right)$. Then with high probability every set of $k \leq \beta n$ vertices of G spans fewer than $c_2 k$ edges.

Proof. The probability in $G(n, p = c_1/n)$ that there exists a vertex subset violating the required property is at most

$$\sum_{i \leq \beta n} \binom{n}{i}\binom{\binom{i}{2}}{c_2 i} \cdot p^{c_2 i} \leq \sum_{i \leq \beta n} \left(\frac{en}{i}\right)^i \cdot \left(\frac{eip}{2c_2}\right)^{c_2 i} = \sum_{i \leq \beta n} \left[\frac{en}{i} \cdot \left(\frac{ec_1 i}{2c_2 n}\right)^{c_2}\right]^i$$

$$= \sum_{i \leq \beta n} \left[\frac{e^{c_2+1}c_1^{c_2}}{(2c_2)^{c_2}} \cdot \left(\frac{i}{n}\right)^{c_2-1}\right]^i.$$

Denote the i-th summand of the last sum by a_i. Then, if $i \leq n^{1/2}$ we get: $a_i \leq \left(O(1)n^{-\frac{c_2-1}{2}}\right)^i$, implying $\sum_{1 \leq i \leq n^{1/2}} a_i = o(1)$. For $n^{1/2} \leq i \leq \beta n$, we have, recalling the expression for β:

$$a_i \leq \left[\frac{e^{c_2+1}c_1^{c_2}}{(2c_2)^{c_2}} \cdot \left(\frac{c_2}{5c_1}\right)^{c_2}\right]^i = \left(e \cdot \left(\frac{e}{10}\right)^{c_2}\right)^i = o(n^{-1}).$$

It follows that $\sum_{i \leq \beta n} a_i = o(1)$, and the desired property of the random graphs holds with high probability. $\square$

One can also cap easily the typical maximum degree in (a nearly spanning subgraph of) a random graph.

Proposition 6.3. *For every $C > 0$ and all sufficiently small $\delta > 0$ the following holds. Let G be a random graph drawn from the probability distribution $G\left(n, \frac{C}{n}\right)$. Then with high probability every set of $\frac{\delta}{\ln \frac{1}{\delta}} n$ vertices of G touches fewer than δn edges.*

We omit the straightforward proof.

Observe that if $G = (V, E)$ satisfies the conclusion of the above proposition, then by deleting $\frac{\delta}{\ln \frac{1}{\delta}} n$ vertices of highest degrees in G, one obtains a spanning subgraph $G' = (V', E')$ on $|V'| = \left(1 - \frac{\delta}{\ln \frac{1}{\delta}}\right) n$ vertices and with $|E'| \geq |E| - \delta n$ edges, in which all degrees are at most $2\ln(1/\delta)$. (Otherwise, all deleted vertices are of degree at least $2\ln(1/\delta)$, forming a subset touching at least δn edges – a contradiction.)

We now formulate the main result of this section.

Theorem 6.4 ([34]). *Let $c_1 > c_2 > 1$, $0 < \beta < 1$, $\Delta > 0$. Let $G = (V, E)$ be a graph on n vertices, satisfying:*

1. $\frac{|E|}{|V|} \geq c_1$;

2. *every vertex subset $U \subset V$ of size $|U| \leq \beta n$ spans fewer than $c_2 |U|$ edges;*

3. $\Delta(G) \leq \Delta$.

Then G contains an induced subgraph $G^* = (V^*, E^*)$ on at least βn vertices which is an α-expander, for $\alpha = \frac{c_1 - c_2}{\Delta \cdot \left\lceil \log \frac{1}{\beta} \right\rceil}$.

Putting it informally, every locally sparse graph G of bounded maximum degree contains a linearly sized expander G^*. In our terminology, the first two conditions above say precisely that G is a (c_1, c_2, β)-graph. They are spelled out in full in the statement above so as to make it self-contained.

Proof of Theorem 6.4 (sketch). The proof proceeds in rounds. We describe the first round here, the rest is fairly similar.

Choose a constant $0 < \delta \ll c_1 - c_2$. Let $H = (W, F)$ be a minimal by inclusion non-empty induced subgraph of G, satisfying $|F|/|W| \geq c_1$. (Such a subgraph exists, as G itself meets this condition.) Due to the local sparseness assumption, we have $|W| > \beta n$. Also, every subset $U \subseteq W$ touches at least $c_1 |U|$ edges of H. Otherwise, deleting U is easily seen to produce a smaller induced subgraph H', still meeting the requirement stated in the definition of H – a contradiction. (We could add that H is connected due to its minimality, but we do not need this property here.)

Let us analyze now the expansion properties of H. If a subset $U \subset W$ has at most βn vertices, then it spans at most $c_2 |U|$ edges in G, and thus in H, and yet touches at least $c_1 |U|$ edges. This shows that at least $(c_1 - c_2)|U|$ edges of H leave U. Recalling the maximum degree assumption, we conclude that U has at least $(c_1 - c_2)|U|/\Delta$ neighbors outside of it in H.

At this point of the proof we have assured that H is a $(\beta n, \frac{c_1 - c_2}{\Delta})$-expander. We can complete the proof by appealing to the general (and somewhat inexplicit) statement of Theorem 2.7. We prefer however, just like in the original paper [34], to provide a self-contained argument, delivering also a better (and explicit) estimate for the order of the resulting subgraph and for its expansion strength.

Consider now medium-sized subsets U in H. Let $\beta n < |U| \leq |W|/2$. Such U touches at least $c_1 |U|$ edges of U. If every such U spans at most $(c_1 - \delta)|U|$ edges, we get it expanding to $\delta |U|/\Delta$ vertices outside, thus indicating that H is a desired expander. Otherwise there is a subset U, $\beta n < |U| \leq |W|/2$, spanning at least $(c_1 - \delta)|U|$ edges. We can now switch our attention to the induced graph $H[U] = G[U]$, whose order is at most half that of the original graph, yet we did not give much in terms of its density, compared to the assumption on G – it is now at least $c_1 - \delta$.

Iterating this argument at most $O(\log(1/\beta))$ times, we eventually arrive at the desired linearly sized α-expander inside G. □

The assumption about bounded maximum degree $\Delta(G) = O(1)$ is used in the proof only to trade edge expansion for vertex expansion.

The proof of Theorem 6.4 is inherently non-algorithmic, as it uses the existence of a minimal by inclusion subgraph H of prescribed density, and such a subgraph is hard to find efficiently. It turns out that with some loss in constants, one can provide an algorithmic proof of the theorem. The argument for such a proof is based on the fact that the proof of (the first part of) the Cheeger inequality (4.1) can be made constructive in the sense that, given a graph $G = (V, E)$ on n vertices, one can find in time polynomial in n a subset $U \subset V$, $\mathrm{vol}_G(U) \leq \mathrm{vol}_G(V)/2$, satisfying $e_G(U, V \setminus U) \leq \sqrt{2\mu(G)} \cdot \mathrm{vol}_G(U)$. (See [2], or [13, Ch. 2], or [14], or [25, Sect. 4.5].) This yields the following algorithmic result.

Theorem 6.5 ([34]). *Let $c_1 > c_2 > 1$, $0 < \beta < 1$, $\Delta > 0$. There exist a constant $\alpha = \alpha(c_1, c_2, \beta, \Delta) > 0$ and an algorithm, that, given an n-vertex graph $G = (V, E)$ with $\Delta(G) \leq \Delta$ and $|E|/|V| \geq c_1$, finds in time polynomial in n a subset $U \subset V$ of size $|U| \leq \beta n$, spanning at least $c_2 |U|$ edges in G, or an induced α-expander $G^* = (V^*, E^*) \subseteq G$ on at least βn vertices.*

See Section 2 of [34] for the proof.

Theorem 6.4 can be used to argue that a supercritical random graph $G(n, c/n)$, $c > 1$, contains **whp** an induced expander of linear size. This can be quite handy as it allows to extend immediately known properties of expanders to sparse random graphs.

Corollary 6.6. *For every $\epsilon > 0$ there exist $\alpha, \beta > 0$ such that a random graph $G \sim G\left(n, \frac{1+\epsilon}{n}\right)$ contains with high probability an induced bounded degree α-expander on at least βn vertices.*

Proof. Due to the standard monotonicity arguments we can assume that ϵ is small enough where necessary.

We will utilize several (very standard) facts about supercritical random graphs. It is known (see, e.g., [26, Ch. 5]) that **whp** $G \sim G\left(n, \frac{1+\epsilon}{n}\right)$ contains a connected component $C_1 = (V_1, E_1)$ (the so called giant component) satisfying:

$$
\begin{aligned}
|V_1| &= 2\epsilon(1 + o_\epsilon(1))n, \\
\frac{|E_1|}{|V_1|} &= 1 + (1 + o_\epsilon(1))\frac{\epsilon^2}{3}.
\end{aligned}
$$

Also, by Proposition 6.3 **whp** every $\frac{\epsilon^3}{10 \ln \frac{1}{\epsilon}} n$ vertices of C_1 touch at most $\frac{\epsilon^3}{3} n$ edges. Deleting $\frac{\epsilon^3}{10 \ln \frac{1}{\epsilon}} n$ vertices of highest degrees from C_1, one gets a graph $G_0 = (V_0, E_0)$ of maximum degree $\Delta(G_0) \leq 7 \ln \frac{1}{\epsilon}$. In addition,

$$|V_0| \geq \left(2\epsilon(1 + o_\epsilon(1)) - \frac{\epsilon^3}{10 \ln \frac{1}{\epsilon}} \right) n = 2\epsilon(1 + o_\epsilon(1))n \,,$$

$$|E_0| \geq |E_1| - \frac{\epsilon^3}{3}n \geq |V_1| \left(1 + \frac{\epsilon^2}{3} + o(\epsilon^2) \right) - \frac{\epsilon^3}{3}n$$

$$\geq |V_0| \left(1 + \frac{\epsilon^2}{3} + o(\epsilon^2) \right) - \frac{\epsilon^3}{3}n \geq |V_0| \left(1 + \frac{\epsilon^2}{7} \right).$$

Finally, applying Proposition 6.2 with $c_1 = 1 + \epsilon$, $c_2 = 1 + \frac{\epsilon^2}{10}$, we get that **whp** every $k \leq \beta n$ vertices of G_0 (with $\beta = \beta(\epsilon)$ from Proposition 6.2) span fewer than $(1 + \epsilon^2/10)k$ edges. The conditions are set to call Theorem 6.4 and to apply it to G_0; we conclude that, given the above likely events, G_0 contains a linearly sized α-expander. $\qquad\square$

We remark here that in order to get this qualitative result, one does not really need to apply the heavy machinery of random graphs — it is enough actually to argue from the "first principles". Indeed, given the likely existence of the giant component C_1 in the supercritical regime, one can prove easily (for example, through sprinkling) that its density is typically above 1.

Let us give yet another illustrative example of how Theorem 6.4 can be utilized. This example comes from the realm of positional games. (The reader can consult monograph [24] for a general background on the subject.) In a *Maker–Breaker* game two players, called Maker and Breaker, claim alternately free edges of the complete graph K_n, with Maker moving first. Maker claims one edge at a time, while Breaker claims $b \geq 1$ edges (or all remaining fewer than b edges if this is the last round of the game). The integer parameter b is the so-called *game bias*. Maker wins the game if the graph of his edges in the end possesses a given monotone graph theoretic property, Breaker wins otherwise, with draw being impossible. In the *long cycle game* Maker's goal is to create as long as possible cycle. For $b \geq n/2$, Bednarska and Pikhurko proved [9] that Breaker has a strategy to force Maker's graph being acyclic. On the other hand, if $b = b(n)$ is such that Maker ends up with at least n edges, his graph contains a cycle (of some length) in the end. We can argue, using expanders, that already for $b = (1 - \epsilon)n/2$, Maker can force a linearly long cycle in his graph.

Theorem 6.7 ([35]). *For every $\epsilon > 0$ there exists $c > 0$ such that for all sufficiently large n, when playing the b-biased Maker–Breaker game on $E(K_n)$ with $b \leq (1 - \epsilon)\frac{n}{2}$, Maker has a strategy to create a cycle of length at least cn.*

Proof (sketch). Maker plays randomly (i.e., by taking a random free edge) during the first $(1 + \epsilon/2)n$ rounds of the game. Observe that at any moment during these rounds, the board still has $\Theta(n^2)$ free edges, and thus the probability of any edge to be chosen by Maker at a given round is $O(n^{-2})$. This allows us to apply standard union-bound type calculations to claim that with positive probability Maker, playing against any strategy of Breaker, can create a graph M_0 on n vertices with the following properties:

- M_0 has $\left(1 + \frac{\epsilon}{2} \right) n$ edges;

- every $k \leq \delta n$ vertices span at most $\left(1 + \frac{\epsilon}{8} \right) k$ edges;

- every δn vertices touch at most $\frac{\epsilon n}{4}$ edges, for some $\delta = \delta(\epsilon) > 0$. Since the game analyzed is a perfect information game with no chance moves, it follows that in fact Maker has a (deterministic) strategy to create a graph M_0 with the above stated properties. Take such M_0 and delete δn vertices of highest degrees. The obtained graph M_1 has $(1 - \delta)n$ vertices, at least $\left(1 + \frac{\epsilon}{4}\right)n$ edges, maximum degree $\Delta(M_1) \leq \frac{\epsilon}{2\delta}$, and every $k \leq \delta n$ vertices span at most $\left(1 + \frac{\epsilon}{8}\right)k$ edges. Applying Theorem 6.4 shows that such M_1, being a part of Maker's graph by the end of the game, contains a linearly sized α-expander, for some $\alpha = \alpha(\epsilon) > 0$. As we will argue in the next section (see Theorem 7.4), Maker's graph contains then a linearly long cycle.   □

7 Long paths and cycles

Now we start investigating extremal properties of expanders directly. The first objects to explore are paths and cycles. As we explained in Section 4, it is unrealistic to expect appearance of many substructures in every expanding graphs. Luckily paths and cycles (the latter ones possibly with some restrictions on their lengths) are not in this excluded category, and indeed, as we will argue in this section, every α-expander on n vertices contains cycles whose length is linear in n.

We first describe the most basic – and very handy – tool for arguing about long paths and cycles in expanding graphs, which is the Depth First Search algorithm (DFS).

7.1 DFS algorithm

The *Depth First Search* is a well known graph exploration algorithm, usually applied to discover connected components of an input graph. As it turns out, due to its nature (of always pushing deeper, as its name suggests), this algorithm is particularly suitable for finding long paths and cycles in graphs. Some illustrative examples of its applications in the context of extremal graphs and random graphs include [1, 3, 8, 11, 36, 40, 49].

Recall that the DFS (Depth First Search) is a graph search algorithm that visits all vertices of a graph and eventually discovers the structure of the connected components of G. The algorithm receives as an input a graph $G = (V, E)$; it is also assumed that an order σ on the vertices of G is given, and the algorithm prioritizes vertices according to σ. The algorithm's output is a spanning forest F of G, whose connected components are identical to those of G. Each connected component C of F is a tree rooted at vertex r, the first vertex of C according to σ.

The algorithm maintains three sets of vertices, letting S be the set of vertices whose exploration is complete, T be the set of unvisited vertices, and $U = V \setminus (S \cup T)$, where the vertices of U are kept in a stack (the last in, first out data structure). It initializes with $S = U = \emptyset$ and $T = V$, and runs till $U \cup T = \emptyset$. At each round of the algorithm, if the set U is non-empty, the algorithm queries T for neighbors of the last vertex v that has been added to U, scanning T according to σ. If v has a neighbor u in T, the algorithm deletes u from T and inserts it into U. If v does not have a neighbor in T, then v is popped out of U and is moved to S. If U is empty, the algorithm chooses the first vertex of T according to σ, deletes it from T and pushes it into U.

The following are basic properties of the DFS algorithm:

(P1) at each round of the algorithm one vertex moves, either from T to U, or from U to S;

(P2) at any stage of the algorithm, it has been revealed already that the graph G has no edges between the current set S and the current set T;

(P3) the set U always spans a path;

(P4) let F be a forest produced by the DFS algorithm applied to G. If $(u,v) \in E(G) \backslash E(F)$, then one of $\{u,v\}$ is a predecessor of the other in F (along the unique path to the root).

Properties **(P1)**, **(P2)** are immediate upon a brief reflection on the description of the DFS algorithm. For Property **(P3)**, observe that when a vertex u is added to U, it happens because u is a neighbor of the last vertex v in U; thus u augments the path spanned by U, of which v is the last vertex; moving the last vertex v of U over to S deletes the last vertex of the path spanned by U, and hence still leaves the updated U spanning a path. Property **(P4)**, applied for example in [3], is perhaps less obvious, and therefore we supply its short proof here. Since u,v are connected by an edge, they obviously belong to the same connected component C of G (which is also a connected component of F, vertex-wise). Assume that v was discovered by the algorithm (and moved to U) earlier than u. If u is not under v in the corresponding rooted tree component T of F, then the algorithm has completed exploring v and moved it over to S before getting to explore u (and to move it to U) — obviously a contradiction.

7.2 Long paths

We now exploit the features of the DFS algorithm to derive the existence of long paths in expanding graphs. The following simple to prove statement is stated explicitly in [33] (see Proposition 2.1 there).

Proposition 7.1. *Let k, ℓ be positive integers. Assume that $G = (V, E)$ is a graph on more than k vertices, in which every vertex subset S of size $|S| = k$ satisfies: $|N_G(S)| \geq \ell$. Then G contains a path of length ℓ.*

Proof. Run the DFS algorithm on G, with σ being an arbitrary ordering of V. Look at the moment during the algorithm execution when the size of the set S of already processed vertices becomes exactly equal to k (there is such a moment due to Property **(P1)** above, as the vertices of G move into S one by one, till eventually all of them land there). By Property **(P2)**, the current set S has no neighbors in the current set T, and thus $N_G(S) \subseteq U$, implying $|U| \geq \ell$. The last move of the algorithm was to shift a vertex from U to S, so before this move U was one vertex larger. The set U always spans a path in G, by Property **(P3)**. Hence G contains a path of length ℓ. $\qquad\square$

It thus follows that if G is α-expander on n vertices, then G contains a path of length at least $\alpha n/2$ (apply Proposition 7.1 with $k = n/2$, $\ell = \alpha n/2$). As we have seen, with the right technology (DFS in this case) the proof of this result becomes essentially a one-liner. The estimate delivered by this argument is fairly tight, also in terms of the dependence on α – if G is taken to be the complete bipartite graph with sides of sizes αn and $(1 - \alpha)n$, then a longest path in G has $2\alpha n$ vertices.

We now describe an application of Proposition 7.1 to size Ramsey numbers. (See, e.g., [35] or [17] for the relevant background.) For graphs G, H and a positive integer r, we write $G \to (H)_r$ if every r-coloring of the edges of G produces a monochromatic copy of H. The *r-color size Ramsey number* $\hat{R}(H, r)$ is defined as the minimal possible number of edges in a graph G, for which $G \to (H)_r$. Let us discuss briefly the multi-color size Ramsey number $\hat{R}(P_n, r)$ of the path P_n on n vertices. Beck in his groundbreaking paper [7] proved that $\hat{R}(P_n, 2) = O(n)$; his argument can easily be extended to show that $\hat{R}(P_n, r) = O_r(n)$ for any fixed $r \geq 2$. If so, the relevant question now is what is the hidden dependence on r in the above bound. Here we argue that $\hat{R}(P_n, r) = O(r^2 \log r)n$.

The following proposition is tailored to handle size Ramsey numbers of long paths.

Proposition 7.2. *Let $d, r > 0$. Let $G = (V, E)$ be a graph on $|V| = N$ vertices with average degree at least d. Assume that for an integer $0 < n < N$, every vertex subset $W \subset V$ of cardinality at most $2n$ spans at most $(d/8r)|W|$ edges. Then every edge subset $E_0 \subset E$ with $|E_0| \geq |E|/r$ contains a path of length n.*

Proof. Denote $G_0 = (V, E_0)$. Clearly the average degree of G_0 is at least d/r. Find a subgraph $G_1 = (V_1, E_1) \subseteq G_0$ of minimum degree at least $d/2r$. Since $|E_1| \geq (d/4r)|V_1|$, we have $|V_1| > 2n$ by the proposition's assumption. Let now $U \subset V_1$ be an arbitrary set of n vertices in G_1. Then the set $U \cup N_{G_1}(U)$ spans all edges of G_1 touching U, and there are at least $|U|(d/2r)/2 = nd/4r$ of them. Hence $|U \cup N_{G_1}(U)| \geq 2n$. We derive: $|N_{G_1}(U)| \geq n$. Applying Proposition 7.1 to G_1 with $k = \ell = n$, the proposition follows. $\square$

Note that the assumption in the proposition is fairly similar to the definition of locally sparse graphs as in Section 6.

We can now prove

Theorem 7.3. *There exists an absolute constant $C_0 > 0$ such that for any integer $r \geq 2$ and sufficiently large n, $\hat{R}(P_n, r) \leq C_0 r^2 \log r \cdot n$.*

This result can be read out from [35] (Theorem 4 there). An alternative proof has recently been offered by Dudek and Prałat [17], it also relies on a DFS-based argument. The estimate of Theorem 7.3 is close to being tight – Dudek and Prałat proved in a prior paper [16] that $\hat{R}(P_n, r) = \Omega(r^2)n$; a somewhat better absolute constant for the lower bound has been obtained by the author [35].

Proof (sketch). Let $N = 25rn$, and consider the binomial random graph $G(N, p)$ with $p = \frac{Cr \log r}{N}$, where $C > 0$ is sufficiently large constant. Then with high probability $G \sim G(N, p)$ has average degree $d = (1 + o(1))Np = (1 + o(1))Cr \log r$. In order to bound the typical local density of G, observe that for a subset $W \subset [N]$ of cardinality $|W| = i \leq 2n$, the number $e_G(W)$ of edges spanned by W in G is a binomial random variable with parameters $\binom{i}{2}$ and p. Recalling the easy estimate $\Pr[\mathrm{Bin}(n, p) \geq k] \leq \binom{n}{k}p^k \leq (enp/k)^k$, we can write:

$$\Pr\left[e_G(W) \geq \frac{Ci \log r}{9}\right] \leq \left[\frac{\frac{ei^2}{2} \cdot \frac{Cr \log r}{N}}{\frac{Ci \log r}{9}}\right]^{\frac{Ci \log r}{9}} = \left(\frac{9er}{2} \cdot \frac{i}{N}\right)^{\frac{Ci \log r}{9}} = \left(\frac{9e}{50} \cdot \frac{i}{n}\right)^{\frac{Ci \log r}{9}}.$$

Summing over all $i \leq 2n$ and over all choices of $W \subset [N]$ with $|W| = i$, performing standard asymptotic manipulations, and recalling that $C > 0$ can be chosen to be sufficiently large, we derive that with high probability every such W spans at most $(d/8r)|W|$ edges.

Let now $f : E(G) \to [r]$ be an r-coloring of $E(G)$. Take E_0 to be the majority color, clearly $|E_0| \geq |E|/r$. Applying Proposition 7.2 to G and E_0, we conclude that E_0 contains a path of length n. It follows that every r-coloring of $E(G)$ contains a monochromatic path of length n. Hence $\hat{R}(P_n, r) \leq |E(G)| = O(r^2 \log r)n$. $\square$

7.3 Long cycles

With more effort (still based on the DFS algorithm and its properties) one can prove that every α-expander on n vertices contains a cycle of length linear in n. Notice that here, unlike in the case of paths, we cannot expect to get a cycle of any prescribed length, but rather a cycle at least as long as the lower bound given by the next theorem.

Theorem 7.4 ([35]). *Let $k > 0, \ell \geq 2$ be integers. Let $G = (V, E)$ be a graph on more than k vertices, satisfying:*

$$|N_G(W)| \geq \ell, \quad \text{for every } W \subset V, \ \frac{k}{2} \leq |W| \leq k.$$

Then G contains a cycle of length at least $\ell + 1$.

Proof (sketch). Observe first that G has a connected component C on more than k vertices. Let T be the tree obtained by applying the DFS algorithm to C, under an arbitrary order σ of its vertices, and let r be its root (which is the first vertex of C under σ). Then $|V(T)| = |V(C)| > k$. Now we find a vertex v and a subset X of its children in T such that the subtrees of T rooted at the vertices of X have between $k/2$ and k vertices in total. (This can be done by going from r down the tree till we find a vertex v whose subtree T_v has more than k vertices, but the subtrees T_x of its children in T are all less than k in their sizes; the required set X can then be recruited from the children of v in T.) Let $W = \bigcup_{x \in X} V(T_x)$. Denote by P the path in T from the root r to v. Then $k/2 \leq |W| \leq k$, and by Property **(P4)** of the DFS algorithm we have: $N_G(W) \subseteq V(P)$. Let $v^* \in V(P)$ be the farthest from v vertex in $N_G(W) \cap V(P) = N_G(W)$. Clearly its distance from v is at least $|N_G(W)| - 1 \geq \ell - 1$, by the theorem's assumption. Let w be a neighbor of v^* in W. Then the cycle, formed by the path in T from w to v^* (which includes the part of P from v to v^*) and the edge (w, v^*), has length at least $\ell + 1$. $\qquad \square$

Assuming that G is an α-expander on $n = \Omega(1/\alpha)$ vertices and applying Theorem 7.4 to G, we derive the existence of a cycle of length more than $\alpha n/4$ in G. The order of magnitude in n obtained is obviously optimal; moreover, the order of dependence on α in this bound is optimal as well, due to the same example of $K_{\alpha n, (1-\alpha)n}$.

7.4 Cycle lengths

Theorem 7.4 argues that if G is an α-expander on n vertices, then G has a cycle of length $\Omega(\alpha)n$. Also, due to Proposition 3.1, G has cycles as short as $O_\alpha(\log n)$. What can be said then about cycle lengths in G between the two extremes? Does G have linearly many in n cycle lengths? How well can one approximate an integer ℓ within the set of cycle lengths in G? The answer is given by the following theorem:

Theorem 7.5 ([21]). *For every $\alpha > 0$ there exist A, C, n_0 such that for every α-expander $G = (V, E)$ on $n \geq n_0$ vertices and every $\ell \in [C \ln n, n/C]$, the graph G has a cycle whose length is between ℓ and $\ell + A$.*

Hence the set $L(G)$ of cycle lengths in an α-expander G on n vertices is well spread and approximates every number ℓ in the range $[C \ln n, n/C]$ within an additive constant. In particular, G has linearly many cycle lengths.

The proof of Theorem 7.5 is somewhat involved. We outline instead the proof of a weaker statement, assuring that the set $L(G)$ of cycle lengths in an α-expander G on n

vertices has cardinality linear in n. The proof borrows some ideas from prior papers on cycle lengths, such as [52].

We start with the following lemma.

Lemma 7.6. *Let $\epsilon > 0$. There are constants $C_1 = C_1(\alpha, \epsilon)$ and $C_2 = C_2(\alpha, \epsilon) > 0$ such that for every $v \in V$ the graph G contains a spanning tree T rooted at v with levels $L_0 = \{v\}, L_1, \ldots, L_{k_0}, \ldots, L_{k_1}$ such that $k_0 = O(\log n), k_1 \leq k_0 + C_1$,*

$$\left| \bigcup_{i=k_0}^{k_1} L_i \right| \geq (1 - \epsilon)n \,,$$

and the degrees of all $u \in \bigcup_{i=k_0}^{k_1} L_i$ in T are at most C_2.

The lemma guarantees that one can find a (spanning) tree T in G with few consecutive "thick" layers L_i, $i = O(\log n)$, containing nearly all vertices of G and such that the degrees in T of vertices in these thick layers are all bounded.

Let now T be the tree guaranteed by Lemma 7.6 with $\epsilon = \alpha/4$ and arbitrary v. Denote $W = \bigcup_{i=k_0}^{k_1} L_i$. Observe that every $U \subset W$, $|U| = n/2$, has at least $\alpha|U| - (n - |W|) \geq \alpha n/4$ neighbors in W. Hence W spans a path P of length at least $\alpha n/4$ by Proposition 7.1.

For each $u \in L_{k_0}$ let T_u be the full subtree of T rooted at u. Due to the bounded degree assumption on T we have $|T_u| \leq C_2^{C_1}$.

For each $u \in L_{k_0}$ with $T_u \cap V(P) \neq \emptyset$ (there are at least $\frac{|V(P)|}{C_2^{C_1}} = \Theta(n)$ of those) choose a highest (closest to v) vertex $x \in T_u \cap V(P)$. Let X be the set of chosen vertices. Now, choose a maximum subset $X_0 \subseteq X$ such that the distance along P between any two vertices of X_0 is at least $C_1 + 1$. We have still $|X_0| = \Theta(n)$.

Let T' be the minimal subtree of T whose set of leaves is X_0, let r be its root. Due to the minimality T' branches at r. Let $A \subset X_0$ be the set of vertices of X_0 in one of the branches, and let $B = X_0 - A$. We can assume that $|B| \geq |A|$, implying $|B| \geq |X_0|/2$. Let a be an arbitrary vertex of A. At least $|B|/2 \geq |X_0|/4$ vertices of B are on the same side of a in P, denote this set by B_0.

For every $b \in B_0$ we get a cycle C_b in G as follows: first, the path from r to a in T', then we move from a to b along P, and finally take the path from b to r in T'. (These three pieces do not intersect internally, as a and b were chosen as highest vertices of P in their corresponding subtrees T_u, $u \in L_{k_0}$.) It is easy to check that the lengths of the cycles C_b are all distinct. Altogether we get $|B_0| = \Theta(n)$ different cycle lengths as promised.

8 Minors in expanding graphs

Minors is one of the most central notions in modern graph theory. It is thus only natural to expect to see meaningful research connecting expanding graphs and minors. And indeed, there have been several papers, addressing this subject directly or indirectly. We will mention some of them in this section, along with a description of new results due to Rajko Nenadov and the author.

First we recall the definition of a minor. Let $G = (V, E)$, $H = (U, F)$ be graphs with $U = \{u_1, \ldots, u_k\}$. We say that G contains H as a *minor* if there is a collection $(U_1, \ldots, U_k)$ of pairwise disjoint vertex subsets (supernodes) in V such that each U_i spans a connected subgraph in G, and whenever $(u_i, u_j) \in F$, the graph G has an edge between U_i and U_j. (Then contracting each U_i to a single vertex produces a copy of (a supergraph of) H.)

Notice that if G contains a minor of H then $|V(G)| \geq |V(H)|$ and $|E(G)| \geq |E(H)|$; these trivial bounds provide an obvious but meaningful benchmark for minor embedding statements.

As we discussed in Section 5, there is a very close connection between expanders and graphs without small separators. This connection enables to extend known results about embedding minors in graphs without small separators to claims about embedding minors in expanding graphs. In particular, it follows from the result of Plotkin, Rao and Smith [48] that an α-expanding graph on n vertices, $\alpha > 0$ a constant, contains a minor of the complete graph $K_{c\sqrt{n/\log n}}$. (In fact, the complete minor delivered by the result of [48] is shallow, i.e., each supernode U_i has diameter $O(\log n)$ in G.) An even stronger result has been announced by Kawarabayashi and Reed, who claimed in [29] that (translating it to the language of expanders) an α-expanding graph on n vertices contains a minor of $K_{c\sqrt{n}}$. Also, Kleinberg and Rubinfeld proved in [30] that an α-expander on n vertices of maximum degree Δ contains all graphs with $O(n/\log^\kappa n)$ vertices and edges as minors, for $\kappa = \kappa(\alpha, \Delta) > 1$. Phrasing it differently, a bounded degree α-expander is *minor universal* for all graphs with $O(n/\log^\kappa n)$ vertices and edges. This is optimal up to polylogarithmic factors due to the above stated trivial bound, as there exist n-vertex expanders with $\Theta(n)$ edges. (Formally, the number of vertices n is another bottleneck here.) It is worth mentioning here that the bounded maximum degree assumption $\Delta(G) = \Delta = O(1)$ is essential for the argument of Kleinberg and Rubinfeld, as the latter relies heavily on mixing properties of random walks; this issue has been briefly touched upon in Section 3.

We now present recent results, along with outlines of their proofs.

8.1 Large minors in expanding graphs

Our aim here is to present a universality-type result, guaranteeing the existence of a minor of *every* graph with bounded (as a function of n) number of vertices and edges in every α-expanding graph. This theorem is obtained jointly with Rajko Nenadov.

Theorem 8.1. *For every $\alpha > 0$ there exist c, n_0 such that the following holds. Let G be a graph on $n \geq n_0$ vertices, and let H be a graph with at most $cn/\log n$ vertices and edges. Then G has a separator of size at most αn, or G contains H as a minor.*

(We bound the number of vertices in H as well so as to avoid the pathological case of H having $O(n/\log n)$ edges, but many more vertices, say, more than n, thus making the embedding of H as a minor in G impossible.)

The guarantee $|V(H)| + |E(H)| = O(n/\log n)$ of the above theorem is asymptotically optimal up to multiplicative constants due to the following argument. Let G be an expanding graph of n vertices with logarithmic girth girth$(G) = \Theta(\log n)$, say, the celebrated Lubotzky-Phillips-Sarnak graph [43], and let H be a collection of k vertex-disjoint triangles. If H is a minor of G, then the image of every triangle in a minor embedding of H in G contains a cycle, and these cycles are disjoint for distinct triangles in H. It follows that $k \cdot \text{girth}(G) \leq n$, implying $k = O(n/\log n)$. Hence the largest collection of disjoint triangles that can be embedded in G as a minor counts $O(n/\log n)$ vertices.

Proof. The proof borrows substantially from the ideas of Plotkin, Rao and Smith [48]. Just as in many arguments of this sort, we start by assuming for convenience (of the proof) that the target graph $H = (U, F)$ has maximum degree (at most) three. This is possible as by splitting the vertices of H and edges emanating from them, we can construct a graph

$H' = (U', F')$ with $\Delta(H') \leq 3$ and $|U'| = O(|U| + |F|)$ so that H is a minor of H'. Since minor containment is transitive, we can embed instead H' in G as a minor.

Rephrasing Corollary 3.2, we get the following handy lemma:

Lemma 8.2. *For every $\alpha > 0$ there exists $C_0 > 0$ such that every graph G on n vertices has diameter at most $C_0 \log n$, or contains a subset $U \subset V(G)$ satisfying: $|U| \leq n/2$ and $|N_G(U)| \leq \alpha n$.*

The proof of Theorem 8.1 is described as an algorithmic procedure that, given G and H with $|V(H)| = k \leq \frac{\alpha n}{6C_0 \log n}$ and $\Delta(H) \leq 3$ (where C_0 is the constant from Lemma 8.2), outputs a small separator in G or a minor embedding of H in G. In fact, this procedure can be easily turned into a polynomial time algorithm performing this task. Let us assume $V(H) = [k]$. The algorithm maintains and updates a partition $V = A \cup B \cup C$ of the vertex set of G, where the set A will eventually contain a large and non-expanding set, witnessing the existence of a small separator in G, or the set B will contain the desired minor embedding of H; the set C will serve as a vertex reservoir. We will maintain $|N_G(A, C)| \leq \alpha|A|$. We will also maintain and update a subset $I_0 \subset [k]$, which will describe the current induced subgraph of H, minor embedded in $G[B]$. Accordingly, there will be disjoint subsets $B_i \subseteq B$, $i \in I_0$, each spanning a connected subgraph in G; we commit ourselves to having always: $|B_i| \leq 2C_0 \log n$.

We initialize $A = B = \emptyset$, $C = V$, $I_0 = \emptyset$.

The procedure will repeat the following loop. Choose an arbitrary $i \in [k] \setminus I_0$. Let X be the set of neighbors of i in I_0, $|X| \leq 3$. For $j \in X$, choose an arbitrary $v_j \in N_G(B_j, C)$. If for some $j \in X$ there is no such neighbor, we dump B_j into A and update accordingly: $A := A \cup B_j$, $B := B \setminus B_j$, $I_0 := I_0 \setminus \{j\}$. Otherwise, we apply Lemma 8.2 to the induced subgraph $G[C]$. In case it outputs a set U, $|U| \leq |C|/2$, $|N_G(U, C \setminus U)| \leq \alpha|U|$, we move U over to A and update: $A := A \cup U$, $C := C \setminus U$. In the complementary case, where $G[C]$ has logarithmic diameter, we find a subset $Y \subset C$ of size $|Y| \leq 2 \cdot \operatorname{diam}(G[C]) \leq 2C_0 \log n$, such that $G[Y]$ is connected and contains vertices $v_j, j \in X$ (choose one of v_j's as the pivot vertex and connect it by paths of length at most $C_0 \log n$ to other v_j's). We then update: $B := B \cup Y$, $C := C \setminus Y$, $B_i := Y$, $I_0 := I_0 \cup \{i\}$. (In case $X = \emptyset$, we can simply take Y to be an arbitrary vertex of C and perform the same update.) The set B_i is connected and has an edge connecting it to B_j, $j \in X$.

Since we always add to the current A a piece U satisfying $|N_G(U, C)| \leq \alpha|U|$, and the set C only shrinks as the algorithm proceeds, we observe that indeed at any point of the execution the current set A satisfies: $|N_G(A, C)| \leq \alpha|A|$. Also, the set B is not too large – it is always composed of sets B_i, $i \in I_0$, and thus has at most $k \cdot 2C_0 \log n = \alpha n/3$ vertices.

The above procedure eventually reaches the situation when A exceeds $n/3$ or $I_0 = [k]$. In the former case, it is easy to see that the first moment the size of A goes above $n/3$, it satisfies: $|A| \leq 2n/3$. Also, $|N_G(A)| \leq |B| + |N_G(A, C)| \leq \alpha n/3 + \alpha|A| \leq \alpha n/3 + 2\alpha n/3 = \alpha n$. In this case, the set $N_G(A)$ witnesses the existence of a small separator in G, separating between A and $V(G) \setminus (A \cup N_G(A))$. In case where the stopping criterion has been reached due to $I_0 = [k]$, the set B spans a minor of H in G, where the image of $u_i \in V(H)$ is the (connected) set B_i. $\square$

Corollary 8.3. *Let G be an α-expander on n vertices. Then G is minor universal for the class of all graphs H with at most $cn/\log n$ vertices and edges, for some $c = c(\alpha) > 0$.*

Let us add few remarks here. First, as we have indicated, the proof presented is algorithmic and allows to find in time polynomial in n a minor embedding of a given graph H

with $|V(H)|+|E(H)| \leq cn/\log n$. Secondly, the minor embedding produced is shallow, i.e., the image of every vertex $u \in V(H)$ in the embedding is a connected set U in G, spanning a graph of diameter logarithmic in n. Working out the details of the proof, we see that the constant c in Theorem 8.1, and thus in Corollary 8.3 depends quadratically on α. Finally, notice that the argument above can be trivially adapted to the case where $\alpha = \alpha(n)$ is vanishing.

8.2 Complete minors in expanding graphs

Theorem 8.1 guarantees that every expanding graph G on n vertices contains every graph H with $O(n/\log n)$ vertices and edges as a minor. As we argued in the previous subsection, this estimate on the size of H is optimal. For some particular types of H however we can do better. For example, due to Proposition 7.1 an expanding n-vertex graph G contains a path of length linear in n. One can also prove easily that such G contains a linearly sized star as a minor. (First find a linearly long path P in G, then argue that due to expansion one can find in G a matching M of linear size with every edge intersecting $V(P)$ in exactly one vertex; contracting $V(P)$ to a single vertex produces the required star.)

A particularly prominent class of graphs to try and find as minors is cliques. How large a clique minor can one hope to find in a weak expander? Denote by $ccl(G)$ the clique contraction number of G, which is the order of a largest clique minor to be found in G. Recall that if H is a minor of G, then $|E(H)| \leq |E(G)|$. Since weak expanders on n vertices can have linearly few in n edges, the best bound one can hope to get is $ccl(G) = \Theta(\sqrt{n})$. Theorem 8.1 comes quite close and guarantees that a graph G on n vertices without separators of small linear size has $ccl(G) = \Omega\left(\sqrt{n/\log n}\right)$.

The above mentioned result of Kawarabayashi and Reed [29] improves this estimate, proving that a graph G on n vertices without sublinear separators satisfies: $ccl(G) = \Omega(\sqrt{n})$. In fact, their result is much more general as it proves the following alternative: for any $h(n)$, an n-vertex graph G has a minor of the complete graph K_h, or a separator of order $O(h\sqrt{n})$.

Here we present a proof of the following theorem, a joint work with Rajko Nenadov.

Theorem 8.4. *For every $\alpha > 0$ there exist c, n_0 such that the following holds. Let G be a graph on $n \geq n_0$ vertices and maximum degree d. Assume that the edge isoperimetric number $i(G)$ satisfies: $i(G) \geq \alpha d$. Then $ccl(G) \geq c\sqrt{n}$.*

Here the edge isoperimetric number $i(G)$ of a graph $G = (V, E)$ is defined as:

$$i(G) = \min_{\substack{U \subset V \\ |U| \leq |V|/2}} \frac{e_G(U, V \setminus U)}{|U|}.$$

This quantity is fairly similar to the Cheeger constant $h(G)$ (to the extent that it is sometimes called the Cheeger constant of G); they are in fact exactly equal up to the factor d for the case of G being a d-regular graph.

The above theorem is superseded by the much more general result of Kawarabayashi and Reed. However, the proof of the latter is quite involved, and the full version of the conference paper [29] is still waiting to be published. We thus believe that presenting a reasonably short proof of Theorem 8.4 – whose outline will come shortly – certainly has merit.

Theorem 8.4 implies in particular that an α-expanding graph on n vertices with bounded degrees has a complete minor of the order of magnitude $\sqrt{n}$. The assumption about edge expansion being comparable to the maximum degree is an artifact of the proof and its techniques (specifically, the use of random walks), and can perhaps be lifted.

Proof. The proof is somewhat similar to the proof of Theorem 8.1, and uses some ideas from the arguments of Plotkin, Rao and Smith [48], and of Krivelevich and Sudakov [39]. Basically, we construct a complete minor of K_k in G, for an appropriately chosen k, a supernode by a supernode. Each time, when embedding the next supernode B_i in G, we need to make sure that B_i both spans a connected graph in G, and contains a neighbor of every presently embedded supernode B_j; due to expansion we can assume that the set of neighbors U_j of B_j is reasonably large. We will aim to do it economically, i.e., to get a relatively small set B_i, so as to allow enough room for embedding of k supernodes. A fairly natural idea for finding such a set, already suggested in [39], is to use random walks, and to take B_i to be the trace of a long enough random walk W, hoping it will hit each of the neighborhoods U_j. Our hopes are supported by the expansion properties of G, enabling to argue that W behaves similarly to a random set of the same size in terms of its hitting properties. In reality, it turns out to be more beneficial to take B_i to be the trace of W along with shortest paths from W to each U_j, arguing that typically the addition of these shortest paths does not add much to the size of W.

Now we get to the actual proof. We start it by introducing a basic tool based on random walks. Recall that a *lazy random walk* on a graph $G = (V, E)$ with the vertex set $V = \{1, \ldots, n\}$ is a Markov chain whose matrix of transition probabilities $P = P(G) = (p_{i,j})$ is defined by

$$p_{i,j} = \begin{cases} \frac{1}{2\deg_G(i)}, & \text{if } \{i, j\} \in E(G) \\ \frac{1}{2}, & \text{if } i = j, \\ 0, & \text{otherwise.} \end{cases}$$

This Markov chain has the stationary distribution π defined as $\pi(i) = \deg_G(i)/2e(G)$. As usual, for a subset $A \subseteq [n]$, we write $\pi(A) = \sum_{i \in A} \pi(i)$.

The following lemma gives an upper bound on the probability that a lazy random walk avoids a given subset $U \subseteq V(G)$.

Lemma 8.5. *Let G be a graph with n vertices and maximum degree $d = d(n)$. Then for any $U \subseteq V(G)$ the probability that a lazy random walk on G, which starts from the stationary distribution π and makes ℓ steps, does not visit U is at most*

$$\exp\left(-\frac{i(G)^3}{8d^3} \cdot \frac{|U|\ell}{n}\right).$$

We now describe the proof of Lemma 8.5.

Let $1 = \lambda_1 \geq \lambda_2 \geq \ldots \geq \lambda_n$ be the eigenvalues of the transition matrix P. The *spectral gap* of P (or of G) is defined as $\eta(G) = 1 - \lambda_2$. The following result of Mossel et al. [46] (more precisely, the first case of [46, Theorem 5.4]) relates the spectral gap to the probability that a random walk does not leave a specific subset. We state its version tailored for our needs.

Theorem 8.6 ([46]). *Let G be a connected graph with n vertices and let $\eta(G)$ be the spectral gap of the transition matrix $P = P(G)$. Then the probability that a lazy random walk of* *length ℓ, starting from a vertex chosen according to the stationary distribution π, does not leave a non-empty subset $A \subseteq V(G)$ is at most*

$$\pi(A)(1 - \eta(G)(1 - \pi(A)))^\ell.$$

In order to bound the spectral gap of G, we use the celebrated result of Jerrum and Sinclair [28, Lemma 3.3], relating the spectral gap of $P(G)$ to its *conductance* $\Phi(G)$, defined as

$$\Phi(G) = \min_{\substack{S \subseteq V \\ 0 < \pi(\bar{S}) \leq 1/2}} \frac{\sum_{i \in S, j \notin S} \pi(i)p_{i,j}}{\pi(S)}.$$

Lemma 8.7 ([28]). *Let $G = (V, E)$ be a connected graph. Then the second eigenvalue λ_2 of the transition matrix P satisfies: $1 - \lambda_2 \geq \Phi(G)^2/2$.*

Finally, it is easy to verify that $\Phi(G) \geq i(G)/2d$, where as before $i(G)$ is the edge isoperimetric number of G, and d is its maximum degree.

Proof of Lemma 8.5. Consider a non-empty subset $U \subset V(G)$. Theorem 8.6 states that a lazy random walk never leaves the set $A = V(G) \setminus U$ (i.e., never visits U) with probability at most

$$\pi(A)(1 - \eta(G)(1 - \pi(A)))^\ell = (1 - \pi(U))(1 - \eta(G)\pi(U))^\ell$$
$$\leq \exp(-\eta(G)\ell\pi(U)) \leq \exp\left(-\frac{i(G)^2}{8d^2} \cdot \ell\pi(U)\right), \qquad (8.1)$$

where in the last inequality we used Lemma 8.7 and $\Phi(G) \geq i(G)/2d$. From the trivial bound $2e(G) \leq dn$ and $\delta(G) \geq i(G)$ we get

$$\pi(U) \geq \frac{|U|i(G)}{2e(G)} \geq |U|\frac{i(G)}{dn},$$

which after plugging into (8.1) gives the desired probability that a random walk misses U. $\qquad\qquad\square$

The next lemma implements our promise to utilize (extensions of) random walks as connected hitting sets.

Lemma 8.8. *For every $\beta > 0$ there exist positive $C = \Theta(1/\beta^3)$ and n_0 such that the following holds. Let G be a graph with $n \geq n_0$ vertices, maximum degree d, and $i(G) \geq \beta d$. Given k and s such that $ks \geq 2n$, and subsets $U_1, \ldots, U_k \subseteq V(G)$, where each U_j is of size $|U_j| \geq s$, there exists a connected set $Y \subseteq V(G)$ of size at most*

$$|Y| \leq C \cdot \frac{n}{s} \ln\left(\frac{ks}{n}\right)$$

which intersects every U_j.

We mention in passing that for many pairs $(k(n), s(n))$, by choosing subsets $U_i \subset [n]$, $|U_i| = s$, at random one can see that there is a familiy $\{U_i\}_{i=1}^k$, whose covering number has order of magnitude $(n/s)\log(ks/n)$. Thus Lemma 8.8 delivers a nearly optimal promise of the size of a hitting set, with an additional – and potentially important – benefit of this set spanning a connected subgraph in G.

Proof. Let

$$\ell = \frac{8}{\beta^3} \cdot \frac{n}{s} \ln\left(\frac{ks}{n}\right) ,$$

and consider a lazy random walk W in G which starts from the stationary distribution and makes ℓ steps. The desired connected subset Y is now constructed by taking the union of W with a shortest path between U_j and W, for each $j \in [k]$. We argue that with positive probability Y is of required size.

For $1 \le j \le k$, let X_j be the random variable measuring the distance from U_j to W in G. Then the set Y has expected size at most $\ell + \sum_{j=1}^k E[X_j]$.

In order to estimate the expectation of X_j, for a positive integer z write $p_{j,z} = Pr[X_j = z]$ and $p_{j,\ge z} = Pr[X_j \ge z]$. Trivially $p_{j,z} = p_{j,\ge z} - p_{j,\ge z+1}$. Hence

$$E[X_j] = \sum_{z \ge 1} z p_{j,z} = \sum_{z \ge 1} z \left(p_{j,\ge z} - p_{j,\ge z+1}\right) \le \sum_{z \ge 1} p_{j,\ge z} .$$

Let $U_{j,z}$ be the set of vertices of G at distance at most z from U_j. The graph G is easily seen to be a β-expander, and similarly to Proposition 3.1 we have: $|U_{j,z}| \ge \min\left\{s(1+\beta)^z, \frac{n}{2}\right\}$. Applying Lemma 8.5 to $U_{j,z}$, we obtain for $z \le \log(n/2)/\log(1+\beta)$:

$$p_{j,\ge z} \le \exp\left\{-\frac{\beta^3(1+\beta)^z s \ell}{8n}\right\} .$$

Plugging in this estimate, recalling the value of ℓ, and doing fairly straightforward arithmetic, we can derive:

$$E[X_j] = \sum_{z \ge 1} p_{j,\ge z} = O\left(\left(\frac{n}{ks}\right)^{1+\beta}\right) .$$

Hence

$$E[|Y|] = \ell + \sum_{j=1}^k E[X_j] = \ell + O\left(k\left(\frac{n}{ks}\right)^{1+\beta}\right) = O(\ell) .$$

$\square$

We are now ready to complete the proof of Theorem 8.4. Let C be a constant given by Lemma 8.8 with $\beta = \alpha/4$, and set

$$b = \sqrt{\frac{8C}{\alpha} \cdot n} \qquad \text{and} \qquad k = \frac{\alpha n}{6b} = \Theta(\alpha^3)\sqrt{n} .$$

The argument here is somewhat similar to the proof of Theorem 8.1. We maintain and update a partition $V(G) = A \cup B \cup C$ so that always $e_G(A, C) < \beta d|A|$. There is also a set of indices $I_0 \subseteq [k]$ (corresponding to the set of supernodes $B_i, i \in I_0$), such that the subsets $B_i \subset B$ are pairwise disjoint, span connected subgraphs in G, and $|B_i| = b$. Moreover, if $i_1 \ne i_2 \in I_0$, then G has an edge between B_{i_1} and B_{i_2}.

Initially $A = B = \emptyset$, $C = V$, $I_0 = \emptyset$. As long as $|A| \le n/3$ or $I_0 \subsetneq [k]$, we repeat the following loop. If there exists $i \in I_0$ such that $e_G(B_i, C) < \beta d|B_i|$, we move B_i to A and update: $A := A \cup B_i$, $B := B \setminus B_i$, $I_0 := I_0 \setminus \{i\}$. Otherwise, recalling that $\Delta(G) \le d$, we derive that $|N_G(B_i, C)| \ge \beta b$ for all $i \in I_0$.

Now look into the induced subgraph $G[C]$. If there is a subset $U \subset C$, $|U| \le |C|/2$ with $e_G(U, C \setminus U) < \beta d|U|$, we move U to A and update $A := A \cup U$, $C := C \setminus U$. Otherwise, the subgraph $G[C]$ satisfies $i(G[C]) \geq \beta d$. We now apply to it Lemma 8.8 with $s = \beta b$ and the sets $U_j = N_G(B_j, C)$, $j \in I_0$ (adding dummy sets U_j to get k' sets U_j altogether, all of size at least s, so that $2n < k's < 3n$) to get a connected set Y of size $|Y| \leq C\frac{n}{s}\ln 3 < b$. Finally we extend Y to a connected set Y' of exactly b vertices in C, choose an arbitrary $i \in [k] \setminus I_0$ and update: $B := B \cup Y'$, $C := C \setminus Y'$, $B_i := Y$, $I_0 := I_0 \cup \{i\}$.

If the procedure terminated due to $I_0 = [k]$, then we have found the required K_k-minor. Suppose that this is not the case. We have $|B| < bk = \alpha n/6$. In the last step the set A has become larger than $n/3$, but still $|A| \leq 2n/3$. This implies $\min(|A|, |V \setminus A|) \geq n/3$, and hence $e_G(A, V \setminus A) \geq i(G)n/3 \geq \alpha n d/3$. On the other hand, $e_G(A, B) \leq d|B| \leq \alpha n d/6$, and $e_G(A, C) < \alpha d/4 \cdot |A| \leq \alpha d n/6$. Altogether $e_G(A, V \setminus A) < e_G(A, B) + e_G(A, C) < \alpha n d/6 + \alpha n d/6 = \alpha n d/3$. The obtained contradiction shows that the algorithm always runs to a successful end, outputting a minor of K_k in G. □

Let us mention that [37] presents a result about large complete minors in expanding graph with growing degrees. Since the result's statement is somewhat involved and assumes stronger edge expansion of small sets, we decided not to state it here, referring the interested reader to [37] for details.

8.3 Large minors in random graphs

We conclude this section by presenting consequences of the above results for random graphs.

As we argued in Section 6 (see Corollary 6.6 there), a supercritical random graph $G \sim G(n, c/n)$, $c > 1$, typically contains an expanding subgraph G^* of bounded degree on linearly many vertices. Applying Corollary 8.3 and Theorem 8.4 to G^*, we derive the following corollaries.

Corollary 8.9. *For every $c > 1$ there exists $\delta > 0$ such that a random graph $G \sim G(n, c/n)$ is with high probability minor universal for the class of all graphs H with at most $\delta n/\log n$ vertices and edges.*

Corollary 8.10. *For every $c > 1$ there exists $\delta > 0$ such that a random graph $G \sim G(n, c/n)$ contains with high probability a minor of the complete graph on $\lfloor \delta \sqrt{n} \rfloor$ vertices.*

Both corollaries provide an asymptotically optimal order of magnitude. For Corollary 8.9, notice that $G \sim G(n, c/n)$ contains typically $o(n/\log n)$ cycles of length at most $c_0 \log n$, for some $c_0 = c_0(c) > 0$ small enough (straightforward first moment argument), and thus we can repeat the argument from Section 8.1. For Corollary 8.10, observe that $G \sim G(n, c/n)$ has typically $O(n)$ edges.

Corollary 8.10 reproves a result of Fountoulakis, Kühn and Osthus [18], obtained by applying direct ad hoc methods (and by working quite hard one may add).

Finally, recall that for $d \geq 3$ a random d-regular graph $G_{n,d}$ is an α-expander for $\alpha = \alpha(d) > 0$ (see Section 4 for discussion). We thus obtain:

Corollary 8.11. *For every integer $d \geq 3$ there exists $\delta > 0$ such that a random graph $G \sim G_{n,d}$ contains with high probability a minor of the complete graph on $\lfloor \delta \sqrt{n} \rfloor$ vertices.*

This has been proven in another paper of Fountoulakis, Kühn and Osthus [19], through the use of the so-called configuration model, and a quite substantial contiguity result about random graphs. The argument presented here provides an alternative, and perhaps more conceptual, way to derive this result.

Acknowledgement. The author wishes to thank the anonymous referee for their careful reading and helpful remarks. He is also very grateful to Rajko Nenadov for his cooperation in parts of the research presented in this survey. Finally, the author thanks Limor Friedman, Rajko Nenadov and Wojciech Samotij for their input and remarks.

References

[1] M. Ajtai, J. Komlós and E. Szemerédi, *The longest path in a random graph*, Combinatorica 1 (1981), 1–12.

[2] N. Alon, *Eigenvalues and expanders*, Combinatorica 6 (1986), 83–96.

[3] N. Alon, M. Krivelevich and P. Seymour, *Long cycles in critical graphs*, Journal of Graph Theory 35 (2000), 193–196.

[4] N. Alon and V.D. Milman, λ_1, *isoperimetric inequalities for graphs, and superconcentrators*, Journal of Combinatorial Theory, Series B 38 (1985), 73–88.

[5] N. Alon, P. Seymour and R. Thomas, *A separator theorem for nonplanar graphs*, Journal of the American Mathematical Society 3 (1990), 801–808.

[6] N. Alon and J.H. Spencer, The probabilistic method, 4th edition, Wiley, New York, 2016.

[7] J. Beck, *On size Ramsey number of paths, trees, and circuits. I*, Journal of Graph Theory 7 (1983), 115–129.

[8] I. Ben-Eliezer, M. Krivelevich and B. Sudakov, *The size Ramsey number of a directed path*, Journal of Combinatorial Theory, Series B 102 (2012), 743–755.

[9] M. Bednarska and O. Pikhurko, *Biased positional games on matroids*, European Journal of Combinatorics 26 (2005), 271–285.

[10] B. Bollobás, *The isoperimetric number of random regular graphs*, European Journal of Combinatorics 9 (1988), 241–244.

[11] S. Brandt, H. Broersma, R. Diestel and M. Kriesell, *Global connectivity and expansion: long cycles in f-connected graphs*, Combinatorica 26 (2006), 17–36.

[12] A.E. Brouwer and W.H. Haemers, Spectra of graphs, Springer, New York, 2012.

[13] F.R.K. Chung, Spectral graph theory, CBMS Regional Conference Series in Mathematics 92, American Mathematical Society, Providence, 1997.

[14] F. Chung, *Four proofs for the Cheeger inequality and graph partition algorithms*, AMS/IP Studies in Advanced Mathematics 48 (2010), 331–349.

[15] J. Dodziuk, *Difference equations, isoperimetric inequality and transience of certain random walks*, Transactions of the American Mathematical Society 284 (1984), 787–794.

[16] A. Dudek and P. Prałat, *On some multicolor Ramsey properties of random graphs*, SIAM Journal on Discrete Mathematics 31 (2017), 2079–2092.

[17] A. Dudek and P. Prałat, *Note on the multicolour size-Ramsey number for paths*, Electronic Journal of Combinatorics 25 (2018), P 3.35.

[18] N. Fountoulakis, D. Kühn and D. Osthus, *The order of the largest complete minor in a random graph*, Random Structures & Algorithms 33 (2008), 127–141.

[19] N. Fountoulakis, D. Kühn and D. Osthus, *Minors in random regular graphs*, Random Structures & Algorithms 35 (2009), 444–463.

[20] J. Friedman, *A proof of Alon's second eigenvalue conjecture and related problems*, Memoirs of the American Mathematical Society, 2008.

[21] L. Friedman, M. Krivelevich and R. Nenadov, *Cycle lengths in expanding graphs*, manuscript.

[22] A. Frieze and M. Karoński, Introduction to Random Graphs, Cambridge University Press, 2015.

[23] J.R. Gilbert, J.P. Hutchinson and R.E. Tarjan, *A separator theorem for graphs of bounded genus*, Journal of Algorithms 5 (1984), 391–407.

[24] D. Hefetz, M. Krivelevich, M. Stojaković and T. Szabó, Positional Games, Birkhäuser, Basel, 2014.

[25] S. Hoory, N. Linial and A. Wigderson, *Expander graphs and their applications*, Bulletin of the American Mathematical Society 43 (2006), 439–561.

[26] S. Janson, T. Łuczak and A. Ruciński, Random Graphs, Wiley, New York, 2000.

[27] M. Jerrum and A. Sinclair, *Conductance and the rapid mixing property for Markov chains: the approximation of the permanent resolved*, Proceedings of the 20th Annual ACM Symposium on Theory of Computing (STOC'88), 1988, 235–244.

[28] M. Jerrum and A. Sinclair, *Approximate counting, uniform generation and rapidly mixing Markov chains*, Information and Computation 82 (1989), 93–133.

[29] K. Kawarabayashi and B. Reed, *A separator theorem in minor-closed classes*, Proceedings of the 51st Annual Symposium on Foundations of Computer Science (FOCS'10), 2010, 153–162.

[30] J. Kleinberg and R. Rubinfeld, *Short paths in expander graphs*, Proceedings of the 37th Annual Symposium on Foundations of Computer Science (FOCS'96), 1996, 86–95.

[31] J. Komlós and E. Szemerédi, *Topological cliques in graphs*, Combinatorics, Probability and Computing 3 (1994), 247–256.

[32] J. Komlós and E. Szemerédi, *Topological cliques in graphs II*, Combinatorics, Probability and Computing 5 (1996), 79–90.

[33] M. Krivelevich, Long paths and Hamiltonicity in random graphs, in: *Random Graphs, Geometry and Asymptotic Structure*, editors N. Fountoulakis and D. Hefetz, London Mathematical Society Student Texts 84, Cambridge University Press, 2016, pp. 4–27.

[34] M. Krivelevich, *Finding and using expanders in locally sparse graphs*, SIAM Journal on Discrete Mathematics 32 (2018), 611–623.

[35] M. Krivelevich, *Long cycles in locally expanding graphs, with applications*, Combinatorica, to appear.

[36] M. Krivelevich, C. Lee and B. Sudakov, *Long paths and cycles in random subgraphs of graphs with large minimum degree*, Random Structures & Algorithms 46 (2015), 320–345.

[37] M. Krivelevich and R. Nenadov, *Complete minors in graphs without sparse cuts*, manuscript.

[38] M. Krivelevich and B. Sudakov, Pseudo-random graphs, in: *More sets, graphs and numbers*, editors E. Györi, G. O. H. Katona and L. Lovász, Bolyai Society Mathematical Studies 15, 2006, pp. 199–262.

[39] M. Krivelevich and B. Sudakov, *Minors in expanding graphs*, Geometric and Functional Analysis 19 (2009), 294–331.

[40] M. Krivelevich and B. Sudakov, *The phase transition in random graphs — a simple proof*, Random Structures & Algorithms 43 (2013), 131–138.

[41] D.A. Levin, Y. Peres and E.L. Wilmer, Markov chains and mixing times, American Mathematical Society, Providence, 2009.

[42] R.J. Lipton and R.E. Tarjan, *A separator theorem for planar graphs*, SIAM Journal on Applied Mathematics 36 (1979), 177–189.

[43] A. Lubotzky, R. Phillips and P. Sarnak, *Ramanujan graphs*, Combinatorica 8 (1988), 261–277.

[44] R. Montgomery, *Logarithmically small minors and topological minors*, Journal of the London Mathematical Society 91 (2015), 71–88.

[45] G. Moshkovitz and A. Shapira, *Decomposing a graph into expanding subgraphs*, Random Structures & Algorithms 52 (2018), 158–178.

[46] E. Mossel, R. O'Donnell, O. Regev, J. Steif, and B. Sudakov, *Non-interactive correlation distillation, inhomogeneous Markov chains and the reverse Bonami-Beckner inequality*, Israel Journal of Mathematics 154 (2006), 299–336.

[47] M.S. Pinsker, *On the complexity of a concentrator*, Proceedings of the 7th International Teletraffic Conference, 1973, 318/1–318/4.

[48] S. Plotkin, S. Rao and W. Smith, *Shallow excluded minors and improved graph decompositions*, Proceedings of the 5th Annual ACM-SIAM Symposium on Discrete Algorithms (SODA'94), 1994, 462–470.

[49] O. Riordan, *Long cycles in random subgraphs of graphs with large minimum degree*, Random Structures & Algorithms 45 (2014), 764–767.

[50] P. Sarnak, *What is . . . an expander?*, Notices of the American Mathematical Society 51 (2004), 762–763.

[51] A. Shapira and B. Sudakov, *Small complete minors above the extremal edge density*, Combinatorica 35 (2015), 75–94.

[52] B. Sudakov and J. Verstraëte, *Cycle lengths in sparse graphs*, Combinatorica 28 (2008), 357–372.

[53] N.C. Wormald, Models of random regular graphs, in: *Surveys in Combinatorics*, editors J. Lamb and D. Preece, London Mathematical Society Lecture Note Series 267, Cambridge University Press, 1999, pp. 239–298.

School of Mathematical Sciences,
Raymond and Beverly Sackler Faculty of Exact Sciences,
Tel Aviv University,
Tel Aviv, 6997801,
Israel
krivelev@tauex.tau.ac.il

Supersingular isogeny graphs in cryptography

Kristin E. Lauter and Christophe Petit

Abstract

We describe recent applications of expander graphs in cryptography, particularly supersingular isogeny graphs. The security of these cryptographic constructions relies on the assumption that computing paths in these graphs is practically infeasible, even with a quantum computer. One of these cryptographic constructions is currently considered for standardization by NIST. We also recall a related construction based on Lubotzky–Philips–Sarnak's celebrated Ramanujan graphs. We describe an efficient path-finding algorithm for these graphs which was motivated by the cryptographic application, and we mention connections to the problem of optimal quantum circuit synthesis.

1 Introduction

Most of our current security infrastructures will become completely insecure once quantum computers are built. Their security relies on the computational hardness of mathematical problems such as factoring and discrete logarithm problems in various abelian groups. These problems are indeed intractable using today's computers but can be solved in polynomial time on quantum computers. Post-quantum cryptography aims at developing new security protocols that will remain secure even after quantum computers are built. The biggest cybersecurity agencies in the world, including GCHQ and the NSA, have recommended a move towards post-quantum cryptography, and the new generation of cryptographic standards will aim at post-quantum security. The American National Institute for Standards and Technologies (NIST) have invited all academics to contribute to a future post-quantum cryptography standard with new algorithms and analysis [PQC].

Supersingular Isogeny Graphs were proposed for use in cryptography in 2006 by Charles, Goren, and Lauter [CGL06]. Supersingular isogeny graphs are examples of Ramanujan graphs, i.e. optimal expander graphs. This means that relatively *short* walks on the graph approximate the uniform distribution, i.e. walks of length approximately equal to the logarithm of the graph size. Walks on expander graphs are often used as a good source of randomness in computer science, and the reason for using *Ramanujan* graphs is to keep the path length short. But the reason these graphs are important for cryptography is that *finding paths* in these graphs, i.e. *routing,* is believed to be hard: there are no known subexponential algorithms to solve this problem, either classically or on a quantum computer. For this reason, systems based on the hardness of problems on Supersingular Isogeny Graphs are currently under consideration for standardization in the NIST Post-Quantum Cryptography (PQC) Competition. Other submissions to the competition include systems based on the hardness of finding short vectors in lattices (lattice-based cryptography), hardness of decoding random linear error-correcting codes (code-based cryptography), and hardness of solving multivariate systems of polynomial equations.

[CGL06] proposed a general construction for cryptographic hash functions based on the hardness of finding a walk on a graph between two given vertices of a given length (a similar construction for Cayley graphs was proposed in [TZ93]). The path-finding problem is the following: given fixed starting and ending vertices representing the start and end points of a path on the graph of a fixed length, find a path of that length between them. A hash function can be defined by using the input to the function as directions for walking around the graph: the output is the label for the ending vertex of the walk. Finding collisions for the hash function is equivalent to finding cycles in the graph, and finding pre-images is equivalent to path-finding in the graph. Backtracking is not allowed in the walks by definition, to avoid trivial collisions.

In [CGL06], two concrete examples of families of optimal expander graphs (Ramanujan graphs) were proposed, the Lubotzky–Phillips–Sarnak (LPS) graphs [LPS88], and the Supersingular Isogeny Graphs (Pizer) [Piz98], where the path finding problem was supposed to be hard. This work was presented at the 2005 and 2006 NIST Hash Function workshops, but the LPS hash function was quickly attacked and broken in two papers in 2008, a collision attack [TZ08] and a pre-image attack [PLQ08]. The preimage attack gives an algorithm to efficiently find paths in LPS graphs, a problem which had been open for several decades. The PLQ algorithm given in [PLQ08] does not, however, find a path of any given specific length, it just finds a path of some length which is bounded by a multiple of the diameter of the graph. The PLQ path-finding algorithm uses the explicit description of the graph as a Cayley graph in $PSL_2(2, \mathbb{F}_p)$, where vertices are 2×2 matrices with entries in $\mathbb{F}_p$ satisfying certain properties.

In 2011, Jao–De Feo [DFJP14] proposed another cryptographic application of Supersingular Isogeny Graphs, namely a key exchange protocol. Another application, to constructing cryptographic signatures, was proposed by Galbraith–Petit–Silva [GPS17]. The security of these applications relies on various hard problems on Supersingular Isogeny Graphs, but the relationships between these hard problems was not clear until recent papers [PL17, EHLMP18, CFLMP18] clarified and studied the relationships.

Given the PLQ attack on the path-finding problem in LPS graphs, it is important to understand the connection between LPS graphs and Supersingular Isogeny Graphs, and this was studied and presented in [CFLMP18]. Furthermore, a surprising connection between the PLQ path-finding algorithm and quantum circuit design was found by Peter Sarnak, who pointed out the fundamental similarity between the PLQ algorithm and the Ross–Selinger algorithm for quantum gate decompositions [Ros15], which was discovered independently.

In this survey, we introduce Supersingular Isogeny Graphs (Section 2); we cover their cryptographic applications including hash functions, key-exchange, and signatures (Section 3); we explain the underlying hard problems and known relationships between the problems (Sections 4 and 5); and we cover LPS graphs, path-finding algorithms in LPS graphs, connections to quantum circuit design and the relationship between Supersingular Isogeny Graphs and LPS graphs (Section 6).

This survey was written to accompany the first author's Invited Lecture at the 27th British Combinatorial Conference at the University of Birmingham, July 29–August 2, 2019. The article covers the topics of the lecture, and consists of lightly edited sections from our papers on this topic, written together and with other coauthors. We thank our coauthors for permission to include this content here.

The survey covers combinatorics objects related to various number theory and cryptography concepts. We refer to [KL07] for an introduction to cryptography; to [Sil09] for the necessary background on elliptic curves and elliptic curve cryptography; and to [Voi18] for quaternion algebras.

2 Preliminaries

2.1 Definitions and Background on Elliptic Curves

We start by recalling some basic and well-known results about elliptic curves and isogenies. They can all be found in [Sil09].

An elliptic curve is a curve of genus one with a specific base point $\mathcal{O}$ which can be used to define a group law, see for example [Sil09] for details. Cryptographers define an elliptic curve to be the set of projective solutions to a Weierstrass equation, with a group law defined via the chord-and-tangent method relative to the given base point. If E is an elliptic curve defined over a field K and $\mathrm{char}(K) \neq 2,3$, we can write a short Weierstrass equation for E:

$$E : y^2 = x^3 + a \cdot x + b,$$

where $a, b \in K$. Two important quantities related to an elliptic curve are its discriminant Δ and its j-invariant, denoted by j. They are defined as follows.

$$\Delta = 16 \cdot (4 \cdot a^3 + 27 \cdot b^2) \quad \text{and} \quad j = -1728 \cdot \frac{a^3}{\Delta}.$$

Two elliptic curves are isomorphic over $\bar{K}$ if and only if they have the same j-invariant.

Definition 2.1. Let E_0 and E_1 be two elliptic curves. An *isogeny* from E_0 to E_1 is a surjective morphism

$$\phi : E_0 \to E_1,$$

which is a group homomorphism. A morphism between algebraic varieties is a function between the varieties that is given locally by polynomials.

An example of an isogeny is the multiplication-by-m map $[m]$,

$$[m] : E \to E$$
$$P \mapsto m \cdot P.$$

The degree of an isogeny is defined as the degree of the finite extension

$$\bar{K}(E_0)/\phi^*(\bar{K}(E_1)),$$

where $\bar{K}(*)$ is the function field of the curve, and ϕ^* is the map of function fields induced by the isogeny ϕ. By convention, we set

$$\deg([0]) = 0.$$

The degree map is multiplicative under composition of isogenies:

$$\deg(\phi \circ \psi) = \deg(\phi) \cdot \deg(\psi)$$

for all chains $E_0 \xrightarrow{\phi} E_1 \xrightarrow{\psi} E_2$. For an integer $m > 0$, the multiplication-by-m map has degree m^2.

The most natural way to represent an isogeny is as a rational map. In cryptographic protocols the isogenies have very large degrees, hence this is not an efficient representation. However when the isogeny degree is smooth (a product of small primes, possibly not distinct) the isogeny can be represented as a composition of low degree isogenies, each of which given as a rational map.

Theorem 2.2. *[Sil09] Let $E_0 \to E_1$ be an isogeny of degree m. Then, there exists a unique isogeny*

$$\hat{\phi} : E_1 \to E_0$$

such that $\hat{\phi} \circ \phi = [m]$ on E_0, and $\phi \circ \hat{\phi} = [m]$ on E_1. We call $\hat{\phi}$ the dual isogeny to ϕ. We also have that

$$\deg(\hat{\phi}) = \deg(\phi).$$

For an isogeny ϕ, we say ϕ is separable if the field extension $\bar{K}(E_0)/\phi^*(\bar{K}(E_1))$ is separable. We then have the following lemma.

Lemma 2.3. *Let $\phi : E_0 \to E_1$ be a separable isogeny. Then*

$$\deg(\phi) = \# \ker(\phi).$$

In this paper, we only consider separable isogenies and frequently use this convenient fact.

Any point P of order m defines an isogeny ϕ of degree m,

$$\phi : E \to E/\langle P \rangle.$$

We will refer to such an isogeny as a cyclic isogeny (meaning that its kernel is a cyclic subgroup of E). More generally, a finite subgroup G of E generates a unique isogeny of degree $\#G$, up to automorphism. For ℓ prime, we also say that two curves E_0 and E_1 are ℓ-isogenous if there exists an isogeny $\phi : E_0 \to E_1$ of degree ℓ. When m is small or a product of small primes, this isogeny can be computed efficiently using Vélu's formulae [Vél71, DFJP14].

We define $E[\ell]$, the ℓ-torsion subgroup of E, to be the kernel of the multiplication-by-ℓ map. If $\operatorname{char}(K) > 0$ and $\ell \geq 2$ is an integer coprime to $\operatorname{char}(K)$, or if $\operatorname{char}(K) = 0$, then the points of $E[\ell]$ are

$$E[\ell] = \{P \in E(\bar{K}) : \ell \cdot P = \mathcal{O}\} \cong \mathbb{Z}/\ell\mathbb{Z} \times \mathbb{Z}/\ell\mathbb{Z}.$$

For ℓ prime, the ℓ-torsion subgroup contains $\ell + 1$ cyclic subgroups of order ℓ, each of them defining an isogeny of degree ℓ.

If an elliptic curve E is defined over a field of characteristic $p > 0$ and its endomorphism ring over $\bar{K}$ is an order in a quaternion algebra, we say that E is supersingular. Every isomorphism class over $\bar{K}$ of supersingular elliptic curves in characteristic p has a representative defined over $\mathbb{F}_{p^2}$, thus we will often let $K = \mathbb{F}_{p^2}$ (for some fixed prime p).

2.2 Supersingular isogeny graphs

Supersingular isogeny graphs were introduced into cryptography in [CGL06]. To define a supersingular isogeny graph, fix a finite field K of characteristic p, a supersingular elliptic curve E over K, and a prime $\ell \neq p$. Then the corresponding isogeny graph is constructed as follows. The vertices are the $\bar{K}$-isomorphism classes of elliptic curves which are $\bar{K}$-isogenous to E. Each vertex is labeled with the j-invariant of the curve. The edges of the graph correspond to the ℓ-isogenies between the elliptic curves. As the vertices are isomorphism classes of elliptic curves, isogenies that differ by composition with an automorphism of the image are identified as edges of the graph. I.e. if E_0, E_1 are $\bar{K}$-isogenous elliptic curves, $\phi : E_0 \to E_1$ is an ℓ-isogeny and $\epsilon \in \operatorname{Aut}(E_1)$ is an automorphism, then ϕ and $\epsilon \circ \phi$ are identified and correspond to the same edge of the graph.

If $p \equiv 1 \pmod{12}$, there is no non trivial automorphism, hence we can uniquely identify an isogeny with its dual to make it an undirected graph. It is a multigraph in the sense that there can be multiple edges if no extra conditions are imposed on p. Three important properties of these graphs follow from deep theorems in number theory:

1. The graph is connected for any $\ell \neq p$ (special case of [CGL09, Thm 4.1]).

2. A supersingular isogeny graph has roughly $p/12$ vertices. [Sil09, Thm 4.1]

3. Supersingular isogeny graphs are optimal expander graphs, in particular they are Ramanujan. (special case of [CGL09, Thm 4.2]).

Remark 3. In order to avoid trivial collisions in cryptographic hash functions based on isogeny graphs, it is best if the graph has no short cycles. Charles, Goren, and Lauter show in [CGL06] how to ensure that isogeny graphs do not have short cycles by carefully choosing the finite field one works over. For example, they compute that a 2-isogeny graph does not have double edges (i.e. cycles of length 2) when working over $\mathbb{F}_p$ with $p \equiv 1 \bmod 420$. Similarly, we computed that a 3-isogeny graph does not have double edges for $p \equiv 1 \bmod 9240$. Given that $420 = 2^2 \cdot 3 \cdot 5 \cdot 7$ and $9240 = 2^3 \cdot 3 \cdot 5 \cdot 7 \cdot 11$, it follows that neither the 2-isogeny graph nor the 3-isogeny graph has double edges for $p \equiv 1 \bmod 9240$.

3 Applications

3.1 Charles–Goren–Lauter Cryptographic Hash Function

Hash functions are cryptographic algorithms which take bitstrings as input, and return bitstrings of a fixed size. A hash function $h : \mathcal{M} \to \mathcal{H}$ must primarily satisfy three properties:

- Preimage resistance: given $h \in \mathcal{H}$, it must be computationally infeasible to compute $m \in \mathcal{M}$ such that $h = H(m)$.

- Collision resistance: it must be computationally infeasible to compute $m, m' \in \mathcal{M}$ such that $H(m) = H(m')$.

- Second preimage resistance: given $m \in \mathcal{M}$, it must be computationally infeasible to compute $m' \in \mathcal{M}$ such that $H(m) = H(m')$ and $m' \neq m$.

Hash functions are a fundamental primitive in cryptography, used for authentication and integrity purposes. An example of a cryptographic hash function commonly used in practice is the SHA-2 algorithm.

In [CGL06], a cryptographic hash function was defined:

$$h : \{0,1\}^r \to \{0,1\}^s$$

based on the Supersingular Isogeny Graph (SIG) for a fixed prime p of cryptographic size, and a fixed small prime $\ell \neq p$. The hash function processes the input string in blocks which are used as directions for walking around the graph starting from a given fixed vertex. Every step in the walk corresponds to an isogeny of degree ℓ, and these can be computed either using Vélu's formulae or using the modular polynomial of degree ℓ. The output of the hash function is the j-invariant of an elliptic curve over $\mathbb{F}_{p^2}$ which requires $2\log(p)$ bits to represent, so $m = 2\lceil \log(p) \rceil$. For the security of the hash function, it is necessary to avoid the generic *birthday attack*. This attack runs in time proportional to the square root of the size of the graph, which is the *Eichler class number*, roughly $\lfloor p/12 \rfloor$. So in practice, we must pick p so that $\log(p) \approx 256$ for 128-bit security.

The integer r is the length of the bit string input to the hash function. The case $\ell = 2$ is the easiest case to implement and a common choice. In this case the graph is 3-regular and r is precisely the number of steps taken on the walk in the graph, with no backtracking allowed, so the input is processed bit-by-bit. In order to ensure that the walk reaches a sufficiently random vertex in the graph, the number of steps should be roughly $\log(p) \approx 256$. A CGL-hash function is thus specified by giving the primes p, ℓ, the starting point of the walk, and the integers $r \approx 256$, s. (Extra congruence conditions were imposed on p to make it an undirected graph with no small cycles.)

The hard problems stated in [CGL06] corresponded to the important security properties of *collision* and *preimage resistance* for this hash function. For preimage resistance, the problem [CGL06, Problem 3] stated was: given p, ℓ, $r > 0$, and two supersingular j-invariants modulo p, to find a path of length r between them:

Problem 3.1 (Path-finding [CGL06]). *Let p and ℓ be distinct prime numbers, $r > 0$, and E_0 and E_1 two supersingular elliptic curves over $\mathbb{F}_{p^2}$. Find a path of length r in the ℓ-isogeny graph corresponding to a composition of r ℓ-isogenies leading from E_0 to E_1 (i.e. an isogeny of degree ℓ^r from E_0 to E_1).*

It is worth noting that, to break the preimage resistance of the specified hash function, you must find a path of exactly length r, and this is analogous to the situation for breaking the security of the key-exchange protocol below. However, the problem of finding any path between two given vertices in the SIG graphs is also still open. For the LPS graphs, the algorithm presented in [PLQ08] did not find a path of a specific given length, but it was still considered to be a "break" of the hash function.

As for collision resistance, it was shown in [PL17, EHLMP18] that it is also equivalent to the path-finding problem under appropriate heuristic assumptions, when the initial vertex for the hash computation is randomly chosen. On the other hand, an efficient collision algorithm was provided for special initial vertices or when a path to one of those special vertices was known.

3.2 Key Exchange ([DFJP14])

Key exchange is a cryptographic protocol by which two parties, after exchanging messages over a public network, agree on a common secret key (which can later be used for encrypting and authenticating communications). An example is the Diffie–Hellman protocol, which together with its elliptic curve version is in widespread use for example to establish secure ("https") browser connections. While both the original Diffie–Hellman protocol and its elliptic curve variant can be broken easily by quantum computers, the following variant using supersingular isogeny graphs, due to Jao–De Feo [DFJP14], is believed to be resistant to quantum computers.

Let E be a supersingular elliptic curve defined over $\mathbb{F}_{p^2}$, where $p = \ell_A^n \cdot \ell_B^m \pm 1$, ℓ_A and ℓ_B are primes, and $n \approx m$ are approximately equal. We have players A (for Alice) and B (for Bob), representing the two parties who wish to engage in a key-exchange protocol with the goal of establishing a shared secret key by communicating via a (possibly) insecure channel. The two players A and B generate their public parameters by each picking two points P_A, Q_A such that $\langle P_A, Q_A \rangle = E[\ell_A^n]$ (for A), and two points P_B, Q_B such that $\langle P_B, Q_B \rangle = E[\ell_B^m]$ (for B).

Player A then secretly picks two random integers $0 \leq m_A, n_A < \ell_A^n$. These two integers will be player A's secret parameters. A then computes the isogeny ϕ_A

$$E \xrightarrow{\phi_A} E_A := E/\langle [m_A]P_A + [n_A]Q_A \rangle.$$

Player B proceeds in a similar fashion and secretly picks $0 \leq m_B, n_B < \ell_B^m$. Player B then generates the (secret) isogeny

$$E \xrightarrow{\phi_B} E_B := E/\langle [m_B]P_B + [n_B]Q_B \rangle.$$

So far, A and B have constructed the following diagram.

To complete the diamond, we proceed to the exchange part of the protocol. Player A computes the points $\phi_A(P_B)$ and $\phi_A(Q_B)$ and sends $\{\phi_A(P_B), \phi_A(Q_B), E_A\}$ along to player B. Similarly, player B computes and sends $\{\phi_B(P_A), \phi_B(Q_A), E_B\}$ to player A. Both players now have enough information to construct the following diagram,

$$
\begin{array}{ccc}
 & E_A & \\
\nearrow^{\phi_A} & & \searrow^{\phi'_A} \\
E & & E_{AB} \\
\searrow^{\phi_B} & & \nearrow^{\phi'_B} \\
 & E_B &
\end{array}
\qquad (3.1)
$$

where

$$E_{AB} \cong E/\langle [m_A]P_A + [n_A]Q_A, [m_B]P_B + [n_B]Q_B \rangle.$$

Player A can use the knowledge of the secret information m_A and n_A to compute the isogeny ϕ'_B, by quotienting E_B by $\langle [m_A]\phi_B(P_A) + [n_A]\phi_B(Q_A) \rangle$ to obtain E_{AB}. Player B can use the knowledge of the secret information m_B and n_B to compute the isogeny ϕ'_A, by quotienting E_A by $\langle [m_B]\phi_A(P_B) + [n_B]\phi_A(Q_B) \rangle$ to obtain E_{AB}. A separable isogeny is determined by its kernel, and so both ways of going around the diagram from E result in computing the same elliptic curve E_{AB}.

The players then use the j-invariant of the curve E_{AB} as a shared secret.

Remark 4. Given a list of points specifying a kernel, one can explicitly compute the associated isogeny using Vélu's formulas [Vél71]. In principle, this is how the two parties engaging in the key-exchange above can compute ϕ_A, ϕ_B, ϕ'_A, ϕ'_B [Vél71]. However, in practice for cryptographic size subgroups, this approach is not practical, and thus a different approach is taken, based on breaking the isogenies into n (resp. m) steps, each of degree ℓ_A (resp. ℓ_B). This equivalence will be explained below.

The security of the key-exchange protocol is based on the following hardness assumption, which was introduced in [DFJP14] and called the Supersingular Computational Diffie–Hellman (SSCDH) problem.

Problem 3.2 (Supersingular Computational Diffie–Hellman (SSCDH)). *Let p, ℓ_A, ℓ_B, n, m, E, E_A, E_B, E_{AB}, P_A, Q_A, P_B, Q_B be as above.*

Let ϕ_A be an isogeny from E to E_A whose kernel is equal to $\langle [m_A]P_A + [n_A]Q_A \rangle$, and let ϕ_B be an isogeny from E to E_B whose kernel is equal to $\langle [m_B]P_B + [n_B]Q_B \rangle$, where m_A,n_A (respectively m_B,n_B) are integers chosen at random between 0 and ℓ_A^m (respectively ℓ_B^m), and not both divisible by ℓ_A (resp. ℓ_B).

Given the curves E_A, E_B and the points $\phi_A(P_B)$, $\phi_A(Q_B)$, $\phi_B(P_A)$, $\phi_B(Q_A)$, find the j-invariant of

$$E_{AB} \cong E/\langle [m_A]P_A + [n_A]Q_A, [m_B]P_B + [n_B]Q_B \rangle;$$

see diagram (3.1).

3.3 Identification Protocols

An identification protocol is an interactive protocol by which one party (the Prover) can prove to another party (the verifier) that he knows some secret data, without revealing any information on this secret. Such protocols have numerous applications, including to building digital signature schemes.

This section presents an identification protocol from [GPS17] similar to the graph isomorphism zero-knowledge protocol, in which one reveals one of two graph isomorphisms, but never enough information to deduce the secret isomorphism. It relies on the hardness of a problem which is a relaxation of Problem 1, that is, to find an isogeny between two given vertices in the Supersingular Isogeny Graph, without the restriction on the degree of the isogeny or the length of the path:

Problem 3.3. *Let p be a prime number. Let E, E' be supersingular elliptic curves over $\mathbb{F}_{p^2}$, chosen uniformly at random. Find an isogeny $E \to E'$.*

As explained below in Section 4, this problem is heuristically equivalent to the problem of computing the endomorphism ring of a random supersingular elliptic curve (Problem 4.1). Note that the endomorphism ring of any supersingular elliptic curve is isomorphic to a maximal order in the quaternion algebra $B_{p,\infty}$ over $\mathbb{Q}$ ramified only at p and ∞. Also, an isogeny between two curves E_1 and E_2 can be naturally mapped to an ideal with left and right orders isomorphic to $\operatorname{End}(E_1)$ and $\operatorname{End}(E_2)$. We refer to Section 4 for more details.

Even though the endomorphism ring computation problem is believed to be hard for randomly chosen curves, there are some particular curves for which it is easy. For example, when $p \equiv 3 \pmod 4$, we can write down explicitly the endomorphism ring of the curve with j-invariant $j = 1728$. Let $E_0 : y^2 = x^3 + Ax$ over a field $\mathbb{F}_{p^2}$, where $p \equiv 3 \pmod 4$, $j(E_0) = 1728$, and $\#E_0(\mathbb{F}_{p^2}) = (p + 1)^2$. When $p \equiv 3 \pmod 4$, the quaternion algebra $B_{p,\infty}$ ramified at p and ∞ can be canonically represented as

$$B_{p,\infty} \cong \mathbb{Q}\langle \mathbf{i}, \mathbf{j} \rangle = \mathbb{Q} + \mathbb{Q}\mathbf{i} + \mathbb{Q}\mathbf{j} + \mathbb{Q}\mathbf{k},$$

where $\mathbf{i}^2 = -1$, $\mathbf{j}^2 = -p$ and $\mathbf{k} := \mathbf{ij} = -\mathbf{ji}$. The endomorphism ring of E_0 is isomorphic to the maximal order $\mathcal{O}_0$ with $\mathbb{Z}$-basis $\{1, \mathbf{i}, \frac{1+\mathbf{k}}{2}, \frac{\mathbf{i}+\mathbf{j}}{2}\}$. Indeed, there is an isomorphism of quaternion algebras

$$\theta : B_{p,\infty} \to \operatorname{End}(E_0) \otimes \mathbb{Q}$$

sending $(1, \mathbf{i}, \mathbf{j}, \mathbf{k})$ to $(1, \phi, \pi, \pi\phi)$ where $\pi : (x, y) \to (x^p, y^p)$ is the Frobenius endomorphism, and $\phi : (x, y) \to (-x, \iota y)$ with $\iota^2 = -1$.

1. The public key is a pair (E_0, E_1) and the private key is an isogeny $\varphi : E_0 \to E_1$.

2. The prover performs a random walk starting from E_1 of degree L in the graph, obtaining a curve E_2 and an isogeny $\psi : E_1 \to E_2$, and reveals E_2 to the verifier.

3. The verifier challenges the prover with a random bit $b \leftarrow \{0, 1\}$.

4. If $b = 0$, the prover sends ψ to the verifier.

   If $b = 1$, the prover does the following:

   – Compute $\text{End}(E_2)$ and translate the isogeny path between E_0 and E_2 into a corresponding ideal I giving the path in the quaternion algebra.

   – Use the **Find new path** algorithm to compute an "independent" path between $\text{End}(E_0)$ and $\text{End}(E_2)$ in the quaternion algebra, represented by an ideal J.

   – Translate the ideal J to an isogeny path η from E_0 to E_2.

   – Return η to the verifier.

5. The verifier accepts the proof if the answer to the challenge is indeed an isogeny between E_1 and E_2 or between E_0 and E_2, respectively.

Figure 1:   [GPS17] Identification Scheme

To generate the public and private keys, we start at E_0 and take a random isogeny (walk in the graph) $\varphi : E_0 \to E_1$ and, using this knowledge, compute $\text{End}(E_1)$. The public information is E_1. The secret is $\text{End}(E_1)$, or equivalently a path from E_0 to E_1. Under the assumption that computing the endomorphism ring is hard, the secret key cannot be computed from the public key only.

This scheme requires three algorithms, explained in detail in [GPS17].

Translate isogeny path to ideal: Given E_0, $\mathcal{O}_0 = \text{End}(E_0)$ and a chain of isogenies from E_0 to E_1, compute $\mathcal{O}_1 = \text{End}(E_1)$ and a left $\mathcal{O}_0$-ideal I whose right order is $\mathcal{O}_1$.

Find new path: Given a left $\mathcal{O}_0$-ideal I corresponding to an isogeny $E_0 \to E_2$, produce a new left $\mathcal{O}_0$-ideal J corresponding to an "independent" isogeny $E_0 \to E_2$ of powersmooth degree.

Translate ideal to isogeny path: Given E_0, $\mathcal{O}_0$, E_2, I such that I is a left $\mathcal{O}_0$-ideal whose right order is isomorphic to $\text{End}(E_2)$, compute a sequence of prime degree isogenies giving the path from E_0 to E_2.

Figure 1 gives the interaction between the prover and the verifier. The isogenies involved in this protocol are summarised in the following diagram:

Here L is chosen large enough to ensure that $j(E_2)$ is nearly uniformly distributed, and as a product of small primes to ensure that the isogenies can be efficiently computed.

Figure 1 gives a canonical, recoverable identification protocol, but the challenge is only one bit. The two translation algorithms mentioned above in the $b = 1$ case are described in [GPS17]. They rely on the fact that $\text{End}(E_0)$ is known. The algorithms are efficient when the degree of the random walk is powersmooth, and for this reason all isogenies in the protocols are taken to be of powersmooth degree. The powersmooth version of the quaternion isogeny algorithm of Kohel–Lauter–Petit–Tignol is also described and analysed in [GPS17]. The random walks are taken to be sufficiently long such that their output has close to uniform distribution.

The process is repeated to reduce the cheating probability. The security of the protocol relies on the hardness of Problem 3.3:

Theorem 3.4. *[GPS17] Let λ be a security parameter and $t \geq \lambda$. If Problem 3.3 is computationally hard, then the identification scheme obtained from t parallel executions of the protocol in Figure 1 is a non-trivial, recoverable canonical identification scheme that is secure against impersonation under (classical) passive attacks.*

In [DFJP14], another identification scheme was proposed using ideas similar to the key exchange protocol. The advantage of the protocol of [GPS17] is that it relies on a more standard and potentially harder computational problem. Whereas the former gives the attacker auxiliary points, this one relies on the problem of computing the endomorphism ring of a random supersingular elliptic curve with no additional information.

4 Endomorphism Rings of Supersingular Elliptic Curves

Section 3 gave an overview of three cryptographic applications of Supersingular Isogeny Graphs along with the statements of the hard problems they rely on, Problems 1, 2, and 3. In addition, there is the related hard problem of computing the endomorphism ring of a supersingular elliptic curve, which we explain next.

Problem 4.1 (Endomorphism ring computation problem.). *Given a supersingular invariant j, compute the endomorphism ring of $E(j)$.*

The endomorphism ring of a supersingular elliptic curve is isomorphic to a maximal order in the quaternion algebra $B_{p,\infty}$ over $\mathbb{Q}$ ramified only at p and ∞. A quaternion algebra is generated as a $\mathbb{Q}$-module by four elements $\{1, i, j, k\}$ where $i^2 = a$, $j^2 = b$, $ij = -ji$ and $k = ij$ for some integers a, b, and is often denoted by (a, b). We refer to Vignéras [Vig80] for the arithmetic of quaternion algebras and the definitions and properties of the trace, reduced norm, orders and ideals.

Pizer [Piz80] gave the following explicit description of $B_{p,\infty}$ for all p along with a basis for one maximal order.

Proposition 4.2. *[Piz80, p368–369] Let $p > 2$ be a prime. Then we can define $B_{p,\infty}$ and the maximal order $\mathcal{O}_0$ as follows:*

p	(a,b)	$\mathcal{O}_0$
3 mod 4	$(-p,-1)$	$\langle 1, j, \frac{j+k}{2}, \frac{1+i}{2} \rangle$
5 mod 8	$(-p,-2)$	$\langle 1, j, \frac{2-j+k}{4}, \frac{-1+i+j}{2} \rangle$
1 mod 8	$(-p,-q)$	$\langle \frac{1+j}{2}, \frac{i+k}{2}, \frac{j+ck}{q}, k \rangle$

where in the last row $q \equiv 3 \pmod 4$, $(p/q) = -1$ and c is some integer with

$$q | c^2 p + 1.$$

Assuming that the generalized Riemann hypothesis is true, there exists $q = O(\log^2 p)$ satisfying these conditions.

We represent quaternion algebra elements as linear combinations of $1, i, j, k$, where moreover q is minimal in the case $p \equiv 1 \pmod 8$. In all cases the maximal orders $\mathcal{O}_0$ given by Proposition 4.2 contain $\langle 1, i, j, k \rangle$ as a small index subring.

Deuring [Deu41] showed that supersingular elliptic curves over $\overline{\mathbb{F}}_p$ (up to isomorphism) are in one-to-one correspondence with maximal orders of $B_{p,\infty}$ (up to conjugation by an invertible element of $B_{p,\infty}$). More precisely, Deuring's correspondence associates to a supersingular invariant j any maximal order $\mathcal{O}$ such that $\mathcal{O} \cong \operatorname{End}(E)$. Moreover any left ideal I of $\mathcal{O}$ corresponds to an isogeny $\phi_I : E \to E_I$ with kernel

$$\ker \phi_I = \{P \in E \,|\, \alpha(P) = 0, \forall \alpha \in I\}.$$

This is a 1-1 correspondence provided that the degree of ϕ_I is coprime to p. In addition, we can identify the right order of I, $\mathcal{O}_R(\mathcal{I})$ with the endomorphism ring of E_I.

When $p \equiv 3 \pmod 4$ the curve $y^2 = x^3 + x$ is supersingular with invariant $j = 1728$. This curve corresponds to a maximal order $\mathcal{O}_0$ with $\mathbb{Z}$-basis $\{1, i, \frac{1+k}{2}, \frac{i+j}{2}\}$ under Deuring's correspondence, and there is an isomorphism of quaternion algebras

$$\theta : B_{p,\infty} \to \operatorname{End}(E_0) \otimes \mathbb{Q}$$

sending $(1, i, j, k)$ to $(1, \phi, \pi, \pi\phi)$ where $\pi : (x, y) \to (x^p, y^p)$ is the Frobenius endomorphism, and $\phi : (x, y) \to (-x, \iota y)$ with $\iota^2 = -1$. More generally, it is easy to compute j-invariants corresponding to the maximal orders given by Proposition 4.2:

Proposition 4.3. *There is a polynomial time algorithm that given a prime $p > 2$, computes a supersingular invariant $j_0 \in \mathbb{F}_p$ such that $\operatorname{End}(E(j_0)) \cong \mathcal{O}_0$ (where $\mathcal{O}_0$ is as given by Proposition 4.2) together with a map $\phi \in \operatorname{End}(E(j_0))$ such that*

$$\theta : B_{p,\infty} \to \operatorname{End}(E(j_0)) \otimes \mathbb{Q} : (1, i, j, k) \to (1, \phi, \pi, \pi\phi)$$

is an isomorphism of quaternion algebras.

Proof. Consider Algorithm 1 below. Step 1 can be executed in time polynomial in $\log p$ using complex multiplication, as in Bröker's algorithm [Brö09]. The cardinality of $\mathcal{J}$ is equal to the class number of $\mathbb{Q}(\sqrt{-q})$, and this is bounded by q. To compute ϕ in Step 3 one can simply compute all isogenies of degree q using Vélu's formulae and identify the one corresponding to an endomorphism. The map ϕ defines an isomorphism of quaternion algebras

$$\theta : B_{p,\infty} \to \operatorname{End}(E(j_0)) \otimes \mathbb{Q} : (1, i, j, k) \to (1, \phi, \pi, \pi\phi).$$

To perform the check in Step 4, one applies θ to the numerators of $\mathcal{O}_0$ basis elements, and check whether the resulting maps annihilate the D torsion, where D is the denominator. $\square$

In general we are interested in *constructing* Deuring's correspondence for arbitrary maximal orders and supersingular j invariants. This could a priori have three different meanings, given by Problems 4.4, 4.5 and 4.6 below.

Algorithm 1 Computing Deuring correspondence for special orders

---

Require: A prime p.

Ensure: A supersingular invariant $j_0 \in \mathbb{F}_p$ such that $\mathcal{O}_0 \cong \text{End}(E(j_0))$, and an endomorphism $\phi \in \text{End}(E(j_0))$ such that $n(\phi) = q$ and $\text{tr}(\phi) = 0$.

1: Compute $\mathcal{J}$ a set of supersingular invariants j such that $E(j)$ has complex multiplication by R_D, the integer ring of $\mathbb{Q}(\sqrt{-q})$.

2: **for** $j \in \mathcal{J}$ **do**

3:     Compute ϕ an endomorphism of degree q of $E(j)$.

4:     **if** $\text{End}(E(j)) \cong \mathcal{O}_0$ **then**

5:         **return** j and ϕ.

6:     **end if**

7: **end for**

Problem 4.4 (Deuring Correspondence List.). *Construct a list of all pairs $(j, \mathcal{O})$ where j is a supersingular invariant and $\mathcal{O}$ is isomorphic to the endomorphism ring of $E(j)$.*

This problem was considered by Cerviño in [Cer04] and [LM04] who presented similar algorithms which compute representation numbers for *all* supersingular elliptic curves and *all* maximal orders of $B_{p,\infty}$, and then compares the two lists of representation numbers to realize Deuring's correspondence. Since representation numbers of size up to $O(p^{1/2})$ may be needed to distinguish any pair of orders, the algorithm runs in time at least $O(p^2)$ times a polynomial function of $\log p$.

As there are roughly $p/12$ supersingular invariants and they require $O(\log p)$ bits to represent, any algorithm for Problem 4.4 will at best run in a time $O(p \log p)$. We can hope for more efficient algorithms if we are only interested in constructing the correspondence for a given order or curve.

Problem 4.5 (Constructive Deuring Correspondence.). *Given a maximal order $\mathcal{O} \subset B_{p,\infty}$, return a supersingular j invariant such that the endomorphism ring of $E(j)$ is isomorphic to $\mathcal{O}$.*

Problem 4.6 (Inverse Deuring Correspondence.). *Given a supersingular invariant j, compute a maximal order $\mathcal{O} \in B_{p,\infty}$ such that the endomorphism ring of $E(j)$ is isomorphic to $\mathcal{O}$.*

The j-invariant is naturally represented as an element of $\mathbb{F}_{p^2}$, and it is unique up to Galois conjugation. The maximal order is unique up to conjugation by an invertible quaternion element, and it can be described by a $\mathbb{Z}$-basis, namely four elements $1, \omega_2, \omega_3, \omega_4 \in B_{p,\infty}$ such that $\mathcal{O} = \mathbb{Z} + \omega_2\mathbb{Z} + \omega_3\mathbb{Z} + \omega_4\mathbb{Z}$. Choosing a Hermite basis makes this description unique.

The endomorphism ring can be returned as four rational maps that form a $\mathbb{Z}$-basis with respect to scalar multiplication (in fact 3 maps, since one of these maps can always be chosen equal to the identity map). The maps themselves can usually not be returned in their canonical expression as rational maps, as in general this representation will require a space larger than the degree, and the degrees can be as big as p.

Various representations of the maps are a priori possible. Any valid representation should be *concise* and *useful*, in the sense that it must require a space polynomial in $\log p$ to store, and it must allow the evaluation of the maps at arbitrary elliptic curve points in a time polynomial in both $\log p$ and the space required to store those points. To the best of our knowledge these two conditions are sufficient for all potential applications of Problem 4.1. When its degree is a smooth number, an endomorphism can be efficiently represented as a composition of small degree isogenies.

A first approximation to an exponential time algorithm to solve Problem 4.1 was provided by Kohel in his PhD thesis [Koh96]. The resulting algorithm explores a tree in an ℓ-isogeny graph (for some small integer ℓ) until a collision is found, corresponding to an endomorphism. The expected cost of this procedure is $O(\sqrt{p})$ times a polynomial in $\log p$. Repeating this procedure a few times, possibly with different values of ℓ, we obtain a set of endomorphisms which generate a subring of the whole endomorphism ring, and Kohel's algorithm relies on the assumption that they actually generate the whole endomorphism ring. Unfortunately it is not true in general that they generate the whole endomorphism ring, as has been shown recently in [BCEMP18].

The endomorphism ring computation problem was also considered in [DG16] for curves defined over $\mathbb{F}_p$. The identification protocol and signature schemes developed in [GPS17] explicitly rely on its potential hardness for security. We remark that the problem of computing endomorphism rings in the ordinary case is completely different, and Kohel's thesis does provide an algorithm in that case which computes the endomorphism ring.

The algorithm uses the structure of the isogeny graph in the ordinary case, which is completely different to the supersingular case considered here, and has the shape of a "volcano" (a central circle from which many regular trees depart), unlike the supersingular graph which has optimal expansion properties. Other approaches to computing the endomorphism ring in the ordinary case have been suggested, but these rely on the action of a commutative class group on the nodes of the graph and the commutative nature of the endomorphism ring of ordinary curves, and it is not clear how they could be adapted to supersingular curves.

We observe that Problems 4.6 and 4.1 take the same input, and their outputs are also "equal" in the sense they are isomorphic. For this reason the two problems have sometimes been referred to interchangeably. We stress, however, that being isomorphic does not a priori guarantee that the isomorphism is efficiently computable, the same way as discrete logarithms can be computed in the additive group $\mathbb{Z}/(p-1)\mathbb{Z}$ but not in the multiplicative group $\mathbb{F}_p^*$. In particular, a solution to Problem 4.6 does not *a priori* provide a *useful* description of the endomorphism ring so that one can for example evaluate endomorphisms at given points. Similarly, a solution to Problem 4.1 does not *a priori* provide a $\mathbb{Z}$-basis for an order in $B_{p,\infty}$, and this is necessary for example to apply the algorithms of [KLPT14].

It turns out that the two problems are equivalent and [PL17] provides efficient algorithms to go from a representation of the endomorphism ring as a $\mathbb{Z}$-basis over $\mathbb{Q}$ to a representation as rational maps and conversely.

5   Relationships between Hard Problems

5.1   Endomorphism Rings and Path Finding

It was proved in [PL17] and [EHLMP18] that hardness of the endomorphism ring computation problem is equivalent to the path-finding problem in the ℓ-isogeny graph with varying path length.

Proposition 5.1 ([PL17]). *Assume there exists an efficient algorithm for the endomorphism ring computation problem. Then there is an efficient algorithm to solve the path-finding problem in the ℓ-isogeny graph.*

Proof. The reduction of this problem to the endomorphism ring computation problem is given in Algorithm 2. Besides two black box calls to an algorithm for the endomorphism ring computation problem, it uses other efficient algorithms, including [PL17, Algorithm 2] to translate a description of an endomorphism ring as rational maps into a description of a maximal order in $B_{p,\infty}$, both the ℓ power and the powersmooth versions of the quaternion isogeny algorithm [KLPT14], and the translation algorithm from ideals to isogenies.   □

Algorithm 2 Reduction from preimage resistance to endomorphism ring computation

---

Require: Two supersingular invariants $j_s, j_t \in \mathbb{F}_{p^2}$.
Ensure: A sequence of j invariants $j_s = j_0, j_1, \ldots, j_e = j_t$ such that for any i there exists an isogeny of degree ℓ from $E(j_i)$ to $E(j_{i+1})$.
1: Compute $\mathrm{End}(E(j_s))$ and $\mathrm{End}(E(j_t))$.
2: Compute $\mathcal{O}_s \approx \mathrm{End}(E(j_s))$ and $\mathcal{O}_t \approx \mathrm{End}(E(j_t))$ with [PL17, Algorithm 2].
3: Compute $\mathcal{O}_0$-left ideals I_s and I_t with right orders respectively $\mathcal{O}_s$ and $\mathcal{O}_t$.
4: Compute $\mathcal{O}_0$-left ideals J_s and J_t with norm ℓ^e for some e, in the same classes as I_s and I_t respectively.
5: **for** $J \in \{J_s, J_t\}$ and corresponding $E \in \{E(j_s), E(j_t)\}$ **do**
6:     Compute a sequence of ideals $J_i = \mathcal{O}_0 q + \mathcal{O}_0 \ell^i$ for $i = 0, \ldots, e$
7:     **for all** i **do**
8:         Compute K_i with powersmooth norm in the same class as I_i.
9:         Translate K_i into an isogeny $\varphi_i : E_0 \to E_i$.
10:     **end for**
11:     Deduce a sequence $(j_0, j(E_1), j(E_2), \ldots, j_e = j(E))$.
12: **end for**
13: **return** $(j(E_s), \ldots, j_0, \ldots, j(E_t))$ the concatenation of both paths.

The reverse direction may look easier a priori. Applying the path-finding algorithm until it returns two paths which are not equal gives a non scalar endomorphism of the curve. Four linearly independent endomorphisms give a full rank subring of the endomorphism ring; and heuristically one expects that a few of such maps will be sufficient to generate the whole ring. To compute the endomorphism ring one would therefore call the collision finding algorithms multiple times until the resulting maps generate the full endomorphism rings.

Proposition 5.2 ([PL17]). *Assume there exists an efficient path-finding algorithm in the ℓ-isogeny graph. Then under plausible heuristic assumptions there is an efficient algorithm to solve the endomorphism ring computation problem.*

Proof. The reduction algorithm is given by Algorithm 3 below. Note that in Step 7 the discriminant can be computed from the Gram matrix, which can be efficiently computed. Heuristically, one expects that the loop will be executed at most $O(\log p)$ times. Indeed let us assume that after adding some elements to the subring we have a subring of index N. Then we can heuristically expect any new randomly generated endomorphism to lie in this subring with a probability only $1/N$. Moreover when it does not lie in the subring, the element will decrease the index by a non trivial integer factor of N.   □

Algorithm 3 Reduction from endomorphism ring computation to collision resistance

---

Require: A supersingular invariant $j \in \mathbb{F}_{p^2}$.
Ensure: The endomorphism ring of $E(j)$.
 1: Let $\mathcal{R} = \langle 1 \rangle \subset \mathrm{End}(E(j))$.
 2: **repeat**
 3:     Perform a random walk in the graph, leading to a new vertex j'.
 4:     Apply a collision finding algorithm on j', leading to an endomorphism of $E(j')$.
 5:     Deduce an endomorphism ϕ of $E(j)$ by concatenating paths.
 6:     Set $\mathcal{R} \leftarrow \langle \mathcal{R}, \phi \rangle$.
 7:     Compute the discriminant of $\mathcal{R}$.
 8: **until** $\mathrm{disc}(\mathcal{R}) = 4p^2$.
 9: **return** a $\mathbb{Z}$ basis for $\mathcal{R}$.

5.2   Path Finding and Key Exchange

In [DFJP14], De Feo–Jao–Plût proposed a set of five hard problems related to the
security of the Key Exchange protocol. It is natural to ask what is the relation between
the problems stated in [DFJP14] and the path-finding problem on Supersingular Isogeny
Graphs proposed in [CGL06]. The security of the Key Exchange is related to the hardness
of the Path Finding Problem (Problem 3.1 above) as follows.

Given an instance of the key-exchange protocol to be attacked, we *know* that there exists
a path of length n between E and E_A, and the hard problem is to find it. In this case n is
roughly half the diameter of the graph, since $p = \ell_A^n \ell_B^m \pm 1$ and $n \approx m$, and the diameter of
these graphs, both LPS and SIG graphs, has been extensively studied. It is known that the
diameter of the graphs is roughly $\log(p)$ (it is $c\log(p)$, where c is a constant between 1 and
2, (see for example [Sar18])). That means that if r is greater than $c\log(p)$, then given two
vertices, it is likely that a path of length r between them may exist. The fact that walks of
length greater than $c\log(p)$ approximate the uniform distribution very closely means that
you are not likely to miss any significant fraction of the vertices with paths of that length,
because that would constitute a bias. Also, if $r \gg \log(p)$ then there may be many paths of
length r. However, if r is much less than $\log(p)$, such as $\frac{1}{2}\log(p)$, there may be *no path* of
such a short length between two given vertices. See [LP15] for a discussion of the "sharp
cutoff" property of Ramanujan graphs. Thus if an algorithm exists to find a path between
two vertices which are so close together, then it is very likely to be the path which was
constructed in the key exchange.

The set-up for the key-exchange requires $p = \ell_A^n \ell_B^m \pm 1$, where n and m are roughly the
same size, and ℓ_A and ℓ_B are very small, such as $\ell_A = 2$ and $\ell_B = 3$. It follows that n and
m are both approximately half the diameter of the graph (which is roughly $\log(p)$). So it
is unlikely we can find paths of length n or m between two random vertices. If a path of
length n exists and Algorithm A finds a path, then it is very likely to be the one which was
constructed in the key exchange. If not, then Algorithm A can be repeated any constant
number of times.

Theorem 5.3. *[CFLMP18] Assume as for the Key Exchange set-up that $p = \ell_A^n \cdot \ell_B^m + 1$ is a
prime of cryptographic size, i.e. $\log(p) \geq 256$, ℓ_A and ℓ_B are small primes, such as $\ell_A = 2$
and $\ell_B = 3$, and $n \approx m$ are approximately equal. Given an algorithm to solve Problem
3.1 (Path-finding), it can be used to solve Problem 3.2 (Key Exchange) with overwhelming*

probability. The failure probability is roughly

$$\frac{\ell_A^n + \ell_A^{n-1}}{p} \approx \frac{\sqrt{p}}{p}.$$

Proof. Given an algorithm (Algorithm A) to solve Problem 3.1, we can use this to solve Problem 3.2 as follows. Given E and E_A, use Algorithm A to find the path of length n between these two vertices in the ℓ_A-isogeny graph. Lemma 4.4 in [CFLMP18] shows how to produce a point R_A which generates the ℓ_A^n-isogeny between E and E_A. Repeat this to produce the point R_B which generates the ℓ_B^m-isogeny between E and E_B in the ℓ_B-isogeny graph. Because the subgroups generated by R_A and R_B have smooth order, it is easy to write R_A in the form $[m_A]P_A + [n_A]Q_A$ and R_B in the form $[m_B]P_B + [n_B]Q_B$. Using the knowledge of m_A, n_A, m_B, n_B, we can construct E_{AB} and recover the j-invariant of E_{AB}, allowing us to solve Problem 3.2.

The reason for the qualification "with overwhelming probability" in the statement of the theorem is that it is possible that there are multiple paths of the same length between two vertices in the graph. If there are multiple paths of length n (or m) between the two vertices, it suffices to repeat Algorithm A to find another path. This approach is sufficient to break the Key Exchange if there are only a small number of paths to try. But we argue that, with overwhelming probability, there are *no* other paths of length n (or m) in the Key Exchange setting:

In the SIG corresponding to (p, ℓ_A), the vertices E and E_A are a distance of n apart. Starting from the vertex E and considering all (non-backtracking) paths of length n, the number of possible endpoints is at most $\ell_A^n + \ell_A^{n-1}$ (no backtracking is allowed), so there are $\ell_A + 1$ choices for the first step and ℓ_A choices for each of the remaining $n - 1$ steps. Considering that the number of vertices in the graph is roughly $\lfloor p/12 \rfloor$, then the probability that a given vertex such as E_A will be the endpoint of one of the walks of length n is roughly

$$\frac{\ell_A^n + \ell_A^{n-1}}{p} \approx \frac{\sqrt{p}}{p} \leq 2^{-128}.$$

This estimate does not use the Ramanujan property of the SIG graphs. While a generic random graph could potentially have a topology which creates a bias towards some subset of the nodes, Ramanujan graphs cannot, as shown in [LP15, Theorem 3.5]. □

Note that, to break the preimage resistance of the specified hash function, you must find a path of exactly length r, and this is analogous to the situation for breaking the security of the key-exchange protocol. However, the problem of finding *any* path between two given vertices in the SIG graphs is also still open.

6 Other Graphs and Related Problems

Supersingular Isogeny Graphs (SIG) were proposed in [CGL06] for use in cryptography to construct cryptographic hash functions, as an example of the general construction relying on graphs with optimal expansion properties and no known algorithm for Path Finding (Problem 3.1). The other graphs proposed in [CGL06] were the LPS graphs, and then in a subsequent paper [CGL09], the dimension-2 analogue of Supersingular Isogeny Graphs, described as superspecial orders in a quaternion algebra over a totally real field. LPS graphs were introduced by Lubotzky, Philips and Sarnak in [LPS88]. They form a family of Ramanujan graphs, and can be described explicitly as a Cayley graph based on a group with a specific given set of generators. Although the LPS graphs had been known for 30 years, no algorithm was known for path-finding in these graphs. As a result of the CGL proposal to construct cryptographic hash functions on LPS graphs, first a collision attack [TZ08] and then a path-finding algorithm [PLQ08] were found. For the LPS graphs, the algorithm presented in [PLQ08] did not find a path of a specific given length, but it was still considered to be a "break" of the hash function. In this section, we describe LPS graphs, the path-finding algorithm from [PLQ08], and the relationship between LPS graphs and SIGs.

6.1 LPS graphs

To describe the LPS graphs we first specify two distinct primes p and ℓ congruent to 1 modulo 4 such that ℓ is a quadratic residue modulo p. We think of p and ℓ as being respectively "large" and "small" primes, and in fact in our complexity estimates we will assume $\ell = O(1)$. By abuse of notation we will often identify elements of $\mathbb{F}_p$ and integers in $\{0, 1, \ldots, p-1\}$. For any ring R, we write $\mathrm{GL}_2(R)$ and $\mathrm{PSL}_2(R)$ respectively for the general linear group and the projective special linear group of rank 2 over R. We define $G := \mathrm{PSL}_2(\mathbb{F}_p)$. We write I for the identity matrix in any of those matrix groups.

We denote by $B = \mathbb{Q}[i, j]$ the quaternion algebra over the rationals generated by two elements i and j such that $i^2 = j^2 = -1$ and $k := ij = -ji$. For any $q = a + bi + cj + dk \in B$, the conjugate of q is

$$\bar{q} := a - bi - cj - dk$$

and the norm of q is

$$n(q) = q\bar{q} = a^2 + b^2 + c^2 + d^2.$$

Note that the norm is multiplicative: for any $q_1, q_2 \in B$, we have $n(q_1 q_2) = n(q_1)n(q_2)$.

By abuse of notation we also use the symbol i for the imaginary unit, and we denote the Gaussian integers by $\mathbb{Z}[i]$. The map $\sigma : \mathrm{GL}_2(\mathbb{Z}[i]) \to B$ defined by

$$m = \begin{pmatrix} a+bi & c+di \\ -c+di & a-bi \end{pmatrix} \;\to\; \sigma(m) = a + bi + cj + dk$$

is an isomorphism of algebras. Note that we have $\det m = n(\sigma(m))$, and that both σ and its inverse are efficiently computable. For any matrix $m = \begin{pmatrix} a+bi & c+di \\ -c+di & a-bi \end{pmatrix} \in \mathrm{GL}_2(\mathbb{Z}[i])$ we define its conjugate matrix by $\overline{m} := \begin{pmatrix} a-bi & -c-di \\ c-di & a+bi \end{pmatrix}$. Clearly, we have $\overline{\overline{m}} = m$ and $\sigma(\overline{m}) = \overline{\sigma(m)}$.

Following [TZ08], for any integer $e \geq 1$ we define E_e as the set of 4-tuples $(a, b, c, d) \in \mathbb{Z}^4$ such that

$$\begin{cases} a^2 + b^2 + c^2 + d^2 = \ell^e, \\ a > 0, \; a = 1 \bmod 2, \\ b = c = d = 0 \bmod 2. \end{cases}$$

Up to multiplication by a unit there are $\ell + 1$ elements of norm ℓ in B; these correspond to elements of E_1. We also denote by Σ the set of matrices corresponding to E_1, and we define

$$\Omega := \left\{ \begin{pmatrix} a+bi & c+di \\ -c+di & a-bi \end{pmatrix} \;\middle|\; (a, b, c, d) \in E_e \text{ for some integer } e > 0 \right\}.$$

Note that Σ is symmetric in the sense that for any $s \in \Sigma$, there exists $s' = \bar{s} \in \Sigma$ with $ss' = \ell I$.

Let $\iota \in \mathbb{F}_p$ be such that $\iota^2 = -1$. Reduction modulo p extends to a group homomorphism $\phi : \mathrm{GL}_2(\mathbb{Z}[i]) \to G$ defined by

$$\phi \begin{pmatrix} a + bi & c + di \\ -c + di & a - bi \end{pmatrix} = \begin{pmatrix} a + b\iota & c + d\iota \\ -c + d\iota & a - b\iota \end{pmatrix}.$$

Let $S = \{\phi(s) | s \in \Sigma\}$ be the set of images of elements in Σ under the homomorphism ϕ. The LPS graph for parameters p and ℓ is the Cayley graph constructed from the group G and the generator set S; in other words it is a graph whose vertices correspond to the elements of G, and such that there is an edge between two vertices corresponding to g_1 and g_2 if and only if there is an element $s \in S$ such that $g_2 = g_1 s$. This graph is an undirected $(\ell + 1)$-regular graph. For any fixed ℓ and increasing p, LPS graphs form a family of Ramanujan graphs, in other words they are optimal expander graphs [LPS88].

An important observation for the path-finding algorithms below is that matrices in Ω admit essentially unique factorizations in the elements of Σ:

Lemma 6.1 ([TZ08], citing [LPS88,Dav03]). *Any matrix in Ω can be expressed in a unique way as a product*

$$M = \pm \ell^r M_1 M_2 \ldots M_e$$

where $\log_\ell(\det(M)) = e + 2r$ *and* $M_i \in \Sigma$ *and* $M_i M_{i+1} \neq \ell I$ *for* $i = 1, \ldots, e - 1$.

6.2 Path Finding in LPS Graphs

Here we describe the algorithm from [PLQ08] and [CP18] to solve the path-finding problem in LPS graphs: i.e. to compute paths between any two vertices in LPS graphs. Let all notation be as above, and let m be a matrix in G that we want to write as a short product of elements from the set S.

The algorithm of Petit et al. [PLQ08] first decomposes the matrix m as a product

$$m = \lambda \cdot D_1 \cdot s \cdot D_2 \cdot s \cdot D_3 \cdot s \cdot D_4 \tag{6.1}$$

where $\lambda \in \mathbb{F}_p^*$, the factors D_i are diagonal matrices with a non zero square determinant, and s is a particular (arbitrary) generator in the set S. As the equation is over the projective special linear group, each diagonal matrix can be normalized as $D_i = \left(\begin{smallmatrix} 1 & 0 \\ 0 & \alpha_i \end{smallmatrix} \right)$ where α_i is a non-zero quadratic residue. Equation 6.1 then amounts to a small polynomial system with four equations and five variables. The algorithm given in [PLQ08] to solve this system picks random solutions until all α_i are square, and it is heuristically expected to need 16 trials on average.

Petit et al. [PLQ08] also provide an algorithm to factor any diagonal matrix with square determinant in a short product of the generators S. This algorithm extends a previous algorithm from Tillich and Zémor [TZ08] for the identity matrix. First the diagonal matrix is lifted into an element of Ω, then a factorization of this element as a product of the elements of Σ is computed. The factorization of the input diagonal matrix in the elements of $S = \phi(\Sigma)$ follows by the group homomorphism ϕ. We now give some details on the first and second step.

Let $m = \left(\begin{smallmatrix} A + B\iota & 0 \\ 0 & A - B\iota \end{smallmatrix} \right) \in G$ be a diagonal matrix with $\det m = A^2 + B^2$ a square. The lifting step consists in finding $e \in \mathbb{N}$ and $\lambda, w, x, y, z \in \mathbb{Z}$ with

$$\begin{cases} (A\lambda + wp)^2 + (B\lambda + xp)^2 + 4p^2(y^2 + z^2) = \ell^e, \\ A\lambda + wp = 1 \bmod 2, \\ B\lambda + xp = 0 \bmod 2. \end{cases} \tag{6.2}$$

After fixing e large enough, Petit et al. solve the norm equation modulo p; as $A^2 + B^2$ is a square this gives two possible values for λ and one is picked randomly. Next, the norm equation is considered modulo p^2: this gives a bilinear equation in w and x, and a random solution is selected. At this point, the norm equation is considered over the integers, and it gives

$$4(y^2 + z^2) = n$$

where

$$n := \left(\ell^e - (A\lambda + wp)^2 - (B\lambda + xp)^2 \right) / p^2 \in \mathbb{Z} \tag{6.3}$$

for λ, x, z chosen as before. This equation has a solution whenever $n/4$ is an integer, and all prime factors of n congruent to 3 modulo 4 appear an even number of times in the factorization of n. To avoid a costly factorization step, Petit et al. suggest to pick random solutions for (w, x) until n is 4 times a prime congruent to 1 modulo 4. Suitable values for (y, z) are then computed with Cornacchia's algorithm. Taking e larger than $\log 8p^4 \approx 4 \log p$ ensures that n is positive in this algorithm.

Once the integers e, λ, w, x, y, z are found they give a matrix

$$\tilde{m} = \begin{pmatrix} (A+wp)+(B+xp)i & 2yp+2zpi \\ -2yp+2zpi & (A+wp)-(B+xp)i \end{pmatrix} \in \Omega$$

reducing to m modulo p. Remember that by Lemma 6.1 factorization is essentially unique in Ω. Tillich and Zémor showed in [TZ08] how to recover all the factors successively: for any $s \in \Sigma$ we have $\tilde{m} = \tilde{m}'s$ with $\tilde{m}' \in \Omega$ if and only if $\tilde{m}s^{-1} \in \Omega$.

6.3 Relationship to Quantum Computation

LPS graphs have recently been suggested for quantum circuit design. In this setting LPS generators correspond to elementary quantum gates, and they are combined to approximate arbitrary quantum gates on a single qbit. Interestingly, an algorithm similar to the diagonal algorithm from [PLQ08] was independently developed later in the context of quantum circuit design [Ros15]. It was later transposed to the cryptographic setting [Sar18], resulting in an algorithm very similar to the Petit–Quisquater–Lauter algorithm in the diagonal case.

6.4 Relationship between LPS and Pizer Graphs

Given the attacks on the LPS path-finding problem, it is natural to investigate whether this approach is relevant to the path-finding problem in Supersingular Isogeny Graphs (SIG). The recent paper [CFLMP18] examines the LPS and Pizer graphs from a number theoretic perspective, highlighting the similarities and differences between the constructions. Both can be thought of as graphs on a double coset space (arising from quaternion algebras). The Ramanujan property of the graphs follows from examining them as double coset spaces. The objects that underly both constructions are similar but different input choices lead to different graphs.

Both the LPS and Pizer (SIG) graphs considered in [CGL06] can be thought of as graphs on

$$\Gamma \backslash \mathrm{PGL}_2(\mathbb{Q}_l) / \mathrm{PGL}_2(\mathbb{Z}_l), \tag{6.4}$$

where Γ is a discrete cocompact subgroup, where Γ is obtained from a quaternion algebra B. Different input choices for the construction lead to different graphs. For the LPS graph Γ can be varied to get an infinite family of Ramanujan graphs, while for Pizer graphs, B varies. In the LPS case, we always work in the Hamiltonian quaternion algebra. For this particular choice of algebra we can rewrite the graph as a Cayley graph. This explicit description is key for breaking the LPS hash function. For the Pizer graphs we do not have such a description. On the Pizer side the graphs may, via Strong Approximation, be viewed as graphs on adèlic double cosets which are in turn the class group of an order of B that is related to the cocompact subgroup Γ. From here one obtains an isomorphism with supersingular isogeny graphs.

Currently there is no known way to use the path-finding algorithm for LPS graphs to attack the hard problems on Supersingular Isogeny Graphs.

Acknowledgements

To prepare this survey to accompany the first author's Invited Lecture at the 27th British Combinatorial Conference, we have pulled together various sections of other papers we have written, together and with other coauthors. This survey consists of lightly edited sections taken from the following preprints and papers: [CFLMP18, PL17, CP18, GPS17]. Sections 1, 2, 3.1, 3.2, 5.2, and 6.4 are from [CFLMP18], Sections 4 and 5.1 are from [PL17], Sections 6.1, 6.2, and 6.3 are from [CP18], and Section 3.3 is from [GPS17]. We thank our coauthors for permission to include this content here.

References

[AAM18] Gora Adj, Omran Ahmadi, and Alfred Menezes, *On isogeny graphs of supersingular elliptic curves over finite fields*, Cryptology ePrint Archive, Report 2018/132, 2018, https://eprint.iacr.org/2018/132.

[Alo86] N. Alon, *Eigenvalues and expanders*, Combinatorica **6** (1986), no. 2, 83–96, Theory of computing (Singer Island, Fla., 1984).

[BCEMP18] Efrat Bank, Catalina Camacho-Navarro, Kirsten Eisentraeger, Travis Morrison, and Jennifer Park, *Cycles in the Supersingular l-Isogeny Graph and Corresponding Endomorphisms*. Preprint 2018. https://arxiv.org/pdf/1804.04063.pdf

[Brö09] Reinier Bröker, Constructing supersingular elliptic curves. *J. Comb. Number Theory*, 1(3):269–273, 2009.

[CP18] Eduardo Carvalho Pinto and Christophe Petit, *Better path-finding algorithms in LPS Ramanujan graphs* Journal of Mathematical Cryptology, September 2018 DOI: 10.1515/jmc-2017-0051

[Cer04] J. M. Cerviño, Supersingular elliptic curves and maximal quaternionic orders. In *Mathematisches Institut, Georg-August-Universität Göttingen: Seminars Summer Term 2004*, pages 53–60. Universitätsdrucke Göttingen, Göttingen, 2004.

[CGL06] Denis X. Charles, Eyal Z. Goren, and Kristin E. Lauter, *Cryptographic hash functions from expander graphs*, J. Cryptology **22** (2009), no. 1, 93–113, available at https://eprint.iacr.org/2006/021.pdf.

[CGL09] Denis X. Charles, Eyal Z. Goren, and Kristin E. Lauter, *Families of Ramanujan graphs and quaternion algebras*, Groups and symmetries, CRM Proc. Lecture Notes, vol. 47, Amer. Math. Soc., Providence, RI, 2009, pp. 53–80.

[Che10] Gaëtan Chenevier, *Lecture notes*, 2010, http://gaetan.chenevier.perso.math. cnrs.fr/coursIHP/chenevier_lecture6.pdf

[CFLMP18] Anamaria Costache, Brooke Feigon, Kristin Lauter, Maike Massierer, and Anna Puskas, *Ramanujan graphs in cryptography*, to appear in *Research Directions in Number Theory: Women in Numbers IV* AWM Springer Series.

[Dav03] Giuliana Davidoff and Peter Sarnak and Alain Valette, *Elementary Number Theory, Group Theory, and Ramanujan Graphs*, London mathematical Society, Cambridge University Press, 2003.

[DFJP14] Luca De Feo, David Jao, and Jérôme Plût, *Towards quantum-resistant cryptosystems from supersingular elliptic curve isogenies*, J. Math. Cryptol. **8** (2014), no. 3, 209–247.

[DG16] Christina Delfs and Steven D. Galbraith. Computing isogenies between supersingular elliptic curves over $\mathbb{F}_p$. *Des. Codes Cryptogr.*, 78(2):425–440, 2016.

[Del71] Pierre Deligne, *Formes modulaires et représentations l-adiques*, Séminaire Bourbaki. Vol. 1968/69, vol. 179, Lecture Notes in Math., no. 355, Springer, Berlin, 1971, pp. 139–172.

[Del74] Pierre Deligne, *La conjecture de Weil. I*, Publications Mathématiques de l'Institut des Hautes Études Scientifiques **43** (1974), no. 1, 273–307.

[Deu41] Max Deuring. *Die Typen der Multiplikatorenringe elliptischer Funktionenkörper.* Abh. Math. Sem. Univ. Hamburg, 14(1):197–272, 1941.

[EHLMP18] K. Eisentraeger, S. Hallgren, K. Lauter, T. Morrison, C. Petit, *Supersingular isogeny graphs and endomorphism rings: reductions and solutions*, Advances in Cryptology - EUROCRYPT 2018 Proceedings. Lecture Notes in Computer Science vol. 10822 Springer Verlag, p. 329–368.

[Gal99] Steven D. Galbraith. Constructing isogenies between elliptic curves over finite fields. *LMS J. Comput. Math.*, 2:118–138, 1999.

[GPS17] Steven D. Galbraith, Christophe Petit, and Javier Silva. Identification protocols and signature schemes based on supersingular isogeny problems. In Tsuyoshi Takagi and Thomas Peyrin, editors, *Advances in Cryptology – ASIACRYPT 2017*, pages 3–33, Cham, 2017. Springer International Publishing.

[Gel75] Stephen S Gelbart, *Automorphic Forms on Adele Groups*, no. 83, Princeton University Press, 1975.

[Iha66] Yasutaka Ihara, *Discrete subgroups of* PL(2, $k_\wp$), Algebraic Groups and Discontinuous Subgroups (Proc. Sympos. Pure Math., Boulder, Colo., 1965), Amer. Math. Soc., Providence, R.I., 1966, pp. 272–278.

[JMV05] David Jao, Stephen D Miller, and Ramarathnam Venkatesan, *Do all elliptic curves of the same order have the same difficulty of discrete log?*, International Conference on the Theory and Application of Cryptology and Information Security, Springer, 2005, pp. 21–40.

[KL07] Jonathan Katz and Yehuda Lindell, *Introduction to Modern Cryptography*, Chapman & Hall/Crc Cryptography and Network Security Series), 2007.

[Koh96] David Kohel. *Endomorphism Rings of Elliptic Curves over Finite Fields*. PhD thesis, University of California, Berkeley, 1996.

[KLPT14] David Kohel, Kristin Lauter, Christophe Petit, and Jean-Pierre Tignol. On the quaternion ℓ-isogeny path problem. *LMS Journal of Computation and Mathematics*, 17:418–432, 2014.

[LM04] Kristin Lauter and Ken McMurdy. *Explicit generators of endomorphism rings of supersingular elliptic curves*. Preprint, 2004. https://phobos.ramapo.edu/~kmcmurdy/research/ss_endomorphisms.pdf

[Li96] Wen-Ch'ing Winnie Li, *A survey of Ramanujan graphs*, Arithmetic, geometry and coding theory (Luminy, 1993), de Gruyter, Berlin, 1996, pp. 127–143.

[LP15] Eyal Lubetzky and Yuval Peres, *Cutoff on all Ramanujan graphs*, Geometric and Functional Analysis **26** (2016), no. 4, 1190–1216.

[LPS88] Alexander Lubotzky, Richard L. Phillips, and Peter Sarnak, *Ramanujan graphs*, Combinatorica **8** (1988), no. 3, 261–277.

[Lub10] Alexander Lubotzky, *Discrete groups, expanding graphs and invariant measures*, Modern Birkhäuser Classics, Birkhäuser Verlag, Basel, 2010, With an appendix by Jonathan D. Rogawski, Reprint of the 1994 edition.

[Mes86] J.-F. Mestre, *La méthode des graphes. Exemples et applications*, Proceedings of the international conference on class numbers and fundamental units of algebraic number fields (Katata, 1986), Nagoya Univ., Nagoya, 1986, pp. 217–242.

[PLQ08] Christophe Petit, Kristin Lauter, and Jean-Jacques Quisquater, *Full cryptanalysis of LPS and Morgenstern hash functions*, Security and Cryptography for Networks (Berlin, Heidelberg) (Rafail Ostrovsky, Roberto De Prisco, and Ivan Visconti, eds.), Springer Berlin Heidelberg, 2008, pp. 263–277.

[PL17] Christophe Petit and Kristin Lauter, Hard and easy problems for supersingular isogeny graphs. Cryptology ePrint Archive, Report 2017/962, 2017. https://eprint.iacr.org/2017/962.

[Piz76] Arnold Pizer, *The representability of modular forms by theta series*, Journal of the Mathematical Society of Japan **28** (1976), no. 4, 689–698.

[Piz80] Arnold Pizer, *An algorithm for computing modular forms on* $\Gamma 0(N)$, Journal of algebra **64** (1980), no. 2, 340–390.

[Piz98] Arnold Pizer, *Ramanujan graphs*, Computational perspectives on number theory (Chicago, IL, 1995), AMS/IP Stud. Adv. Math., vol. 7, Amer. Math. Soc., Providence, RI, 1998, pp. 159–178.

[PQC] *Post-Quantum Cryptography Standardization*, https://csrc.nist.gov/Projects/Post-Quantum-Cryptography/Post-Quantum-Cryptography-Standardization, Accessed: 2018-04-14.

[Ros15] Neil J. Ross, *Optimal ancilla-free Clifford+V approximation of z-rotations*, Quantum Information & Computation, Volume 15 Issue 11-12, September 2015, 932-950.

[Sar18] Naser T. Sardari, *Diameter of Ramanujan graphs and random Cayley graphs*, (2018). Combinatorica, 1–20.

[Sil09] J. H. Silverman, *The Arithmetic of Elliptic Curves*, second ed., Graduate Texts in Mathematics, vol. 106, Springer, Berlin–Heidelberg–New York, 2009.

[TZ93] Jean-Pierre Tillich and Gilles Zémor, *Group-theoretic hash functions*, (1993). Proceedings of the First French-Israeli Workshop on Algebraic Coding, 90–110.

[TZ08] Jean-Pierre Tillich and Gilles Zémor, *Collisions for the LPS expander graph hash function*, Advances in Cryptology – EUROCRYPT 2008 (Nigel Smart, ed.), Springer, 2008, pp. 254–269.

[Vél71] Jacques Vélu, *Isogénies entre courbes elliptiques*, C. R. Acad. Sci. Paris Sér. A-B **273** (1971), A238–A241.

[Vig80] Marie-France Vignéras, *Arithmétique des Algèbres de Quaternions*, Lecture Notes in Mathematics, vol. 800, Springer, Berlin, 1980.

[Voi18] John Voight, *Quaternion Algebras*, 2018, https://math.dartmouth.edu/~jvoight/quat-book.pdf, retrieved January 09, 2019.

Microsoft Research, One Microsoft Way, Redmond, WA 98052 USA
klauter@microsoft.com

School of Computer Science, University of Birmingham
University Rd W, Birmingham B15 2TT United Kingdom
christophe.f.petit@gmail.com

Delta-matroids for graph theorists

Iain Moffatt

Abstract

What happens if you try to develop matroid theory, but start with topological graph theory? This survey provides an introduction to delta-matroids. We aim to illustrate the two-way interaction between graph theory and delta-matroid theory that enriches both subjects. Along the way we shall see intimate connections between delta-matroids and, amongst others, circle graphs, Eulerian circuits, embedded graphs, matchings, pivot-minors, (skew-)symmetric matrices, and vertex-minors.

1 Introduction

"I lectured on matroids at the first formal conference on them [...] in 1964. To me that was the year of the Coming of the Matroids. Then and there the theory of matroids was proclaimed to the mathematical world. And outside the halls of lecture there arose the repeated cry: 'What the hell is a matroid?'"

— W.T. Tutte [1]

Since that 1964 conference, matroids have become a mainstay of combinatorics (and a regular topic of BCC talks [2, 8, 39, 51, 62, 66, 70, 75, 77]). However, our interest here is in a generalisation of matroids called *delta-matroids*. Delta-matroids, introduced in the mid-1980s, are not nearly so well-known, even among matroid theorists. Here, inspired by Tutte's felicitious phrasing, I aim to answer the question 'what the hell is a delta-matroid?'.

This survey is intended to introduce delta-matroids to readers familiar with graph theory. *No prior knowledge of matroids is assumed.* Delta-matroids were introduced in the mid-1980s, independently, by Bouchet in [9]; Chandrasekaran and Kabadi, under the name of *pseudo-matroids*, in [25]; and Dress and Havel, under the name of *metroids*, in [32]. (Here we follow the terminology and notation of Bouchet.) Our focus here is on how delta-matroids relate to graph theory, and we shall see connections between them and circle graphs, Eulerian circuits, embedded graphs, matchings, pivot-minors, (skew-)symmetric matrices, and vertex-minors. In particular, our aim is to illustrate the two-way interaction between graph theory and delta-matroid theory that enriches both subjects.

The emphasis here is on providing an accessible introduction to delta-matroids that conveys the 'flavour' of the subject. It does not provide a comprehensive account of delta-matroids. In particular, many beautiful results have not been included here, even when they are closely related to those that have been. For example, delta-matroids have applications in theoretical computer science, but here we totally ignore this aspect of delta-matroid theory (although Sections 2.4.6 and 2.4.7 hint at why they appear in that area). A graph-theoretic topic that we do not touch on is applications of delta-matroid to graph polynomials, including the Tutte [73], Bollobás-Riordan [6,7], interlace [4,5], Penrose [1,36,63], and transition [43], polynomials (see, for example, [22,29,30,46,56,57]). Also, delta-matroids have close connections with several other generalisations of matroids, and other combinatorial structures (see the remark at the end of Section 2.2). Indeed, some delta-matroid

---

[1]This extract is from Tutte's article, *The Coming of the Matroids*, [75]. It appeared in this *Surveys in combinatorics* series, and is associated with his talk at the 1999 BCC held at the University of Kent at Canterbury.

results are better understood in terms of more general structures or generalised matroids (such as isotropic systems, jump systems, or multimatroids). We do not discuss these generalisations here: asking a reader to absorb the definition of *one* generalisation of a matroid at a time is quite enough!

A number of exercises can be found throughout the text. These exercises are intended to assist with the digestion of definitions and results, and, as such, they are not hard and mostly require only a few minutes of thought. A similar comment holds for the examples and figures. We provide sketches of some proofs, but not all. At the end, there is a list of frequently used notation.

2 What is a delta-matroid?

2.1 A warm up

Rather than diving straight into the definitions, let us start with an example that shows we have been working with delta-matroids since our undergraduate days.

Suppose we have a finite-dimensional vector space V, and two bases $X = \{x_1, \ldots, x_n\}$ and $Y = \{y_1, \ldots, y_n\}$ of V. From our first courses in linear algebra we know, for each i: (i) there is some y_j such that $(X \backslash \{x_i\}) \cup \{y_j\}$ is a basis for V; and (ii) there is some x_j such that $(X \cup \{y_i\}) \backslash \{x_j\}$ is a basis for V. Knowing that the sets X and Y are of the same size, we can conveniently use the *symmetric difference*, $X \bigtriangleup Y := (X \cup Y) \backslash (X \cap Y)$, to express these two properties as

$$(\forall u \in X \bigtriangleup Y)\, (\exists v \in X \bigtriangleup Y)\, (X \bigtriangleup \{u, v\} \in \mathcal{F}), \tag{2.1}$$

where $\mathcal{F}$ is the set of all bases of V.

Matroids and delta-matroids are mathematical structures that satisfy the exchange property in (2.1): a *delta-matroid* is a pair $(E, \mathcal{F})$ where E is a set, and $\mathcal{F}$ is a collection of subsets of E that satisfies (2.1) for all $X, Y \in \mathcal{F}$. If every set in $\mathcal{F}$ has the same size, then the delta-matroid $(E, \mathcal{F})$ is said to be a *matroid*.

Thus the set $\mathcal{F}$ of all bases of the vector space V satisfies (2.1) for all $X, Y \in \mathcal{F}$ and so the pair $(V, \mathcal{F})$ (where V is regarded as a set) forms a *delta-matroid*. Moreover, since every member of $\mathcal{F}$ has the same size, which need not be the case for delta-matroids in general, this delta-matroid is a *matroid*.

2.2 The definition

Here we assume all sets (other than, possibly, fields) are finite, and will do so without further comment. Where there is no potential for confusion, we omit the braces when writing single element sets, for example, writing $X \backslash x$ instead of $X \backslash \{x\}$, or $X \cup x$ instead of $X \cup \{x\}$. The *symmetric difference*, $X \bigtriangleup Y$, of sets X and Y is

$$X \bigtriangleup Y := (X \cup Y) \backslash (X \cap Y).$$

Definition 2.1 (Set system). A *set system* is a pair $D = (E, \mathcal{F})$ where E is a set, and $\mathcal{F}$ is a collection of subsets of E. A set system is *proper* if $\mathcal{F}$ is not empty; it is *trivial* if E is empty.

Example 2.2. Let $E = \{a, b, c\}$,

$$\mathcal{F} = \{\emptyset, \{a\}, \{b\}, \{c\}, \{b, c\}\},$$

and
$$\mathcal{F}' = \{\{a\}, \{b\}, \{a, b\}, \{a, b, c\}\}.$$
Then $D = (E, \mathcal{F})$ and $D' = (E, \mathcal{F}')$ are both set systems.

The Symmetric Exchange Axiom appeared in (2.1).

Definition 2.3 (Symmetric Exchange Axiom). A set system $D = (E, \mathcal{F})$ is said to satisfy the *Symmetric Exchange Axiom* (SEA) if, for all $X, Y \in \mathcal{F}$, if there is an element $u \in X \triangle Y$, then there is an element $v \in X \triangle Y$ such that $X \triangle \{u, v\} \in \mathcal{F}$.

See Figure 1 for an illustration of the Symmetric Exchange Axiom. For ease of reference, here it is in a symbolic form:

$$(\forall\, X, Y \in \mathcal{F})\,(\forall\, u \in X \triangle Y)\,(\exists\, v \in X \triangle Y)\,(X \triangle \{u, v\} \in \mathcal{F}). \qquad \textbf{(SEA)}$$

It is important to notice that the Symmetric Exchange Axiom allows the possibility that $u = v$.

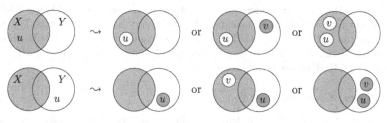

Figure 1: The Symmetric Exchange Axiom, where X, Y, and the shaded parts form feasible sets

Definition 2.4 (Delta-matroid). A *delta-matroid* $D = (E, \mathcal{F})$ is a proper set system that satisfies the Symmetric Exchange Axiom. The set E is called its *ground set*, and the members of $\mathcal{F}$ are called *feasible sets*.

When working with delta-matroids, $E(D)$ is often used to denote the ground set of a delta-matroid D, and $\mathcal{F}(D)$ its collection of feasible sets. Here, although we generally use the letter E for ground sets, for certain classes of delta-matroids we will instead use V. The choice of V or E relates to whether the elements of the ground set correspond most naturally to the vertex set or the edge set of a graph.

Example 2.5. Consider the set systems D and D' in Example 2.2. By examining $\mathcal{F}$, we see that if $X = \{b, c\}$ and $Y = \{a\}$, then $a \in X \triangle Y$, but there there is no $v \in X \triangle Y$ such that $X \triangle \{a, v\} \in \mathcal{F}$. Thus $D = (E, \mathcal{F})$ is *not* a delta-matroid.

On the other hand, it can be checked that $\mathcal{F}'$ satisfies the Symmetric Exchange Axiom and hence $D' = (E, \mathcal{F}')$ is a delta-matroid. Its ground set is $\{a, b, c\}$ and its feasible sets are $\{a\}$, $\{b\}$, $\{a, b\}$, and $\{a, b, c\}$.

Definition 2.6 (Matroid). A delta-matroid is said to be a *matroid* if all of its feasible sets are of the same size.

If a delta-matroid is a matroid, then it is usual to refer to its feasible sets as its *bases*, and to use $\mathcal{B}$, rather than $\mathcal{F}$ to denote its collections of bases.

Remark 5. Introducing matroids as a special type of delta-matroid is somewhat anachronistic. Matroids were introduced by Whitney [78] in the 1930's, while delta-matroids were introduced in the mid-1980's. Furthermore, matroids are much more studied, better known and, better understood than delta-matroids. A reader meeting this topic for the first time should think of a delta-matroid as being a generalisation of a matroid, rather than as a matroid being a special type of delta-matroid, as presented here. Two standard and excellent references for matroid theory are the books [61, 76].

Exercise 2.7. *The standard 'basis definition' of a matroid is as follows: the set system $(E, \mathcal{B})$ is a matroid if (i) $\mathcal{B}$ is non-empty; and (ii) for distinct $A, B \in \mathcal{B}$, if $a \in A \backslash B$, then there exists $b \in B \backslash A$ such that $(A \backslash a) \cup b \in \mathcal{B}$. Verify that this definition of a matroid is equivalent to that given in Definition 2.6.*

Remark 6. The definition of a delta-matroid given here is due to Bouchet and we follow his terminology. As mentioned above, delta-matroids were introduced independently by Bouchet in [9]; Chandrasekaran and Kabadi in [25], under the name of pseudo-matroids; and Dress and Havelin [32], under the name of metroids. Delta-matroids are related to many different matroidal-objects, including the following: Tardos' g-matroids [71], Kung's Pfaffian structures [47], Qi's ditroids [64], Bouchet's symmetric matroids [9], Traldi's transition matroids [72], Bouchet's Isotropic systems [10], jump systems [19], and Bouchet's multimatroids [17]. This list is indicative, not exhaustive.

2.3 Examples of delta-matroids

Having seen the definition of a delta-matroid, we now give a selection of examples of them. Here we provide only constructions and examples, but most, although not all, of these delta-matroids will be discussed in more detail later.

2.3.1 From column spaces Let $\mathbf{A}$ be a matrix with entries in a field $\Bbbk$. Let E be a set of labels for the columns of $\mathbf{A}$, and, define a collection $\mathcal{B}$ of subsets of E by, for each $X \subseteq E$ setting

$$X \in \mathcal{B} \iff X \text{ labels a basis of the column space of } \mathbf{A}.$$

Then the pair $(E, \mathcal{B})$ forms a matroid called the *vector matroid* of $\mathbf{A}$. (This matroid is exactly the example discussed in Section 2.1, and is due to Whitney [78].)

Example 2.8. Working over $\mathbb{R}$, the vector matroid of the matrix

$$\mathbf{A} = \begin{matrix} & \begin{matrix} 1 & 2 & 3 & 4 & 5 \end{matrix} \\ & \begin{bmatrix} 1 & 0 & 0 & 1 & 1 \\ 0 & 0 & 1 & 0 & 1 \end{bmatrix} \end{matrix}$$

has ground set $E = \{1, 2, 3, 4, 5\}$, and its set of bases is

$$\mathcal{B} = \{\{1, 3\}, \{1, 5\}, \{3, 4\}, \{3, 5\}, \{4, 5\}\}.$$

2.4 From (skew-)symmetric matrices

A matrix $\mathbf{A}$ is *symmetric* if $\mathbf{A}^t = \mathbf{A}$, is *skew-symmetric* if $\mathbf{A}^t = -\mathbf{A}$ and the diagonal entries are zero.

Suppose that $\mathbf{A}$ is a symmetric or skew-symmetric matrix over a field $\Bbbk$, and that E labels its rows and columns (in the same order). For $X \subseteq E$, let $\mathbf{A}[X]$ denote the principal submatrix of $\mathbf{A}$ given by the rows and columns indexed by X. Define a collection $\mathcal{F}$ of subsets of E by

$$X \in \mathcal{F} \iff \mathbf{A}[X] \text{ is non-singular,}$$

where $\mathbf{A}[\emptyset]$ is considered to be non-singular. Then the pair $(E, \mathcal{F})$ forms a delta-matroid. (This result is due to Bouchet [12].)

Example 2.9. Working over GF(2), consider the matrices

$$\mathbf{A}_1 = \begin{array}{c} \\ 1 \\ 2 \\ 3 \\ 4 \end{array} \begin{array}{cccc} 1 & 2 & 3 & 4 \\ \begin{bmatrix} 0 & 1 & 1 & 1 \\ 1 & 0 & 1 & 0 \\ 1 & 1 & 0 & 0 \\ 1 & 0 & 0 & 0 \end{bmatrix} \end{array}, \quad \text{and} \quad \mathbf{A}_2 = \begin{array}{c} \\ 1 \\ 2 \\ 3 \\ 4 \end{array} \begin{array}{cccc} 1 & 2 & 3 & 4 \\ \begin{bmatrix} 0 & 1 & 1 & 1 \\ 1 & 0 & 1 & 0 \\ 1 & 1 & 1 & 0 \\ 1 & 0 & 0 & 0 \end{bmatrix} \end{array}.$$

The matrix $\mathbf{A}_1$ gives rise to a delta-matroid $D(\mathbf{A}_1) = (V, \mathcal{F}_1)$ with ground set $V = \{1, 2, 3, 4\}$ and collection of feasible sets

$$\mathcal{F}_1 = \{\emptyset, \{1, 2\}, \{1, 3\}, \{1, 4\}, \{2, 3\}, \{1, 2, 3, 4\}\}.$$

The matrix $\mathbf{A}_2$ gives rise to a delta-matroid $D(\mathbf{A}_2) = (V, \mathcal{F}_2)$ with ground set $V = \{1, 2, 3, 4\}$ and collection of feasible sets

$$\mathcal{F}_2 = \{\emptyset, \{3\}, \{1, 2\}, \{1, 3\}, \{1, 4\}, \{2, 3\}, \{1, 2, 3\}, \{1, 3, 4\}, \{1, 2, 3, 4\}\}.$$

2.4.1   From simple graphs (with vertices forming the ground set) Working over some field $\Bbbk$, the *adjacency matrix over* $\Bbbk$ of a graph G is the matrix whose rows and columns correspond to the vertices of G; and whose (u, v)-entry is the number edges between u and v. When the graph has loops, it is usual to take the (v, v)-entry to be twice the number of vv-edges, however, our convention here is to take it to be equal to the number of vv-edges.

Since adjacency matrices are always symmetric, the previous example provides a way to associate a delta-matroid with a graph: given a graph, form its adjacency matrix over some field, and take the delta-matroid of that matrix. Although this construction works for all graphs and over any field, in this survey we shall consider it only for simple graphs and looped simple graphs over the field of two elements, GF(2). To avoid ambiguity, let us give detailed definitions for these cases.

A *simple graph* is a graph with no loops or multiple edges. A *looped simple graph* is a graph obtained from a simple graph by adding (exactly) one loop to some of its vertices.

Definition 2.10 (Adjacency matrix). The *adjacency matrix*, $\mathbf{A}_G$, of a simple graph or a looped simple graph G is the matrix over GF(2) whose rows and columns correspond to the vertices of G; and where, for $u \neq v$, the (u, v)-entry of $\mathbf{A}_G$ is 1 if the corresponding vertices u and v are adjacent in G, and is 0 otherwise; and the (v, v)-entry of $\mathbf{A}_G$ is 1 if there is a loop at the vertex v, and is 0 otherwise.

Through its adjacency matrix, a delta-matroid $D(\mathbf{A}_G)$ can be associated with a (looped) simple graph G.

Example 2.11. Let G_1 be the simple graph in Figure 2a, G_2 be the looped simple graph in Figure 2b, and let $\mathbf{A}_1$ and $\mathbf{A}_2$ be the matrices in Example 2.9. Then G_1 has adjacency matrix $\mathbf{A}_{G_1} = \mathbf{A}_1$, and G_2 has adjacency matrix $\mathbf{A}_{G_2} = \mathbf{A}_2$. Thus $D(\mathbf{A}_{G_1}) = (V, \mathcal{F}_1)$ and $D(\mathbf{A}_{G_2}) = (V, \mathcal{F}_2)$ both have ground set $V = \{1, 2, 3, 4\}$ and their collections of feasible sets are

$$\mathcal{F}_1 = \{\emptyset, \{1,2\}, \{1,3\}, \{1,4\}, \{2,3\}, \{1,2,3,4\}\},$$

and

$$\mathcal{F}_2 = \{\emptyset, \{3\}, \{1,2\}, \{1,3\}, \{1,4\}, \{2,3\}, \{1,2,3\}, \{1,3,4\}, \{1,2,3,4\}\}.$$

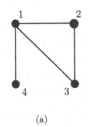 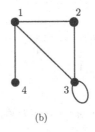 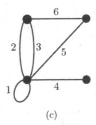

(a)	(b)	(c)

Figure 2: Three graphs

2.4.2 From graphs (with edges forming the ground set) Let $G = (V, E)$ be a connected graph. Define a collection $\mathcal{B}(G)$ of subsets of E by, for each $A \subseteq E$ setting

$$A \in \mathcal{B}(G) \iff (V, A) \text{ a spanning tree of } G.$$

(Recall a subgraph H of G is *spanning* if $V(H) = V(G)$.) Then the pair $(E, \mathcal{B}(G))$ forms a matroid called the *cycle matroid* of G, denoted by $C(G)$. (This is due to Whitney [78].)

Example 2.12. Consider the graph G shown in Figure 2c. Its cycle matroid, $C(G)$, has ground set $E = \{1, 2, 3, 4, 5, 6\}$ and set of bases

$$\mathcal{B} = \{\{2,4,5\}, \{2,4,6\}, \{3,4,5\}, \{3,4,6\}, \{4,5,6\}\}.$$

2.4.3 From graphs in surfaces Let $G = (V, E)$ be a connected graph (cellularly) embedded in a (connected) surface Σ. (Informally, an embedded graph is a graph drawn on a surface in such a way that edges do not intersect, except for where their ends meet at vertices, as in Figure 3. The cellular condition means that each of its faces, i.e. the components of $\Sigma \backslash G$, is homeomorphic to a disc.) Since G and any subgraph H of it can be regarded as a set of curves and points on the surface, we can take a regular neighbourhood $N(H)$ of each subgraph H of G. (Informally, think of $N(H)$ as a surface with boundary that arises by 'thickening up' the drawing of H, as in Figure 4.)

Each regular neighbourhood $N(H)$ of a subgraph H of the embedded graph G has some number of boundary components. We say that H is a *quasi-tree* if $N(H)$ has exactly one boundary component.

Define a collection $\mathcal{F}$ of subsets of E by, for each $A \subseteq E$ setting

$$A \in \mathcal{F} \iff (V, A) \text{ is a quasi-tree.}$$

Then the pair $(E, \mathcal{F})$ forms a delta-matroid. (This result is implicit in Bouchet's paper [13].)

Example 2.13. Let G be the graph in the torus shown in Figure 3. It has an edge set $E = \{1, 2, \ldots, 6\}$. There are exactly nine subset sets A of E for which (V, A) forms a quasi-tree. Figure 4 gives three of these and their corresponding neighbourhoods $N(V, A)$. The pair $(E, \mathcal{F})$ forms a delta-matroid where

$$\mathcal{F} = \{\{2, 4, 5\}, \{2, 4, 6\}, \{3, 4, 5\}, \{3, 4, 6\}, \{4, 5, 6\},$$
$$\{1, 2, 3, 4, 5\}, \{1, 2, 3, 4, 6\}, \{1, 2, 4, 5, 6\}, \{2, 3, 4, 5, 6\}\}.$$

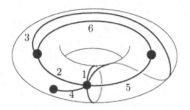

Figure 3: A graph embedded in the torus

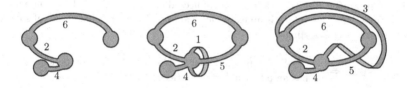

Figure 4: Neighbourhoods of the subgraphs on $\{2, 4, 6\}$, $\{1, 2, 4, 5, 6\}$, and $\{2, 3, 4, 5, 6\}$

Later, we will consider this example in the formalism of ribbon graphs (see Section 5). We will also see that delta-matroids of this type give a topological analogue of cycle matroids.

2.4.4 From Eulerian circuits Let $G = (V, E)$ be a connected 4-regular graph. We are interested in the Eulerian circuits in G. At any vertex v of G there are exactly three possible routes that an Eulerian circuit can take through it. At each vertex, set one choice of route through it as being *forbidden*, and of the other two as *allowed*. Set one allowed route at each vertex as being *preferred*.

With this information, construct a collection $\mathcal{F}$ of subsets of V by, for each $X \subseteq V$, setting

$$X \in \mathcal{F} \iff \begin{array}{l} \text{there is an Eulerian circuit taking only allowed allowed routes} \\ \text{through vertices, and preferred routes at exactly the vertices in } X. \end{array}$$

Then the pair $(V, \mathcal{F})$ forms a delta-matroid. This type of delta-matroid is known as an *Eulerian delta-matroid*. (This result is due to Bouchet [9].)

Example 2.14. Figure 5 shows a 4-regular graph equipped with a set of preferred and forbidden transitions. It has exactly four Eulerian circuits that avoid forbidden transitions. These are given by *abfcde*, which used preferred transitions at 1; *abdefc*, which used preferred transitions at 3; *acfbde*, which used preferred transitions at 1 and 3; *acdbfe*, which used preferred transitions at 1, 2, and 3. Thus we obtain a delta-matroid on $V = \{1, 2, 3\}$ with the collection of feasible sets

$$\mathcal{F} = \{\{1\}, \{3\}, \{1, 3\}, \{1, 2, 3\}\}.$$

Preferred		Forbidden	
vertex	route	vertex	route
1	$\{ae\}, \{bf\}$	1	$\{af\}, \{be\}$
2	$\{cd\}, \{ef\}$	2	$\{ce\}, \{df\}$
3	$\{ac\}, \{bd\}$	3	$\{ad\}, \{bc\}$

Figure 5: A 4-regular graph with preferred and forbidden transitions

2.4.5 From grafts Let $G = (V, E)$ be a connected graph and $T \subseteq V$ be a non-empty set of its vertices. The pair (G, T) is an example of a *graft*. Define a collection $\mathcal{F}$ of subsets of E by, for each $A \subseteq E$ setting

$$A \in \mathcal{F} \iff \quad \begin{array}{l} (V, A) \text{ a spanning forest of } G \text{ in which each component} \\ \text{has an odd number of vertices in } T. \end{array}$$

Then the pair $(E, \mathcal{F})$ forms a delta-matroid, denoted here by $D(G, T)$. (This result is due to Oum [59].)

Example 2.15. The graft (G, T) shown in Figure 6 has a delta-matroid $D(G, T)$ on ground set $E = \{1, 2, 3, 4, 5\}$ and its collection of feasible sets is

$$\mathcal{F} = \{\{3, 5\}, \{4, 5\}, \{1, 2, 3, 5\}, \{1, 2, 4, 5\}, \{1, 3, 4, 5\}, \{2, 3, 4, 5\}\}.$$

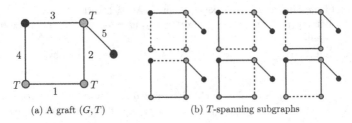

(a) A graft (G, T) (b) T-spanning subgraphs

Figure 6: A graft and its T-spanning subgraphs

2.4.6 From matchings Let $G = (V, E)$ be a simple graph and, for $U \subseteq V$, let $G[U]$ be its induced subgraph on U. A *perfect matching* on G is a subset A of its edges such that each vertex of G is incident with exactly one edge in A. Define a collection $\mathcal{F}$ of subsets of V by, for each $U \subseteq V$,

$$\mathcal{F} := \{U \subseteq V : G[U] \text{ has a perfect matching}\}.$$

Then the pair $(V, \mathcal{F})$ forms a delta-matroid called the *matching delta-matroid of* G. (This is due to Bouchet [14].)

Example 2.16. Let G be the graph with vertex set $V = \{1, 2, 3, 4\}$ given in Figure 2a. Then G has a perfect matching, and so do its restrictions to any edge, and as does the empty graph. Thus with

$$\mathcal{F} = \{\emptyset, \{1, 2\}, \{1, 3\}, \{1, 4\}, \{2, 3\}, \{1, 2, 3, 4\}\}.$$

The pair $(V, \mathcal{F})$ is the matching delta-matroid of G.

2.4.7 From the Greedy Algorithm Suppose we have a proper set system $(E, \mathcal{F})$ and a weight function $w : E \to \mathbb{R}$. We want to find a member of $\mathcal{F}$ of maximum weight, that is, we want to find some $F \in \mathcal{F}$ maximising $w(F) := \sum_{x \in F} w(x)$.

Roughly speaking, the *greedy algorithm* runs though the elements of E from largest to smallest and selects an element if, together with the other previously selected elements, it forms a subset of some $F \in \mathcal{F}$ such that F contains no rejected elements. Otherwise it rejects the element.

Formally, suppose that we have a *separation oracle* telling us for each ordered pair (P, Q), where P and Q are disjoint subsets of E, whether there is some $F \in \mathcal{F}$ containing P and disjoint from Q. If such an F exists, (P, Q) is *separable*. The *greedy algorithm* successively examines each element of E according to an ordering $x_1, x_2, \ldots, x_n$ such that $|w(x_1)| \geq |w(x_2)| \geq \cdots \geq |w(x_n)|$, putting each x_i in either a set A of selected elements or B of rejected elements:

```
A := ∅
B := ∅
for i := 1 to n do
    if w(x_i) ≥ 0 then
        if (A ∪ x_i, B) is separable then
            A := A ∪ x_i
        else
            B := B ∪ x_i
        end if
    else
        if (A, B ∪ x_i) is separable then
            B := B ∪ x_i
        else
            A := A ∪ x_i
        end if
    end if
end for
```

The greedy algorithm *succeeds* if A is a maximum weight member of $\mathcal{F}$, that is, if $w(A) = \max_{F \in \mathcal{F}} w(F)$.

Bouchet [9], and, independently, Chandrasekaran and Kabadi [25] in the equivalent language of pseudomatroids, characterised delta-matroids as the class of set systems for which the greedy algorithm succeeds:

Theorem 2.17. *The greedy algorithm applied to a set system $(E, \mathcal{F})$ succeeds for every weight function $w : E \to \mathbb{R}$ if and only if $(E, \mathcal{F})$ is a delta-matroid.*

3   Delta-matroid essentials

We now give a brief overview of basic delta-matroid constructions and terminology. The definition of a delta-matroid was given in Section 2.2. Isomorphism is defined in the obvious way: two delta-matroids are *isomorphic* if there is a bijection between their ground sets that induces a bijection between their feasible sets. We use equals signs to denote delta-matroids are isomorphic, although we will generally identify isomorphic delta-matroids.

A delta-matroid is said to be *even* if its feasible sets are either all of odd size, or all of even size. Otherwise it is said to be *odd*. We emphasise that the feasible sets of an even delta-matroid may all be of odd size. A delta-matroid it is said to be *normal* if the empty set is feasible.

Example 3.1. The delta-matroids $D(\mathbf{A}_1)$ and $D(\mathbf{A}_2)$ from Example 2.9 are both normal. $D(\mathbf{A}_1)$ is even but $D(\mathbf{A}_2)$ is not. The delta-matroid in Example 2.13 is even, and is not normal.

The feasible sets of a delta-matroid $D = (E, \mathcal{F})$ are graded by their size. Let $\mathcal{F}_{\min}$ denote the collection of all feasible sets in $\mathcal{F}$ of minimum size, and $\mathcal{F}_{\max}$ the collection of all feasible of maximum size. For $k = 0, 1, 2, \ldots$, let $\mathcal{F}_{\min +k}$ denote the collection of all feasible sets in $\mathcal{F}$ that are of size exactly k larger than a minimum sized feasible set. The *width* of a delta-matroid is the difference between the sizes of its largest and smallest feasible sets.

The maximum gap in the collection of sizes of feasible sets of a delta-matroid is two. That is, if a delta-matroid has a feasible set of size k and a larger feasible set, then it has a feasible set of size $k + 1$ or $k + 2$. In particular, this means that for an even delta-matroid, all of $\mathcal{F}_{\min}, \mathcal{F}_{\min +2}, \ldots, \mathcal{F}_{\max}$ are non-empty. For odd delta-matroids, if there is a feasible set of size k and one of size greater than k, then, while there will be a feasible set of size $k + 1$ or $k + 2$, there will not necessarily be both (for example, see the delta-matroids in Theorem 7.11). However, Bouchet proved in [13] that in an odd delta-matroid, there will always be feasible sets of sizes k and $k + 1$, for some k.

Exercise 3.2. *Prove (for example, by induction on $|X \triangle Y|$) that if a delta-matroid has a feasible set X of size k and a larger feasible set, then it has a feasible set Y of size $k + 1$ or $k + 2$.*

When $D = (E, \mathcal{F})$ is a delta-matroid, $D_{\min} := (E, \mathcal{F}_{\min})$ and $D_{\max} := (E, \mathcal{F}_{\max})$ are both matroids. $D_{\min}$ is called the *lower matroid*, and $D_{\max}$ is called the *upper matroid* of D. Bouchet defined these matroids in [13].

Example 3.3. Consider the delta-matroid $D := D(\mathbf{A}_2)$ from Example 2.9. With $E = \{1, 2, 3, 4\}$, we have $D_{\min} = (E, \{\emptyset\})$ and $D_{\max} = (E, \{1, 2, 3, 4\})$. Furthermore, $\mathcal{F}_{\min +2} = \{\{1, 2\}, \{1, 3\}, \{1, 4\}, \{2, 3\}\}$, but the pair $(E, \mathcal{F}_{\min +2})$ does not form a matroid (since the Symmetric Exchange Axiom fails with $X = \{1, 4\}$, $Y = \{2, 3\}$, and $u = 1$).

Exercise 3.4. *Let $D = (E, \mathcal{F})$ be a delta-matroid. Verify that $(E, \mathcal{F}_{\min})$ satisfies the basis definition of a matroid from Exercise 2.7. Conclude that $D_{\min}$ is indeed a matroid.*

A fundamental operation in delta-matroid theory is *twisting* (which is sometimes called *pivoting*). This operation changes a delta-matroid by replacing each feasible set X with its symmetric difference $X \triangle A$, for some fixed set A.

Definition 3.5 (Twist). Let $D = (E, \mathcal{F})$ be a delta-matroid, and $A \subseteq E$. Let

$$\mathcal{F}' := \{X \triangle A : X \in \mathcal{F}\}.$$

Then the *twist* of D by A, denoted $D * A$, is defined as

$$D * A := (D, \mathcal{F}').$$

The *dual* of D, denoted D^*, is defined as $D^* := D * E$.

Bouchet, in [9], showed that the set of delta-matroids is closed under twisting.

Proposition 3.6. *If $D = (E, \mathcal{F})$ is a delta-matroid, then so is $D * A$, for each $A \subseteq E$.*

Example 3.7. If D is the delta-matroid from Example 2.13, then $D * \{3, 4\}$ is the delta-matroid on $\{1, \ldots, 6\}$ with feasible sets

$$\mathcal{F}' = \{\{2, 3, 5\}, \{2, 3, 6\}, \{5\}, \{6\}, \{3, 5, 6\}, \{1, 2, 5\}, \{1, 2, 6\}, \{1, 2, 3, 5, 6\}, \{2, 5, 6\}\}.$$

Exercise 3.8. *Verify the following results about twisting. (1) The twist of a delta-matroid is a delta-matroid (i.e., prove Proposition 3.6). (2) Every delta-matroid is the twist of a normal delta-matroid. (3) The twist of an even delta-matroid is even. (4) $D_{\max}$ is a matroid (use Exercise 3.4). (5) $(D * A) * B = D * (A \triangle B)$.*

We now define deletion and contraction for delta-matroids. In defining these, care must be taken in the special cases when an element is in every feasible set, or does not appear in any feasible set. Such elements are called coloops and loops.

Definition 3.9 (Loop and coloop). Let $D = (E, \mathcal{F})$ be a delta-matroid. Then an element $e \in E$ is a *loop* if it is not in any feasible set of D, and a *coloop* if it is in every feasible set of D.

Example 3.10. In Example 2.12, the element 1 is a loop, and 4 is a coloop.

Definition 3.11 (Deletion). Let $D = (E, \mathcal{F})$ be a delta-matroid, and $e \in E$. Then D *delete* by e, denoted $D \backslash e$, is defined as $D \backslash e := (E \backslash e, \mathcal{F}')$, where

1. when e is not a coloop,

$$\mathcal{F}' = \{X : X \in \mathcal{F} \text{ and } e \notin X\};$$

2. and when e is a coloop,

$$\mathcal{F}' = \{X \backslash e : X \in \mathcal{F} \text{ and } e \in X\}.$$

Thus, in words, if e is not a coloop, the feasible sets of $D\backslash e$ are obtained by restricting to feasible sets of D that do not contain e, and when e is a coloop they are obtained by restricting to the feasible sets of D that do contain e, then removing e from them.

Definition 3.12 (Contraction). Let $D = (E, \mathcal{F})$ be a delta-matroid, and $e \in E$. Then D *contract* by e, denoted D/e, is defined as $D/e := (E\backslash e, \mathcal{F}')$, where

1. when e is not a loop,
$$\mathcal{F}' = \{X\backslash e : X \in \mathcal{F} \text{ and } e \in X\};$$

2. and when e is a loop, $\mathcal{F}' = \mathcal{F}$.

Thus if e is not a loop, the feasible sets of D/e are obtained by restricting to the feasible sets of D that contain E, then removing e from them. When e is a loop D and D/e have the same feasible sets.

Example 3.13. Let $D = (E, \mathcal{F})$ be the delta-matroid from Example 2.13. Then $D\backslash 1$ has ground set $\{2, \ldots, 6\}$ and its collection of feasible sets is

$$\{\{2, 4, 5\}, \{2, 4, 6\}, \{3, 4, 5\}, \{3, 4, 6\}, \{4, 5, 6\}, \{2, 3, 4, 5, 6\}\}.$$

$D/1$ has ground set $\{2, \ldots, 6\}$ and its collection of feasible sets is

$$\{\{2, 3, 4, 5\}, \{2, 3, 4, 6\}, \{2, 4, 5, 6\}\}.$$

$(D/1)\backslash 4$ has ground set $\{2, 3, 5, 6\}$ and its collection of feasible sets is

$$\{\{2, 3, 5\}, \{2, 3, 6\}, \{2, 5, 6\}\}.$$

Exercise 3.14. *Show that if D is a delta-matroid then so are $D\backslash e$ and D/e. (Deletion and contraction are due to Bouchet and Duchamp [20].)*

An important observation is that the notions of deletion and contraction are 'dual' to each other:
$$D/e = (D * e)\backslash e. \tag{3.1}$$
This identity ties up the three delta-matroid operations of deletion, contraction, and twisting in a fundamental way.

Exercise 3.15. *Verify Equation (3.1).*

Observe that when $e \neq f$, the operations of twisting, deleting, and contracting on e, commute with the operations of twisting, deleting, and contracting on f. In particular, this means that for $D = (E, \mathcal{F})$ and $A \subseteq E$, we can define $D\backslash A$ and D/A as the result of deleting, respectively contracting, every element of A in any order.

Definition 3.16 (Minor). A delta-matroid D' is said to be a *minor* of a delta-matroid D if it can be obtained from D through the operations of deletion, contraction and twisting. Furthermore, D' is said to be a *strong-minor* of D if it can be obtained from D through the operations of deletion and contraction (without twisting).

Note that by (3.1), the operation of contraction is redundant in the definition of a minor. We also note that the term 'strong-minor' used here is not a standard term in the literature, but we need to make a distinction between these two types of minor.

Exercise 3.17. *Prove that a delta-matroid is even if and only if it has no minor isomorphic to the delta-matroid $(\{a\}, \{\emptyset, \{a\}\})$. (This result is due to Bouchet [13].)*

4   Graphic matroids

Cycle matroids provide a bridge between graph theory and matroid theory. While there is much to be said about cycle matroids and their role in matroid theory, their importance in terms of the current exposition is that there is a fundamental compatibility between graphs and matroids which means that results in either area can be used to gain insights in the other (Oxley's BCC survey article [62] illustrates this principle well). In Section 5, we shall demonstrate that an analogous connection holds between topological graph theory and delta-matroid theory, and see that many delta-matroid results can be regarded as 'topological' analogues of established matroid results. Our exposition of graphic matroids is tailored towards this aim, and the results mentioned here are standard and can be found in, for example, [61].

The cycle matroid of a connected graph $G = (V, E)$ was described in Section 2.3. The following definition includes the case when G is not connected. Recall that in the context of matroids, a feasible set is called a basis.

Definition 4.1 (Cycle matroid, graphic matroid). Let $G = (V, E)$ be a graph. Let

$$\mathcal{B} := \{F \subseteq E(G) : F \text{ is the edge set of a maximal spanning forest of } G\},$$

Then $C(G) := (E, \mathcal{B})$ is the *cycle matroid* of G.
A matroid is *graphic* if it is isomorphic to the cycle matroid of some graph.

Exercise 4.2. *Verify that the bases of $C(G)$ satisfies the Symmetric Exchange Axiom, and hence that $C(G)$ is a matroid.*

Edge and vertex deletion for graphs is denoted $G\backslash e$ and $G\backslash v$, respectively. Edge contraction is denoted G/e. We allow contraction of loops, and it is defined as the graph resulting from deleting the loop. An edge e of a graph G is a *bridge* if $G\backslash e$ has more components than G.

Exercise 4.3. *Let G be a graph with an edge e. Show that*

1. e is a coloop in $C(G)$ if and only if e is a bridge in G; and

2. e is a loop in $C(G)$ if and only if e is a loop in G.

The usual notion of a *matroid-minor* coincides with strong-minors when the matroid is regarded as a delta-matroid. (Strong-minors are needed as the set of matroids is not closed under twisting.) Graph minors are compatible with matroid-minors, providing a key link between graph and matroid theory.

Theorem 4.4. *Let G be a graph, e be an edge of G, and v be an isolated vertex. Then*

$$C(G\backslash e) = C(G)\backslash e, \quad C(G/e) = C(G)/e, \quad and \quad C(G\backslash v) = C(G).$$

Properties of cycle matroids are intimately linked with properties of plane and planar graphs (a graph is *plane* if it *has* been embedded in the plane, and is *planar* if it *can* be embedded in the plane), as exhibited in the following theorems.

Theorem 4.5. *Let G be a plane graph and G^* be its (geometric) dual. Then*

$$C(G^*) = (C(G))^*.$$

Theorem 4.6. *The following are equivalent for a graph G.*

1. *G is planar*

2. *$C(G)^*$ is graphic*

3. *$C(G)$ has no matroid-minor isomorphic to $C(K_5)$ or $C(K_{3,3})$.*

Of course this theorem should be compared with the Kuratowski-Wagner Theorem which states that a graph G is planar if and only if it has no minor isomorphic to K_5 or $K_{3,3}$.

Exercise 4.7. *By considering different embeddings of a graph consisting of one vertex and two loops, show that, in general, Theorem 4.5 does not hold for non-plane embeddings.*

5 Topological graph theory and delta-matroids

It is often productive to think of matroids as 'generalisations of graphs'. In this section we explain how, analogously, delta-matroids can be thought of as being 'generalisations of graphs in surfaces', a point of view that enriches both fields. The usual passage between graphs and matroids is via cycle matroids, as described in the previous section. The passage between embedded graphs and delta-matroids is via ribbon-graphic delta-matroids. These delta-matroids arise by dropping a hidden topological restriction in the definition of a cycle matroid.

Bouchet first constructed delta-matroids from graphs in surfaces in [13]. His approach was very different, but equivalent, to that presented in this section. He associated a transition system to the medial graph of a graph in a surface and considered the Eulerian delta-matroid that arises from it. We instead approach the subject here through the language of ribbon graphs. The connection between ribbon graph theory and delta-matroid theory, as well as the philosophy that delta-matroid theory generalises topological graph theory, is due to Chun, Moffatt, Noble, and Rueckriemen [29,30]. The equivalence between this approach and Bouchet's is detailed in Section 6, where Eulerian and ribbon-graphic delta-matroids are identified.

5.1 Ribbon graphs

In Section 2.4.3 we saw how a delta-matroid can be associated with a graph in a surface. We now develop this idea. However, to do so it is convenient, and more natural, to work in the language of ribbon graphs, rather than cellularly embedded graphs. This section contains a brief introduction to ribbon graphs. A more comprehensive introduction can be found in [35].

In essence, a ribbon graph is a structure that arises by taking a regular neighbourhood of a graph in a surface, but without throwing away the vertex-edge structure of the graph. See Figure 7. We can think of a ribbon graph informally as 'a graph whose vertices consist of discs, and whose edges consist of ribbons', as in Figure 7c.

Definition 5.1 (Ribbon graph). A *ribbon graph* $G = (V, E)$ is a (possibly non-orientable) surface with boundary represented as the union of two sets of discs, a set V of *vertices*, and a set of *edges* E such that:

1. the vertices and edges intersect in disjoint line segments;

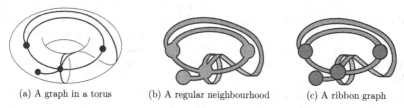

(a) A graph in a torus   (b) A regular neighbourhood   (c) A ribbon graph

Figure 7: Equivalence of graphs in surfaces and ribbon graphs

2. each such line segment lies on the boundary of precisely one vertex and precisely one edge;

3. every edge contains exactly two such line segments.

Ribbon graphs describe exactly cellularly embedded graphs (i.e., graphs embedded on a closed surface such that the faces are all discs). We have discussed how a ribbon graph arises from a cellularly embedded graph (Figure 7). In the other direction, given a ribbon graph, the classification of surfaces with boundary ensures there is a unique way (up to homeomorphism) to embed it in a surface by 'filling in the holes'.

In addition to parameters inherited from graph theory, such as numbers of edges, vertices and components, some topological parameters are associated with ribbon graphs. A ribbon graph is *orientable* if it is orientable as a surface, and is *non-orientable* otherwise. The *genus* of a ribbon graph is its genus as a surface. The *Euler genus*, $\gamma(\mathbb{G})$ of a ribbon graph $\mathbb{G}$ equals its genus if it is non-orientable, and equals twice its genus if it is orientable. A connected ribbon graph is *plane* if it has Euler genus 0 (i.e., if it corresponds to a graph in a sphere).

Ribbon graph equivalence corresponds to cellularly embedded graph equivalence. Two ribbon graphs are *equivalent* if there is a homeomorphism from one to the other (which should be orientation preserving when the ribbon graph is orientable) that sends vertices to vertices, edges to edges, and preserves the cycle order of half-edges at each vertex. We consider ribbon graphs up to this equivalence. Note that ribbon graphs are not embedded in 3-space, and in drawings ribbon graphs, we can 'push' half-twists of edges around the ribbon graph and 'turn vertices over' as illustrated in Figure 8, as well as 'pushing edges through each other'.

Figure 8: Some equivalent drawings of ribbon graphs

Deletion for ribbon graphs is defined in the obvious way:

Definition 5.2 (Deletion). Let $\mathbb{G}$ be ribbon graph, e be an edge of it, and v a vertex. Then $\mathbb{G}$ *delete* e, written $\mathbb{G} \backslash e$ is the ribbon graph obtained from $\mathbb{G}$ by removing the edge e, and $\mathbb{G} \backslash v$ is the ribbon graph obtained from $\mathbb{G}$ by removing the vertex v and all its incident edges.

Contraction for ribbon graphs is more tricky to define. The difficulty is that while we would like to define contraction of an edge e to be the result of merging e and its incident vertices into a single vertex, as we do in the case for graphs, applying this operation to a loop in a ribbon graph can result in an object that is no longer a ribbon graph. To obtain a definition of contraction, we move to the language of arrow presentations, which is due to Chmutov [26].

Definition 5.3 (Arrow presentation). An *arrow presentation* is a set of closed curves, each with a collection of disjoint labelled arrows lying on them, and where each label appears on precisely two arrows.

An arrow presentation is shown in Figure 9a.

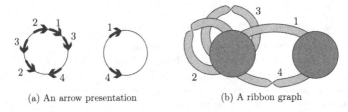

(a) An arrow presentation   (b) A ribbon graph

Figure 9: A ribbon graph and its description as an arrow presentation

Arrow presentations describe ribbon graphs. A ribbon graph G can be formed from an arrow presentation by identifying each closed curve with the boundary of a disc (forming the vertex set of G). Then, for each pair of e-labelled arrows, taking a disc (which will form an edge of G), orienting its boundary, placing two disjoint arrows on its boundary that point in the direction of the orientation, and identifying each e-labelled arrow on this edge. See Figure 9.

Conversely a ribbon graph can be described as an arrow presentation by arbitrarily labelling and orienting the boundary of each edge disc of G. Then on each arc where an edge disc intersects a vertex disc, place an arrow on the vertex disc, labelling the arrow with the label of the edge it meets and directing it consistently with the orientation of the edge disc boundary. The boundaries of the vertex set marked with these labelled arrows give an arrow presentation.

Now suppose that we have a non-loop edge e of a ribbon graph $\mathbb{G}$. Then the natural contraction operation is illustrated in Figure 10a. Figure 10b shows this operation in terms of a 'splicing' operation on arrow presentations. Notice that in terms of arrow presentation this definition is local and does not see if the edge is a loop or not. Thus it can be applied to *any* edge. This gives our definition of contraction.

Definition 5.4 (Contraction). Let $\mathbb{G}$ be ribbon graph with an edge e. Then $\mathbb{G}$ *contract* e, written $\mathbb{G}/e$ is the ribbon graph obtained from $\mathbb{G}$ by the following process: (1) describe $\mathbb{G}$ as an arrow presentation, (2) 'splice' the arrow presentation as indicated in Figure 10b. (That is, delete the two e labelled arrows and the parts of the curves they lie on. Add arcs connecting the two pairs of points that were the tips and tails of the arrow.) (3) The ribbon graph described by this arrow presentation is $\mathbb{G}/e$.

Example 5.5. Figure 11 illustrates the contraction of loops. Notice that the underlying graph of $\mathbb{G}/1$ does not equal the result of contracting the edge 1 in the underlying graph of $\mathbb{G}$, so graph contraction and ribbon graph contraction are not compatible operations.

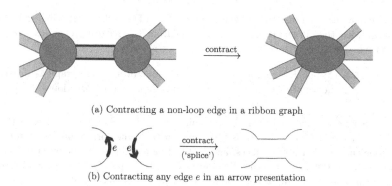

(a) Contracting a non-loop edge in a ribbon graph

(b) Contracting any edge e in an arrow presentation

Figure 10: Descriptions of contraction

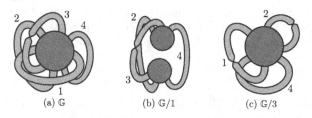

(a) $\mathbb{G}$ (b) $\mathbb{G}/1$ (c) $\mathbb{G}/3$

Figure 11: Contraction for ribbon graphs

Table 1 shows the local effect of deletion and contraction on a ribbon graph.

	non-loop	non-orient. loop	orientable loop	arrow pres.
G				
$G\backslash e$				
G/e				
G^e				

Table 1: Operations on an edge e (highlighted in bold) of a ribbon graph. The ribbon graphs are identical outside of the region shown

Contraction can be defined directly on ribbon graphs as follows. If u_1 and u_2 are the (not necessarily distinct) vertices incident to e, then $\mathbb{G}/e$ denotes the ribbon graph obtained as follows: consider the boundary component(s) of $e \cup u_1 \cup u_2$ as curves on G. For each resulting curve, attach a disc (which will form a vertex of $\mathbb{G}/e$) by identifying its boundary component with the curve. Delete e, u_1 and u_2 from the resulting complex, to get the ribbon graph $\mathbb{G}/e$.

Definition 5.6 (Minor). A ribbon graph $\mathbb{H}$ is a *minor* of a ribbon graph $\mathbb{G}$ if it can be obtained by a sequence of edge deletions, vertex deletions, and contractions.

The other basic operation on ribbon graphs we need here is duality. Recall that the *dual*, G^*, of a graph G in a surface is the graph in the same surface obtained from G by placing one vertex in each of its faces, and embedding an edge of G^* between two of these vertices whenever the faces of G they lie in are adjacent. Edges of G^* are embedded so that they cross the corresponding face boundary (or edge of G) transversally.

Figure 12 shows the construction of a dual, where the plane graphs have been thickened to form ribbon graphs in the plane. We can describe these ribbon graphs as arrow presentations, and Figure 12d shows how the two arrow presentations fit naturally together in the plane with G and G^*. By examining this figure in the locality of an edge (inside the dotted region in the figure) we see that, in terms of arrow presentations, a dual graph can be constructed by using the local change of Figure 13 at each pair of arrows.

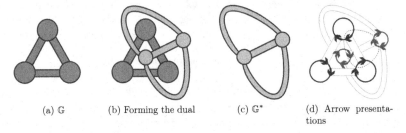

(a) $\mathbb{G}$ (b) Forming the dual   (c) $\mathbb{G}^*$ (d) Arrow presentations

Figure 12: Dual graphs and their arrow presentations

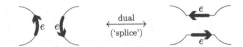

Figure 13: A partial dual in terms of arrow presentations

What we have done is moved from a global construction of the dual G^* of G to a local construction. Since it is local, we can form the edges of a dual graph one at a time. This results in the concept of a *partial dual*, which is due to Chmutov [26]. The partial dual, $\mathbb{G}^A$, of a ribbon graph $\mathbb{G}$ is the result of forming the dual of $\mathbb{G}$, but only at the edges in some set of edges A.

Definition 5.7 (Partial dual). Let $\mathbb{G}$ be ribbon graph with an edge e. Then the *partial dual* of $\mathbb{G}$ with respect to e is the ribbon graph denoted $\mathbb{G}^e$ obtained from $\mathbb{G}$ by the following process: (1) describe $\mathbb{G}$ as an arrow presentation, (2) 'splice' the arrow presentation at the two e-labelled arrows as indicated in Figure 13. (3) The ribbon graph described by this arrow presentation is $\mathbb{G}^e$.

When $e \neq f$ are edges of a ribbon graph $\mathbb{G} = (V, E)$, it is easily seen that $(\mathbb{G}^e)^f = (\mathbb{G}^f)^e$. Thus for $A \subseteq E$, we can define *partial dual* of $\mathbb{G}$ with respect to A, denoted by G^A, to be the ribbon graph obtained from G by forming the partial dual with respect to each edge of A in any order.

Example 5.8. The ribbon graph in Figure 9b can be described by the arrow presentation in Figure 14a. Forming the partial dual with respect to the edges 3 and 4, gives the arrow presentation shown in Figure 14b, which represents the ribbon graph in Figure 11a (so this is $\mathbb{G}^{\{3,4\}}$ when $\mathbb{G}$ is as in Figure 9b).

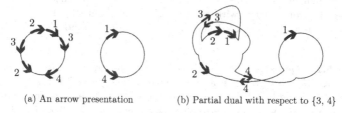

(a) An arrow presentation   (b) Partial dual with respect to $\{3, 4\}$

Figure 14: Forming a partial dual using arrow presentations

Table 1 shows the local effect of forming a partial dual with respect to an edge of a ribbon graph.

It is easy to see that the following properties hold. $\mathbb{G}^* = \mathbb{G}^{E(\mathbb{G})}$, where $\mathbb{G}^*$ is the dual of $\mathbb{G}$; $\mathbb{G}^\emptyset = \mathbb{G}$; $(\mathbb{G}^A)^B = \mathbb{G}^{A \triangle B}$; partial duality acts disjointly on connected components; and $\mathbb{G}^A$ is orientable if and only if $\mathbb{G}$ is. By examining arrow presentations (for example, in Table 1), we immediately see that

$$\mathbb{G}/e = \mathbb{G}^e \backslash e. \tag{5.1}$$

As with contraction, partial duals can be formed without passing through arrow presentations. Let $\mathbb{G} = (V, E)$ be a ribbon graph, $A \subseteq E$, and regard the boundary components of the ribbon subgraph (V, A) as curves on $\mathbb{G}$. Glue a disc to $\mathbb{G}$ along each of these curves by identifying the boundary of the disc with the curve, and remove the interior of all vertices of $\mathbb{G}$. The resulting ribbon graph is $\mathbb{G}^A$.

Exercise 5.9. *Consider the set S of pairs $(\mathbb{G}, e)$ where $\mathbb{G}$ is a ribbon graph and e is one of its edges. Let δ denote the operation $\delta : (\mathbb{G}, e) \mapsto (\mathbb{G}^e, e)$. Let $\tau : (\mathbb{G}, e) \mapsto (\mathbb{G}^{\tau(e)}, e)$ where $\mathbb{G}^{\tau(e)}$ is obtained from $\mathbb{G}$ by adding a 'half-twist' to the edge e (formally, reverse the direction of exactly one e-labelled arrow in an arrow presentation of $\mathbb{G}$). Two ribbon graphs are twisted duals if one can be obtained from the other by a sequence of applications of the operations τ and δ to its edges (see [34]). Verify that the operations τ and δ induce an action of the symmetric group $\langle \delta, \tau \mid \delta^2, \tau^2, (\tau\delta)^3 \rangle$ on S.*

5.2 Ribbon-graphic delta-matroids

Thinking of matroid theory as a generalisation of graph theory, where the passage from a graph G to a matroid is through its cycle matroid $C(G)$, suppose we were set the problem of finding the matroid analogue of topological graph theory. We are thus looking for some matroid analogue of a ribbon graph $\mathbb{G}$. We quickly see that cycle matroids do not provide an effective analogue of ribbon graphs, since they do not see any of their topological information (e.g, the two ribbon graphs that are 2-cycles have the same cycle matroid). To progress let us examine the construction of $C(\mathbb{G})$.

For simplicity, suppose $\mathbb{G}$ is connected. Then the bases of $C(\mathbb{G})$ are the edge sets of the spanning trees of $\mathbb{G}$. A spanning tree of $\mathbb{G}$ can be characterised as a ribbon subgraph that is (1) spanning, (2) has exactly one boundary component, and (3) is of genus 0. With this formulation it is apparent why we are seeing no topological information in $C(\mathbb{G})$ — we are only considering subgraphs of genus 0. We immediately see how to adjust the construction to preserve topological information — drop the genus 0 condition.

This takes us to the concept of a *quasi-tree*, which is a ribbon graph with exactly one boundary component. With this, we can obtain a topological version of a cycle matroid by replacing the words "tree" with "quasi-tree" in its definition. It turns out that this results in a delta-matroid, denoted here by $D(\mathbb{G})$, that is a topological counterpart of a cycle matroid.

Definition 5.10 (Quasi-tree). A *quasi-tree* is a ribbon graph with exactly one boundary component. A ribbon subgraph $\mathbb{H}$ of a connected ribbon graph $\mathbb{G}$ is a *spanning quasi-tree* if $\mathbb{H}$ is a quasi-tree and has the same vertex set as $\mathbb{G}$. By an abuse of notation, if $\mathbb{G}$ is not connected then we say a ribbon subgraph $\mathbb{H}$ is a *spanning quasi-tree* of $\mathbb{G}$ if $\mathbb{H}$ induces a spanning quasi-tree of each connected component of $\mathbb{G}$.

We obtain a topological analogue of a cycle matroid by replacing trees with quasi-trees in Definition 4.1.

Definition 5.11 (Ribbon-graphic delta-matroid). Let $\mathbb{G} = (V, E)$ be a ribbon graph, and let

$$\mathcal{F} := \{F \subseteq E : F \text{ is the edge set of a spanning quasi-tree of } \mathbb{G}\}.$$

We call $D(\mathbb{G}) := (E, \mathcal{F})$ the *delta-matroid of* $\mathbb{G}$.

We say a delta-matroid is *ribbon-graphic* if it is isomorphic to the delta-matroid of some ribbon graph.

Example 5.12. Let $\mathbb{G}$ be the ribbon graph shown in Figure 15a. Its spanning quasi-trees are shown in Figure 15b. From this we see that $D(\mathbb{G}) = (E, \mathcal{F})$ where $E = \{1, 2, 3, 4\}$ and

$$\mathcal{F} = \{\{1\}, \{4\}, \{1, 2\}, \{1, 3\}, \{1, 4\}, \{2, 4\}, \{3, 4\}, \{1, 2, 4\}, \{1, 2, 3, 4\}\}.$$

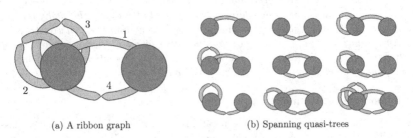

(a) A ribbon graph   (b) Spanning quasi-trees

Figure 15: A ribbon graph and its spanning quasi-trees

Example 5.13. The construction of a delta-matroid given in Section 2.4.3 is just Definition 5.11 phrased in terms of graphs in surfaces. Thus Example 2.13 gives the delta-matroid of the ribbon graph in Figure 7.

In [13], Bouchet proved, using the language of Eulerian circuits in medial graphs, that $D(\mathbb{G})$ is a delta-matroid. Figure 16 sketches a proof in terms of the topology of surfaces. A ribbon graphic proof can be found in [29].

Theorem 5.14. $D(\mathbb{G})$ *as constructed in Definition 5.11 is a delta-matroid.*

Exercise 5.15. *Prove that the feasible sets of $D(\mathbb{G})$ of minimum size are exactly the bases of the cycle matroid of $\mathbb{G}$, and hence $D(\mathbb{G})_{\min} = C(\mathbb{G})$.*

For a ribbon graph $\mathbb{G}$ with k components, it follows by definition that the feasible sets of $D(\mathbb{G})$ are in 1-1 correspondence with the spanning quasi-trees of $\mathbb{G}$. This correspondence can be refined (see [29]) to show that the feasible sets of $D(\mathbb{G})$ with cardinality m are in 1-1 correspondence with the spanning quasi-trees of $\mathbb{G}$ with Euler genus $m - |V| + k$. The following properties of ribbon-graphic delta-matroids follow from this basic result. They were first proved by Bouchet in [13].

Proposition 5.16. *Let $\mathbb{G}$ be a ribbon graph. Then*

1. *the width of $D(\mathbb{G})$ equals the Euler genus $\mathbb{G}$;*

2. *$D(\mathbb{G})$ is even if and only if $\mathbb{G}$ is orientable;*

3. *$D(\mathbb{G})_{\min} = C(\mathbb{G})$;*

4. *$D(\mathbb{G})_{\max} = C(\mathbb{G}^*)^*$;*

5. *$D(\mathbb{G}) = C(\mathbb{G})$ if and only if $\mathbb{G}$ is the disjoint union of plane ribbon graphs.*

Recall from Exercise 4.3 that loops and coloops in cycle matroids correspond to loops and bridges in graphs. The situation in delta-matroids is a little more complicated since a loop in a ribbon graph can have different topological properties: it can be orientable or non-orientable, and *trivial* or *non-trivial*. A loop edge e incident with a vertex v in a ribbon graph is *non-trivial* if there is some cycle C in the ribbon graph such that e and C are met in the cyclic order $eCeC$ when following the boundary of the vertex v. It is *trivial* otherwise. For example, in Figure 11c, the loop 2 is trivial, while loops 1 and 4 are non-trivial. Loops 1 and 2 are non-orientable, and loop 4 is orientable.

Exercise 5.17. *Let $\mathbb{G}$ be a ribbon graph, $D(\mathbb{G}) = (E, \mathcal{F})$, and $e \in E(\mathbb{G})$. Show that*

1. *e is a coloop in $D(\mathbb{G})$ if and only if e is a bridge in $\mathbb{G}$; and*

2. *e is a loop in $D(\mathbb{G})$ if and only if e is a trivial orientable loop in $\mathbb{G}$.*

(This result is from Chun et al. [29].)

In fact, each of the four types of loops in ribbon graphs mentioned above can be recognised in their delta-matroids (see [29]). The corresponding four delta-matroid loop types are often used to define cases in induction arguments for delta-matroids, just as loops and coloops do in the matroid case.

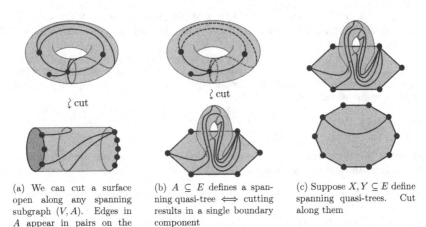

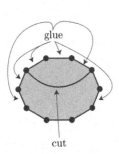

(a) We can cut a surface open along any spanning subgraph (V, A). Edges in A appear in pairs on the boundary, edges not in A are embedded, vertices are on the boundary

(b) $A \subseteq E$ defines a spanning quasi-tree $\Longleftrightarrow$ cutting results in a single boundary component

(c) Suppose $X, Y \subseteq E$ define spanning quasi-trees. Cut along them

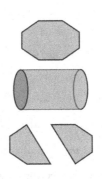

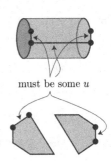

(d) Y can be obtained from X by cutting along edges in $Y \backslash X$ and gluing together edges in $X \backslash Y$ that were previously cut open

(e) Pick $u \in X \triangle Y$. Cutting along $X \triangle u$ gives something with one or two boundary components

(f) If it has two boundary components there must be some $v \in X \triangle Y$ such that cutting or gluing will merge these into a single boundary component, otherwise the property in Figure 16d would fail. Thus $X \triangle \{u, v\}$ is a quasi-tree

Figure 16: A sketch of a proof that the Symmetric Exchange Axiom holds for $D(\mathbb{G})$

5.3   Minors and the interplay with ribbon graphs

From Table 1 it is clear that $\mathbb{G}$ and $\mathbb{G}^e \backslash e (= \mathbb{G}/e)$ have the same numbers of boundary components, as do $\mathbb{G} \backslash e$ and $\mathbb{G}^e$. A consequence of this is that if $\mathbb{H}$ is a spanning quasi-tree of a ribbon graph $\mathbb{G}$, then we can obtain a spanning quasi-tree of its partial dual $\mathbb{G}^A$ by 'toggling' edges in $\mathbb{H}$ that are in A. This sets up a 1-1 correspondence between the spanning quasi-trees of $\mathbb{G}$ and of $\mathbb{G}^A$. Concretely, B is the edge set of a spanning quasi-tree in $\mathbb{G}$ if and only if $B \triangle A$ is the edge set of a spanning quasi-tree in G^A. Phrasing this in terms of delta-matroids gives the following fundamental bridge between delta-matroid theory and ribbon graph theory. The result is from Chun et al. [29].

Theorem 5.18. *Let* $\mathbb{G} = (V, E)$ *be a ribbon graph and* $A \subseteq E$*. Then*

$$D(\mathbb{G}^A) = D(\mathbb{G}) * A.$$

As special case, this theorem completes the classical matroid result stated in Theorem 4.5, that, for *plane* graphs, $C(G^*) = C(G)^*$. Taking $A = E$ in Theorem 5.18 gives that for *any* embedded graph,

$$D(\mathbb{G}^*) = D(\mathbb{G})^*.$$

When $\mathbb{G}$ is plane this identity become the matroid one.

Exercise 5.19. *Using that* $D(\mathbb{G})_{\min} = C(\mathbb{G})$*, deduce from Theorem 5.18 that* $D(\mathbb{G})_{\max} = C(\mathbb{G}^*)^*$.

It was shown in [29] that delta-matroid and ribbon graph deletion and contraction correspond.

Theorem 5.20. *Let* $\mathbb{G}$ *be a ribbon graph, and* $e \in E(\mathbb{G})$*. Then*

$$D(\mathbb{G}\backslash e) = D(\mathbb{G})\backslash e \qquad and \qquad D(\mathbb{G}/e) = D(\mathbb{G})/e.$$

A proof of the deletion result in this theorem can be obtained by considering a ribbon graph locally at an edge e, the three different ways that boundary components can touch this edge, and how the boundary components change under deletion. The contraction result follows from the deletion result, (5.1), and Theorem 5.18.

Theorems 5.18 and 5.20 together give a compatibility between delta-matroid minors and ribbon graph minors:

$$\text{Ribbon graph minors} \quad \xleftrightarrow{\text{compatible}} \quad \text{delta-matroid strong-minors,}$$

$$\begin{array}{c}\text{Ribbon graph minors} \\ \text{and partial duals}\end{array} \quad \xleftrightarrow{\text{compatible}} \quad \text{delta-matroid minors.}$$

This means that we can translate results from one setting to another. Of course, ribbon graphs are not identified with delta-matroids so it may be that translating gives a false or partial result, and, even when the result is true, a new proof may be needed. What is important is that intuition developed in either area can provide intuition in the other.

Exercise 5.21. *Let* $\mathbb{G}_1$ *be the ribbon graph from Figure 17a. Prove that a ribbon graph is orientable if and only if it has no minor equivalent to* $\mathbb{G}_1$*. Formulate a delta-matroid version of this statement, and compare it to the result in Exercise 3.17.*

Let us see how the compatibility between delta-matroids and ribbon graphs can be used in practice. Much of the recent development in ribbon graph theory has been motivated by knot theory. It is a classical and well-known result that alternating knot and link diagrams can be represented by plane graphs. Dasbach et al., in [31], extended this construction to describe *any* (*i.e.*, not only alternating) knot or link diagram as a ribbon graph. Not all ribbon graphs arise from knot and link diagrams. This leads to the problem of characterising the class of ribbon graphs of knots and links. Chmutov showed in [26] that this class consists exactly of ribbon graphs with a plane partial dual (in fact, partial duality was introduced to explain the relationship between the ribbon graphs of knots and links). The following excluded-minor characterisation for this class was given in [52].

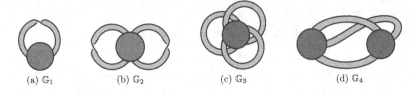

(a) $\mathbb{G}_1$ (b) $\mathbb{G}_2$ (c) $\mathbb{G}_3$ (d) $\mathbb{G}_4$

Figure 17: Ribbon graphs appearing in excluded-minor theorems

Theorem 5.22. *Let* $\mathbb{G}_1$, $\mathbb{G}_3$, *and* $\mathbb{G}_4$ *be the ribbon graphs in Figure 17. Then a ribbon graph* $\mathbb{G}$ *is a partial dual of a plane graph if and only if it has no minor equivalent to* $\mathbb{G}_1$, $\mathbb{G}_3$, *or* $\mathbb{G}_4$.

Let us translate this into delta-matroids. Partial duality corresponds to twisting in delta-matroids (Theorem 5.18), and plane graphs correspond to delta-matroids of width zero (Proposition 5.16(1)), i.e., to matroids. Thus "$\mathbb{G}$ is a partial dual of a plane graph" becomes "D is a twist of a matroid". By Theorem 5.20, ribbon graph minors correspond to strong delta-matroid minors. So "no (ribbon graph) minor equivalent to $\mathbb{G}_1$, $\mathbb{G}_3$, or $\mathbb{G}_4$" becomes "no strong (delta-matroid) minor isomorphic to $D(\mathbb{G}_1)$, $D(\mathbb{G}_3)$, or $D(\mathbb{G}_4)$". Since $\mathbb{G}_3$, or $\mathbb{G}_4$ are partial duals, $D(\mathbb{G}_3)$ and $D(\mathbb{G}_4)$ are twists, so we can rephrase this as "no (delta-matroid) minor isomorphic to $D(\mathbb{G}_1)$ or $D(\mathbb{G}_3)$". Thus we are led to conjecture that "A delta-matroid D is the twist of a matroid if and only if it does not have a minor isomorphic to $D(\mathbb{G}_1)$ or $D(\mathbb{G}_3)$." This turns out to be a result of Duchamp from [33].

Theorem 5.23. *Let* $\mathbb{G}_1$, *and* $\mathbb{G}_3$ *be the ribbon graphs in Figure 17. A delta-matroid* D *is the twist of a matroid if and only if it does not have a minor isomorphic to* $D(\mathbb{G}_1)$ *or* $D(\mathbb{G}_3)$.

Just as in the case of graphs and matroids, sometimes delta-matroid versions of ribbon graph results require an 'extra something', as follows.

Theorem 5.22 was extended to graphs in the real projective plane in [54].

Theorem 5.24. *Let* $\mathbb{G}_2$, $\mathbb{G}_3$, *and* $\mathbb{G}_4$ *be the ribbon graphs in Figure 17. Then a ribbon graph has a partial dual of Euler genus at most one if and only if it has no ribbon graph minor equivalent to* $\mathbb{G}_2$, $\mathbb{G}_3$, *or* $\mathbb{G}_4$.

The direct delta-matroid translation of Theorem 5.24 is "a delta-matroid has a twist of width at most one if and only if it has no minor isomorphic to $D(\mathbb{G}_2)$ or $D(\mathbb{G}_3)$". However, this statement is not true (although it does hold for ribbon-graphic and binary delta-matroids). An additional non-ribbon-graphic delta-matroid needs to be included for the correct result, as was found by Chun et al. in [28].

Theorem 5.25. *Let* $\mathbb{G}_2$ *and* $\mathbb{G}_3$ *be the ribbon graphs in Figure 17. A delta-matroid has a twist of width at most one if and only if it has no minor isomorphic to* $D(\mathbb{G}_2)$ *or* $D(\mathbb{G}_3)$, *or* $(\{1,2,3\},\{\emptyset,\{1\},\{2\},\{3\},\{1,2,3\}\})$.

We have just seen examples of ribbon graph theory informing delta-matroid theory. We now give an example where delta-matroid theory has informed ribbon graph theory.

Proved by Brylawski in [24] and independently by Seymour in [67], the following result says that in a connected matroid M that contains a minor N, it is always possible to delete or contract an element from M to stay connected and keep N as a minor. Results such as this are useful in induction proofs.

Theorem 5.26. *Suppose M is a connected matroid with a connected minor N. If $e \in E(M)\backslash E(N)$, then $M\backslash e$ or M/e is connected with N as a minor.*

Chun, Chun, and Noble, in [27], extended this result to delta-matroids.

Theorem 5.27. *Let D be a connected even delta-matroid with a connected minor D'. If $e \in E(D)\backslash E(D')$, then $D\backslash e$ or D/e is connected with D' as a minor.*

By translating from delta-matroids to ribbon graphs they obtained the following new result about ribbon graphs.

Theorem 5.28. *Let $\mathbb{G}$ be a 2-connected, orientable ribbon graph. If $\mathbb{H}$ is a 2-connected minor of $\mathbb{G}$ and $e \in E(\mathbb{G})\backslash E(\mathbb{H})$, then $\mathbb{G}\backslash e$ or $\mathbb{G}/e$ is 2-connected with $\mathbb{H}$ as a minor.*

Chun, Chun, and Noble were interested in "Splitter Theorems" for delta-matroids in [27]. Their paper includes other, and more impressive, examples of delta-matroid theory informing ribbon graph theory. About one of their ribbon graph results, they wrote: "It is extremely unlikely that we would have established [the result] without the intuition provided by delta-matroids." Describing these results here would require the introduction of a fairly large amount of terminology, so we will settle with the example just seen.

6   Eulerian delta-matroids

In this section we describe a class of delta-matroids arsing from Eulerian circuits, as seen in Section 2.4.4, called *Eulerian* delta-matroids. One of Bouchet's main motivations for introducing delta-matroids was the study of Eulerian circuits through this class.

Our interest here is in the set of Eulerian circuits in a 4-regular graph G. In general, at each vertex there are three ways that an Eulerian circuit can pass though it. Here we want to restrict the set of Eulerian circuits by forbidding, at each vertex, one of these three ways. We then consider the resulting, restricted set of Eulerian circuits.

Let us think how we can record the resulting set of allowed Eulerian circuits. At each vertex there are only two allowed ways an Eulerian circuit may pass through. If we distinguish one of these and call it "preferred" then we can encode each Eulerian circuit by, for each vertex, noting whether or not it follows the preferred route. Thus we can record each allowed Eulerian circuit as a subset U of vertices of G, where we follow the preferred route through a vertex v if and only if $v \in U$. We now formalise this discussion.

Let $G = (V, E)$ be a connected 4-regular graph. Each vertex v of G is incident with exactly four half-edges. (We need to consider half-edges rather than edges as our graphs may have loops.) A *bitransition* at a vertex v is a pairing of its incident half-edges. Each vertex has exactly three bitransitions. A graphical representation of them is given in Figure 18.

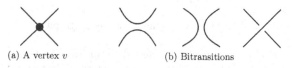

(a) A vertex v (b) Bitransitions

Figure 18: A representation of the bitransitions at v in a 4-regular graph

A *transition system* of the graph G is a choice of bitransition at each of its vertices. Notice that transition systems correspond to circuit coverings of G (by "passing through" each vertex in the way specified by its bitransition). We say that a transition system is *Eulerian* if it corresponds to an Eulerian circuit in G.

Definition 6.1 (Eulerian delta-matroid). Let $G = (V, E)$ be a connected 4-regular graph. At each vertex of G specify one bitransition as *forbidden*, and call the other two *allowed*. Specify one of the two allowed bitransitions at each vertex as being *preferred*. A transition system is *allowed* if it does not contain a forbidden transition. Let T_F denote the transition system consisting of all forbidden bitransitions, and T_P denote the transition system consisting of all preferred bitransitions. Set

$$D(G, T_F, T_P) := (V, \mathcal{F}),$$

where

$$\mathcal{F} = \{U \subseteq V : \text{ there exists an allowed Eulerian transition system of } G$$
$$\text{with preferred bitransition at exactly the vertices of } U\}.$$

A delta-matroid is said to be *Eulerian* if it is isomorphic to $D(G, T_F, T_P)$ for some choice of G, T_F, and T_P.

An example of $D(G, T_F, T_P)$ can be found in Example 2.14, where G is shown in Figure 5 and T_F and T_P are specified by the tables in that figure.

Bouchet [9] proved that $D(G, T_F, T_P)$ is a delta-matroid.

Theorem 6.2. $D(G, T_F, T_P)$, *as constructed in Definition 6.1, is a delta-matroid.*

A direct proof of Theorem 6.2 can be found in [9], where this class of delta-matroids was introduced. Following [29], we see later that Theorem 6.2 follows from a connection between Eulerian and ribbon-graphic delta-matroids.

There are two situations where a set of forbidden bitransitions arises naturally: graphs in surfaces, and directed graphs. Let us start with the case of graphs in surfaces.

Let $G = (V, E)$ be a connected graph embedded in a surface Σ. The *medial graph* G_m of G is the 4-regular graph embedded in Σ obtained by placing a vertex of degree 4 on each edge of G, and then drawing the edges of the medial graph by following the face boundaries of G. See Figure 19.

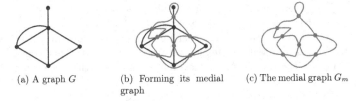

(a) A graph G (b) Forming its medial graph   (c) The medial graph G_m

Figure 19: Constructing a medial graph

We can obtain a set of forbidden and preferred bitransitions for G_m by allowing only bitransitions that pair half-edges that follow a face boundary of G_m through v, and preferring transitions that follow the corresponding edge of G, as illustrated in Figure 20.

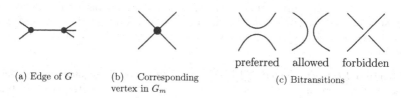

(a) Edge of G (b)   Corresponding   preferred   allowed   forbidden
                       vertex in G_m (c) Bitransitions

Figure 20: Bitransitions in medial graphs

This set of forbidden and preferred bitransitions gives rise to the delta-matroid $D(G_m, T_F, T_P)$. Since the vertices of G_m correspond to the edges of G, this can be regarded as a delta-matroid on the edge set E of G, rather than the vertex set of G_m. Let us denote the resulting delta-matroid on E by $D(G \subset \Sigma)$. This class of delta-matroids was introduced by Bouchet in [13], where they were called the *delta-matroids of maps*.

We can recognise $D(G \subset \Sigma)$ as the delta-matroid of a ribbon graph. For this, let $G = (V, E)$ be a connected graph embedded in a surface Σ, and let $\mathbb{G}$ be its description as a ribbon graph. The edges of G and $\mathbb{G}$ correspond to each other, so the ground sets of $D(G \subset \Sigma)$ and $D(\mathbb{G})$ can be identified. Moreover, by considering Figure 21, it is easy to see that, for each $A \subseteq E$, the boundary components of the ribbon subgraph (V, A) of $\mathbb{G}$ correspond to the allowed circuits in G_m that take the preferred bitransition at the vertices of G_m that are in A. This sets up a 1-1 correspondence between the spanning quasi-trees of $\mathbb{G}$ and the allowed Eulerian transition systems of G_m. It follows that $D(G \subset \Sigma) = D(\mathbb{G})$. We note that Bouchet's results for ribbon-graphic delta-matroids stated in Section 5 were phrased and proved in terms of $D(G \subset \Sigma)$ and transition systems. The connection with ribbon graph theory appeared in [29, 30].

Figure 21: Identifying feasible sets of $D(G \subset \Sigma)$ and $D(\mathbb{G})$

It turns out that *every* Eulerian delta-matroid is the delta-matroid of a ribbon graph. This can be seen by considering a generalisation of partial duality, called twisted duality, that was briefly mentioned in Exercise 5.9. The idea is that every (non-embedded) 4-regular graph F arises as the medial graph of some embedded graph $G \subset \Sigma$, and the twisted duals of $G \subset \Sigma$ give all embedded graphs with a medial graph isomorphic (as a graph) to F. By [34], the forbidden and permitted transitions of the medial graph of one of these must coincide with with the forbidden and permitted transitions of F. A correspondence between Eulerian and ribbon-graphic delta-matroids follows. (The Twisted dual results here are due to Ellis-Monaghan and Moffatt [34]. Further details of the delta-matroid application can be found in [29], and an alternative approach to the result in [13].) This discussion gives the following.

Theorem 6.3. *A delta-matroid D is Eulerian if and only if it is isomorphic to $D(\mathbb{G})$, for some ribbon graph $\mathbb{G}$.*

Theorem 6.3 was proved by Bouchet [13] in the context of delta-matroids of maps $D(G \subset \Sigma)$, the ribbon graph phrasing and approach presented here is from [29].

Taking an apparently different direction for the moment, we consider *Eulerian digraphs*. These are connected digraphs in which the in-degree equals the out-degree at each vertex. We are interested in its *directed Eulerian circuits*, so the circuits must follow the directions of the arcs.

Definition 6.4 (Directed Eulerian delta-matroid). Let $\vec{G}$ be a 4-regular Eulerian digraph. At each vertex there are two bitransitions that agree with the orientation. Take these as the allowed bitransitions, and choose a preferred bitransition at each vertex. Let T_P denote the transition system consisting of all preferred bitransitions. With these choices construct a delta-matroid
$$D(\vec{G}, T_P) := D(G, T_F, T_P).$$

A delta-matroid is said to be *directed Eulerian* if it is isomorphic to $D(\vec{G}, T_P)$ for some $\vec{G}$ and T_P.

From Theorem 6.3 we know every directed Eulerian delta-matroid can be realised as the delta-matroid of a ribbon graph $\mathbb{G}$. However, the directions on the arcs can be used to ensure that we can always construct some such $\mathbb{G}$ that is orientable (see [29] for details). With this we recover the following theorem of Bouchet from [9].

Theorem 6.5. *A delta-matroid D is directed Eulerian if and only if $D = D(\mathbb{G})$, for some orientable ribbon graph $\mathbb{G}$.*

Recalling that $\mathbb{G}$ is orientable if and only if $D(\mathbb{G})$ is even gives the following.

Corollary 6.6. *A delta-matroid is directed Eulerian if and only if it is even and Eulerian.*

As a summary of the identifications of this section,

$$\text{Eulerian delta-matroids} \overset{1\text{-}1}{\longleftrightarrow} \text{delta-matroids of ribbon graphs},$$

$$\text{even Eulerian delta-matroids} \overset{1\text{-}1}{\longleftrightarrow} \text{delta-matroids of orientable ribbon graphs},$$

$$\text{directed Eulerian delta-matroids} \overset{1\text{-}1}{\longleftrightarrow} \text{delta-matroids of orientable ribbon graphs}.$$

Although we have identified ribbon-graphic and Eulerian delta-matroids, it is useful to have both realisations as they provide different insights and applications.

The following natural delta-matroid problems were proposed by Geelen, Iwata, and Murota, in [42]; and Bouchet in [16], respectively. Consider a pair of delta-matroids $D = (E, \mathcal{F})$ and $D' = (E, \mathcal{F}')$ on E. The *partition problem* asks if there is some partition of E into two sets F and F' such that $F \in \mathcal{F}$ and $F' \in \mathcal{F}'$. The *delta-covering problem* is to find feasible sets $F \in \mathcal{F}$ and $F' \in \mathcal{F}'$ maximising $|F \triangle F'|$. The delta-covering problem is clearly a generalisation of the partition problem. These problems originate from the theory of Eulerian circuits. (It is worth noting that the delta-covering problem is a generalisation of the matroid parity problem.)

Let $\vec{G}$ be a 4-regular Eulerian digraph. Two directed Eulerian circuits are *compatible* if they use different bitransitions at each vertex (so the two directed Eulerian circuits 'take different routes' through each vertex). The problem is to determine if $\vec{G}$ admits two compatible directed Eulerian circuits. This is exactly the partition problem when $D = D' = D(\vec{G}, T_P)$, for some T_P.

More generally, we could ask for the construction of compatible directed Eulerian circuits (if they exist), or for the construction of two directed Eulerian circuits with the minimum number of common bitransitions. These are special cases of the delta-covering problem with $D = D' = D(\vec{G}, T_P)$.

Geelen, Iwata, and Murota, in [42], gave an efficient solution to the delta covering problem for a class of delta-matroids known as *linear delta-matroids*. This class includes directed Eulerian delta-matroids (by [16]), and hence gives an efficient algorithm for construction pairs of compatible directed Eulerian circuits in a digraph.

The approach taken in [42] was to reformulate the delta-covering problem as a problem called the *delta-parity problem* (its description is more involved than the delta-covering problem so we omit it here). This problem extends the *parity problem for linearly presented matroids*, an extremely general problem that is known to contain NP-hard problems. Geelen, Iwata, and Murota extended Lovász's Minimax Theorem and efficient solution to the parity problem for linearly presented matroids, [48–50] to solve the delta-matroid problem.

7 Matrices and representability

We revisit the example in Section 2.4. There, given a symmetric or skew-symmetric matrix $\mathbf{A}$ over a field $\Bbbk$, whose rows and columns were labelled (in the same order) by a set E, we formed a delta-matroid $D(\mathbf{A}) := (E, \mathcal{F})$ by taking the labelling set E as the ground set, and, for the collection of feasible sets, we took

$$X \in \mathcal{F} \iff \mathbf{A}[X] \text{ is non-singular.}$$

Recall that $\mathbf{A}[\emptyset]$ is considered to be non-singular, and so $D(\mathbf{A})$ is necessarily normal.

Bouchet proved $D(\mathbf{A})$ is a delta-matroid in [12].

Theorem 7.1. *For every symmetric or skew-symmetric matrix $\mathbf{A}$ over a field $\Bbbk$, the pair $D(\mathbf{A}) := (E, \mathcal{F})$ constructed as above is a normal delta-matroid.*

Remark 7. Bouchet also proved Theorem 7.1 for quasi-symmetric matrices, where $\mathbf{A} = [a_{ij}]$ is *quasi-symmetric* if there is some function $\varepsilon : E \to \{-1, +1\}$ such that $\varepsilon(i)a_{i,j} = \varepsilon(j)a_{j,i}$, for all i, j. (Thus a symmetric matrix is a quasi-symmetric matrix where ε is a constant function.)

The two delta-matroid operations of delete and twist acting on $D(\mathbf{A})$ can be given in terms of operations on $\mathbf{A}$. It is straightforward to see that $D(\mathbf{A})\backslash e$ coincides with the delta-matroid of the matrix obtained from $\mathbf{A}$ by deleting the row and column labelled by e. Thus,

$$D(\mathbf{A})\backslash e = D(\mathbf{A}[E\backslash e]). \tag{7.1}$$

Delta-matroid twisting corresponds to a matrix operation called *pivoting*.

Definition 7.2 (Pivoting for matrices). Let $\mathbf{A}$ be a square matrix over a field $\Bbbk$, whose rows and columns are labelled (in the same order) by a set E. Let $X \subseteq E$. Without loss of generality (reordering if necessary), suppose that X labels the first $|X|$ rows and columns of the matrix. Then $\mathbf{A}$ has a block form

$$\mathbf{A} = \begin{array}{c} \\ X \\ E\backslash X \end{array} \overset{\displaystyle X \qquad E\backslash X}{\left[\begin{array}{c|c} \alpha & \beta \\ \hline \gamma & \delta \end{array} \right]}.$$

Suppose that $\mathbf{A}[X]$ is non-singular. Then the *pivot* of $\mathbf{A}$ with respect to X is the matrix $\mathbf{A} * X$ with block form

$$\mathbf{A} * X = \begin{array}{c} X \\ E \backslash X \end{array} \begin{bmatrix} \begin{array}{c|c} \alpha^{-1} & \alpha^{-1}\beta \\ \hline -\gamma\alpha^{-1} & \delta - \gamma\alpha^{-1}\beta \end{array} \end{bmatrix}.$$

Example 7.3. Working over $\mathrm{GF}(2)$, we have

$$\mathbf{A} = \begin{array}{c} \\ 1 \\ 2 \\ 3 \\ 4 \end{array} \begin{array}{cccc} 1 & 2 & 3 & 4 \\ \begin{bmatrix} 0 & 1 & 1 & 1 \\ 1 & 0 & 1 & 0 \\ \hline 1 & 1 & 1 & 0 \\ 1 & 0 & 0 & 0 \end{bmatrix} \end{array}, \quad \text{and so} \quad \mathbf{A} * \{1,2\} = \begin{array}{c} \\ 1 \\ 2 \\ 3 \\ 4 \end{array} \begin{array}{cccc} 1 & 2 & 3 & 4 \\ \begin{bmatrix} 0 & 1 & 1 & 0 \\ 1 & 0 & 1 & 1 \\ \hline 1 & 1 & 1 & 1 \\ 0 & 1 & 1 & 0 \end{bmatrix} \end{array}.$$

Bouchet, in [12], proved that pivoting in a matrix corresponds to twisting in a delta-matroid.

Theorem 7.4. *Let $\mathbf{A}$ be a symmetric or skew-symmetric matrix over a field $\Bbbk$, whose rows and columns are labelled (in the same order) by a set E. Let $X \subseteq E$, be such that $\mathbf{A}[X]$ is non-singular (or, equivalently, let X be a feasible set of $D(\mathbf{A})$). Then $\mathbf{A} * X$ is a symmetric or skew-symmetric matrix (of the same type as $\mathbf{A}$), and*

$$D(\mathbf{A} * X) = D(\mathbf{A}) * X. \tag{7.2}$$

Using (3.1), we can describe contraction $D(\mathbf{A})/e$ in terms of operations on $\mathbf{A}$ in the case when $\{e\}$ is a feasible set of $D(\mathbf{A})$:

$$D(\mathbf{A})/e = D((\mathbf{A} * e)[E \backslash e]), \quad \text{when } \mathbf{A}[e] \neq [0]. \tag{7.3}$$

While (7.1) gives that for all $X \subseteq E$, $D(\mathbf{A}) \backslash X = D(\mathbf{A}[E \backslash X])$, notice that (7.2) and (7.3) require that X is a feasible set of $D(\mathbf{A})$ (or equivalently, $\mathbf{A}[X]$ is non-singular). Of course, this is not surprising since delta-matroids from matrices are always normal, but the set of normal delta-matroids is not closed under twisting or contracting. What this does mean, however, is that care must be taken when representing delta-matroids by matrices, as we shall see presently.

A normal delta-matroid is *representable* if it can be obtained as the delta-matroid of a matrix. Every delta-matroid is a twist of a normal delta-matroid (just twist by any feasible set), and we say that a delta-matroid is *representable* if one of its twists is the delta-matroid of a matrix.

Definition 7.5 (Representable). Let $D = (E, \mathcal{F})$ be a delta-matroid. We say that D is *representable* over $\Bbbk$, if there exists some $X \subseteq E$ and a symmetric or skew-symmetric matrix $\mathbf{A}$ over a field $\Bbbk$ such that

$$D * X = D(\mathbf{A}).$$

We say that $\mathbf{A}$ is a matrix *representing* D.

Example 7.6. Let $D = (E, \mathcal{F})$ be the delta-matroid with $E = \{1, 2, 3, 4\}$ and

$$\mathcal{F} = \{\{1\}, \{4\}, \{1, 2\}, \{1, 3\}, \{1, 4\}, \{2, 4\}, \{3, 4\}, \{1, 2, 4\}, \{1, 2, 3, 4\}\}.$$

Let

$$\mathbf{A}_1 = \begin{bmatrix} 0 & 1 & 1 & 1 \\ 1 & 0 & 1 & 0 \\ 1 & 1 & 1 & 0 \\ 1 & 0 & 0 & 0 \end{bmatrix}, \quad \text{and} \quad \mathbf{A}_2 = \begin{bmatrix} 0 & 1 & 1 & 0 \\ 1 & 0 & 1 & 1 \\ 1 & 1 & 1 & 1 \\ 0 & 1 & 1 & 0 \end{bmatrix}.$$

Then $D * \{3, 4\} = D(\mathbf{A}_1)$, and $D * \{1, 2, 3, 4\} = D(\mathbf{A}_2)$. Thus $\mathbf{A}_1$ and $\mathbf{A}_2$ are both representing matrices for D.

The definition of representability for delta-matroids requires a choice of a set X to make $D * X$ normal. In general, there are many such sets to choose from (since a necessary and sufficient condition is that X is a feasible set of D), and therefore a delta-matroid D will have many representing matrices. However, it follows readily from the transitivity of twisting and (7.2) that all representing matrices are pivots of one another.

Proposition 7.7. *Working over a fixed field, let $\mathbf{A}_1$ be representing matrix for a delta-matroid D. Then $\mathbf{A}_2$ is a representing matrix for D if and only if $\mathbf{A}_2$ is a pivot of $\mathbf{A}_1$.*

Bouchet and Duchamp proved in [20] that the class of representable delta-matroids is closed under taking minors.

Theorem 7.8. *Let D be a delta-matroid and D' be a minor of it. Then if D is representable by a (skew-)symmetric matrix over $\Bbbk$, so is D'.*

In what will follow, we will mostly focus on representations over the two element field GF(2). Such a representation is called a *binary representation*. Recall that our definition of skew-symmetric matrices requires that the diagonal elements are zero.

Definition 7.9 (Binary). A delta-matroid is *binary* if it is representable over GF(2).

Suppose that we have a delta-matroid $D = (E, \mathcal{F})$ and we know that $D = D(\mathbf{A})$ for some (skew-)symmetric matrix $\mathbf{A}$ over GF(2). Then we know that $\{v\} \in \mathcal{F}$ if and only if $\mathbf{A}[v] = [1]$. This determines the diagonal entries of $\mathbf{A}$. We also know that $\{u, v\} \in \mathcal{F}$ if and only if $\mathbf{A}[\{u, v\}]$ is (skew-)symmetric and non-singular, so, as we know the diagonal entries, the feasible sets of size two determine the off-diagonal entries of $\mathbf{A}$. Specifically, set the (u, v)-entry of $\mathbf{A}$ to be 1 if and only if $\{u\}, \{v\} \in \mathcal{F}$ but $\{u, v\} \notin \mathcal{F}$, or $\{u, v\} \in \mathcal{F}$ but $\{u\}$ and $\{v\}$ are not both in $\mathcal{F}$.

Thus, over GF(2), when $D = D(\mathbf{A})$, its feasible sets of size at most two completely determine the matrix $\mathbf{A}$, and hence they determine D itself. This leads to the following result of Bouchet and Duchamp from [20].

Theorem 7.10. *Let $D = (E, \mathcal{F})$ be a normal set system (i.e., $\emptyset \in \mathcal{F}$). Then there is exactly one binary delta-matroid $D' = (E, \mathcal{F}')$ such that $\mathcal{F}_{\min + k} = \mathcal{F}'_{\min + k}$, for $k = 0, 1, 2$.*

Observe that the construction above gives a way to read off a representing matrix of a binary delta-matroid D: twist by any feasible set X so that $D * X$ is normal. Construct a matrix $\mathbf{A}$ following the above procedure. Then $D * X = D(\mathbf{A})$.

In [20] Bouchet and Duchamp used Theorem 7.10 to show that the minimal non-binary delta-matroids are of width at most four. Equipped with this bound, they obtained the following excluded-minor characterisation of the class of binary delta-matroids.

Theorem 7.11. *A delta-matroid is binary if and only if it has no minor isomorphic to one of the following delta-matroids. The delta-matroids on $\{1,2,3\}$ with collection of feasible sets*

1. $\{\emptyset, \{1,2\}, \{1,3\}, \{2,3\}, \{1,2,3\}\}$,

2. $\{\emptyset, \{1\}, \{2\}, \{3\}, \{1,2\}, \{1,3\}, \{2,3\}\}$,

3. $\{\emptyset, \{2\}, \{3\}, \{1,2\}, \{1,3\}, \{1,2,3\}\}$;

or the delta-matroids on $\{1,2,3,4\}$ with collection of feasible sets

4. $\{\emptyset, \{1,2\}, \{1,3\}, \{1,4\}, \{2,3\}, \{2,4\}, \{3,4\}\}$,

5. $\{\emptyset, \{1,2\}, \{1,4\}, \{2,3\}, \{3,4\}, \{1,2,3,4\}\}$.

A notable application of Theorem 7.11 is the recovery of Tutte's excluded-minor characterisation of binary matroids from [74]. Let D_5 denote the delta-matroid described in Item 5 of the theorem. Then $D_5 * \{1,3\}$ is the only matroid that can be recovered as a twist of any of the delta-matroids in the theorem. The matroid $D_5 * \{1,3\}$ is known as the *uniform matroid* $U_{2,4}$. Thus restricting the theorem to matroids (and technically using Theorem 7.16 to recover Tutte's form), gives Tutte's theorem.

Theorem 7.12. *A matroid is binary if and only if it has no minor isomorphic to $U_{2,4}$.*

It can be checked by either constructing the delta-matroids of all 1-vertex ribbon graphs on at most four edges, or by an appeal to the topology of ribbon graphs (along the lines of Exercise 7.14 below) that none of the delta-matroids in Theorem 7.11 arise from ribbon graphs. It follows that ribbon-graphic delta-matroids are binary, a result of Bouchet from [12] (where it was phrased in terms of Eulerian delta-matroids).

Theorem 7.13. *Every ribbon-graphic delta-matroid is binary.*

Exercise 7.14. *Consider the delta-matroid in Item 2 of Theorem 7.11. By considering what properties the edges of a 1-vertex ribbon graph must have for the feasible sets to form quasi-trees, give an argument that shows that the delta-matroid cannot come from a ribbon graph.*

Knowing that $D(\mathbb{G})$ is binary, it is straightforward to construct a binary representing matrix for it. For this we say that two loops in a ribbon graph that share a vertex are *interlaced* if their ends are met in an alternating order when travelling round the vertex boundary.

Given a ribbon graph $\mathbb{G} = (V, E)$, construct a representing matrix $\mathbf{A}$ as follows. Choose some spanning quasi-tree of $\mathbb{G}$ (for example a maximal spanning forest). Let X be its edge set. Then each component of a the partial dual $\mathbb{G}^X$ has exactly one vertex. Let the (e, e)-entry of $\mathbf{A}$ be 1 if and only if e is non-orientable in $\mathbb{G}^X$. Let both the (e, f)-entry and (f, e)-entry be 1 if e and f are interlaced in $\mathbb{G}^X$, and 0 otherwise. Its easily seen that the feasible sets of size at most 2 in $D(\mathbb{G}^X)$ and $D(\mathbf{A})$ coincide. By Theorems 7.10 and 5.18, it follows that $D(\mathbb{G}) * X = D(\mathbf{A})$.

Example 7.15. Consider Example 5.12 which gives $D(\mathbb{G})$ for the ribbon graph $\mathbb{G}$ in Figure 15a. The set $\{3, 4\}$ is feasible, and $\mathbb{G}^{\{3,4\}}$ is the ribbon graph shown in Figure 11a. The edge 3 is non-orientable. The pairs of interlaced edges are 12, 13, 14, and 23. This gives the matrix

$$\mathbf{A} = \begin{array}{c} \\ 1 \\ 2 \\ 3 \\ 4 \end{array} \begin{array}{c} 1 \quad 2 \quad 3 \quad 4 \\ \begin{bmatrix} 0 & 1 & 1 & 1 \\ 1 & 0 & 1 & 0 \\ 1 & 1 & 1 & 0 \\ 1 & 0 & 0 & 0 \end{bmatrix} \end{array}.$$

$D(\mathbf{A}) * \{3, 4\}$ was computed in Example 7.6, and we see this is exactly $D(\mathbb{G})$ from Example 5.12. Thus $D(\mathbb{G}) = D(\mathbf{A}) * \{3, 4\}$, and so $D(\mathbb{G}) * \{3, 4\} = D(\mathbb{G}^{\{3,4\}}) = D(\mathbf{A})$.

We have previously seen that ribbon graph results can be used to conjecture results about delta-matroids. Sometimes, the analogues of ribbon graph results hold for binary delta-matroids, but not for delta-matroids in general. (An example of this is in [55], where a canonical form for surfaces with boundary was shown to hold on the level of binary delta-matroids, but not in general.) A similar comment holds for the connection between matroids and graphs via cycle matroids.

We close this section with a remark on matroid representability. Every matroid is a delta-matroid, and so Definition 7.5 provides a definition of representability for matroids. A reader familiar with matroid theory might be worried by the fact that this definition of representability is *not* the standard definition of representability from matroid theory. In matroid theory, a matroid M is *representable* over a field $\Bbbk$ if M equals the vector matroid (see Section 2.3.1) of some matrix over $\Bbbk$. Bouchet proved in [12] that the two notions of matroid representability agree.

Theorem 7.16. *A matroid is representable over $\Bbbk$ in the sense of matroid theory if and only if it is representable over $\Bbbk$ in the sense of delta-matroid theory by a skew-symmetric matrix.*

A consequence of this is that since not all matroids are representable in the sense of matroid theory, not all delta-matroids are representable.

8 Simple graphs, pivoting and delta-matroids

This section ties the properties of binary delta-matroids to those of simple graphs and looped simple graphs. It is easy to associate a simple graph with an even binary delta-matroid — consider a representing matrix for a delta-matroid D as being the adjacency matrix of a graph, and associate this graph with the delta-matroid. This construction, however, depends upon a choice of representing matrix for D, and different choices can result in different graphs. We need to understand how the resulting graphs are related. For this we need to consider pivots and related graph operations.

We will see that even binary delta-matroids considered up to twists can be identified with simple graphs considered up to edge pivots. Similarly, binary delta-matroids considered up to twists can be identified with looped simple graphs considered up to elementary pivots. This was first written down by Geelen in [41] (see also [40]) although he notes that the graph-theoretical point-of-view was used by both Bouchet and Cunningham in their discussions with him at the time of that paper.

8.1   Simple graphs, pivots and adjacency matrices

Pivoting is a graph operation related to Kotzig's transformations on Eulerian circuits [45]. It was introduced by Bouchet in the context of isotropic systems [11] and multimatroids [18], and rediscovered by Arratia, Bollobás, and Sorkin when they introduced the interlace polynomial in [3, 4].

Definition 8.1 (Pivoting for graphs). Let G be a simple graph, and uv be an edge. Partition the vertices other than u and v into four classes: (1) vertices adjacent to u but not v, (2) vertices adjacent to v but not u, (3) vertices adjacent to both u and v, (4) vertices adjacent to neither u nor v.

The *pivot* of the edge uv is the graph, $G \wedge uv$, constructed from G as follows. For any vertex pair x, y where x is in one of the classes (1)–(3), and y is in a different class (1)–(3), "toggle" the pair xy in the edge set (so if xy was an edge, make it a non-edge; and if xy was a non-edge, make it an edge). Finally, switch the names of the vertices u and v. See Figure 22.

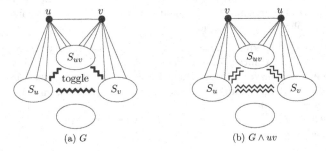

(a) G (b) $G \wedge uv$

Figure 22: Pivoting (edges between the three sets, S_u, S_v, and $S_{u,v}$, are 'toggled', and the names of u and v are switched)

Definition 8.2 (Pivot-minors for graphs). A *pivot-minor* of a graph is any graph that can be obtained from it by edge pivots and vertex deletions.

A related operation is local complementation, first studied by Kotzig in [45]. We use $N_G(v)$ to denote the set of neighbours of a vertex v in the graph G. Note that $v \notin N_G(v)$.

Definition 8.3 (Local complementation). Let G be simple graph. Then the $G * v$ denotes the graph obtained from v by *local complementation* at v. The graph $G * v$ is obtained from G by replacing the induced subgraph on $N_G(v)$ with its complement graph. That is, $G * v$ is obtained from G by 'toggling' the edges and non-edges at vertices in $N_G(v)$.

Definition 8.4 (vertex-minor). A *vertex-minor* of a graph is any graph that can be obtained from it by local complementations and vertex deletions.

Exercise 8.5. *Let uv be an edge of a simple graph G. Verify that, after switching the names of vertices u and v, $G \wedge uv = G * u * v * u = G * v * u * v$. (This is due to Bouchet [11].)*

Recall the adjacency matrix over $\mathrm{GF}(2)$ of a simple graph G is the matrix $\mathbf{A}_G$ whose rows and columns correspond to the vertices of G, and whose (u, v)-entry is 1 if the corresponding vertices u and v are adjacent in G, and is 0 otherwise. The diagonal entries are 0.

Definition 8.6 (Fundamental graph). Let $\mathbf{A}$ be a skew-symmetric matrix over GF(2). A simple graph G is said to be the *fundamental graph* of $\mathbf{A}$ if $\mathbf{A}$ is its adjacency matrix. The graph G is said to be a *fundamental graph* of an even binary delta-matroid D if it is the fundamental graph of some representing matrix of D.

Over GF(2), every simple graph is the fundamental graph of some skew-symmetric matrix, and every skew-symmetric matrix has a fundamental graph, giving a correspondence:

$$\{\text{skew-symmetric matrices over GF(2)}\} \xleftrightarrow{\text{1-1}} \{\text{simple graphs}\}.$$

It is not hard to determine the effect that deleting a vertex and pivoting an edge in a simple graph G has on its adjacency matrix $\mathbf{A}_G$. For a vertex v,

$$\mathbf{A}_{G \setminus v} = \mathbf{A}_G[V \setminus v]. \tag{8.1}$$

Recall that $\mathbf{A}_G[V \setminus v]$ is the matrix obtained from $\mathbf{A}_G$ by deleting the row and column corresponding to v.

For an edge uv of G,

$$\mathbf{A}_G * \{u, v\} = \mathbf{A}_{G \wedge uv}, \tag{8.2}$$

where $\mathbf{A}_G * \{u, v\}$ denotes the pivot of a matrix from Definition 7.2.

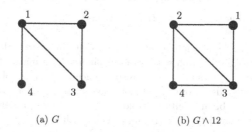

(a) G (b) $G \wedge 12$

Figure 23: A simple graph G and its pivot $G \wedge 12$

Example 8.7. Consider the graph G in Figure 23a. Its pivot $G \wedge 12$ is shown in Figure 23b. The adjacency matrices of these graphs are given below, where it can be seen that $\mathbf{A}_{G \wedge 12} = \mathbf{A}_G * \{1, 2\}$.

$$\mathbf{A}_G = \begin{array}{c} \\ 1 \\ 2 \\ 3 \\ 4 \end{array}\begin{array}{c} \begin{array}{cccc} 1 & 2 & 3 & 4 \end{array} \\ \begin{bmatrix} 0 & 1 & 1 & 1 \\ 1 & 0 & 1 & 0 \\ 1 & 1 & 0 & 0 \\ 1 & 0 & 0 & 0 \end{bmatrix} \end{array}, \quad \text{and} \quad \mathbf{A}_{G \wedge 12} = \begin{array}{c} \\ 1 \\ 2 \\ 3 \\ 4 \end{array}\begin{array}{c} \begin{array}{cccc} 1 & 2 & 3 & 4 \end{array} \\ \begin{bmatrix} 0 & 1 & 1 & 0 \\ 1 & 0 & 1 & 1 \\ 1 & 1 & 0 & 1 \\ 0 & 1 & 1 & 0 \end{bmatrix} \end{array}.$$

Exercise 8.8. *Verify Equations (8.1) and (8.2).*

8.2   Simple graphs and even binary delta-matroid

Let $G = (V, E)$ be a simple graph and $\mathbf{A}_G$ be its adjacency matrix over GF(2). We can form a delta-matroid $D(\mathbf{A}_G)$ from the adjacency matrix. Thus we have a way to associate a delta-matroid $D(\mathbf{A}_G)$ with any simple graph G. Furthermore, since G is simple, $\mathbf{A}_G$ is has zeros on its diagonal, and it follows that every feasible set of $D(\mathbf{A}_G)$ is of even size. Thus $D(\mathbf{A}_G)$ is an even binary delta-matroid.

On the other hand, suppose that $D = (V, \mathcal{F})$ is an even binary delta-matroid. Then we know, for some symmetric or skew-symmetric matrix $\mathbf{A}$ over GF(2) and some $X \subseteq V$, that $D * X = D(\mathbf{A})$. Since $D(\mathbf{A})$ is even and normal, $\mathbf{A}$ must have zeros on its diagonals, and so is the adjacency matrix of some simple graph $G_{D,X}$. This graph is the fundamental graph of the matrix $\mathbf{A}$, and is a fundamental graph of D.

It is straightforward to construct a fundamental graph $G_{D,X}$ of an even binary delta-matroid $D = (V, \mathcal{F})$ directly. Choose a feasible set $X \in \mathcal{F}$. Then $D * X$ is normal. Take V to be the vertex set of $G_{D,X}$, and add an edge uv if and only if $\{u, v\}$ is a feasible set of $D * X$ (or equivalently, if and only if $\{u, v\} \bigtriangleup X$ is a feasible set of D).

Example 8.9. Two simple graphs that are pivots of each other are shown in Figure 23, and their adjacency matrices given in Example 8.7. $D(\mathbf{A}_G) = (E, \mathcal{F})$ and $D(\mathbf{A}_{G \wedge 12}) = (E, \mathcal{F}')$ where $E = \{1, 2, 3, 4\}$,

$$\mathcal{F} = \{\emptyset, \{1, 2\}, \{1, 3\}, \{1, 4\}, \{2, 3\}, \{1, 2, 3, 4\}\},$$

and

$$\mathcal{F}' = \{\emptyset, \{1, 2\}, \{1, 3\}, \{2, 3\}, \{2, 4\}, \{3, 4\}\}.$$

It is readily checked that $D(\mathbf{A}_G) * \{1, 2\} = D(\mathbf{A}_{G \wedge 12})$.

Thus we have a way to associate an even binary delta-matroid with a simple graph, and a way to associate a simple graph with an even binary delta-matroid. By making use of Theorem 7.10, we see this sets up a 1-1 correspondence between simple graphs and *normal* even binary delta-matroids. However, it does not set up a 1-1 correspondence between simple graphs and even binary delta-matroids since, in general, a delta-matroid will have many different fundamental graphs. (Because there is the choice of which $X \subseteq V$ we use to make $D * X$ normal). However, all of the fundamental graphs of the delta-matroid are related through pivots.

Theorem 8.10. *Let D be an even binary delta-matroid, and G be a fundamental graph of D. Then a graph H is also a fundamental graph of D if and only if H can be obtained from G by a sequence of edge pivots.*

Before giving a proof of this lemma, let us say a few words about it. Combining (7.2) and (8.2) gives that, when G is simple with an edge uv, and $\mathbf{A}_G$ is its adjacency matrix over GF(2), then

$$D(\mathbf{A}_{G \wedge uv}) = D(\mathbf{A}_G * \{u, v\}) = D(\mathbf{A}_G) * \{u, v\}. \tag{8.3}$$

As G and H are fundamental graphs of the same delta-matroid, we know that $D(\mathbf{A}_H) = D(\mathbf{A}_G) * X$, for some set X. Thus, in light of (8.3), we need to write $D(\mathbf{A}_G) * X$ as $D(\mathbf{A}_G) * \{u_1, v_1\} * \cdots * \{u_k, v_k\}$, where each pair $\{u_k, v_k\}$ is feasible in the relevant delta-matroid, so that the pivots of the fundamental graph are pivots on an edge.

Proof. First suppose H can be obtained from G by a series of edge pivots, $H = G \wedge u_1v_1 \wedge \cdots \wedge u_kv_k$. Then, by (8.3), $D(\mathbf{A}_H) = D(\mathbf{A}_G) * \{u_1, v_1\} * \cdots * \{u_k, v_k\} = D(\mathbf{A}_G) * \{u_1, v_1, \cdots, u_k, v_k\}$. Thus the matrix $\mathbf{A}_H$ also represents D, so H is a fundamental graph of D.

Conversely, suppose that H is a fundamental graph of D. Then $D(\mathbf{A}_G)$ and $D(\mathbf{A}_H)$ are both twists of D, and so $D(\mathbf{A}_G) * X = D(\mathbf{A}_H)$, for some set X. Moreover, since these two delta-matroids are normal, X must be a feasible set of $D(\mathbf{A}_G)$, and be of even size (since D is an even delta-matroid). If $X = \emptyset$ we are done, so suppose this is not the case. Applying the Symmetric Exchange Axiom to $X \triangle \emptyset$ gives that there is some $\{u_1, v_1\} \subseteq X$ such that $\{u_1, v_1\}$ is a feasible set of $D(\mathbf{A}_G)$. It follows that $D(\mathbf{A}_G) * X = (D(\mathbf{A}_G) * \{u_1, v_1\}) * (X \backslash \{u_1, v_1\})$, so $D(\mathbf{A}_G) * \{u_1, v_1\}$ is normal with $\{u_1, v_1\}$ feasible in $D(\mathbf{A}_G)$, $X \backslash \{u_1, v_1\}$ feasible in $D(\mathbf{A}_G) * \{u_1, v_1\}$, and $(D(\mathbf{A}_G) * \{u_1, v_1\}) * (X \backslash \{u_1, v_1\}) = D(\mathbf{A}_H)$. We can repeat this argument to write $D(\mathbf{A}_H) = D(\mathbf{A}_G) * \{u_1, v_1\} * \cdots * \{u_k, v_k\}$, and, so by (8.3), $D(\mathbf{A}_H) = G \wedge u_1v_1 \wedge \cdots \wedge u_kv_k$. As each $\{u_i, v_i\}$ is feasible in the relevant delta-matroid, this is a sequence of edge pivots, as required. $\qquad \square$

Lemma 8.10 identifies even binary delta-matroids with equivalence classes of simple graphs under pivoting:

$$\{\text{even binary delta-matroids up to twists}\} \overset{1\text{-}1}{\longleftrightarrow} \{\text{simple graphs up to edge pivots}\}. \quad (8.4)$$

Theorem 8.11. *Let D be an even binary delta-matroid, and G be a fundamental graph of D. Then a graph H is a pivot-minor of G if and only if it is a fundamental graph of a minor of D.*

Proof. A pivot-minor of G is obtained by pivoting at edges and deleting vertices. By Theorem 8.10, for an edge uv of G, $G \wedge uv$ is also a fundamental graph of D. For a vertex v of G, the adjacency matrix $\mathbf{A}_{G \backslash v}$ of $G \backslash v$ equals $\mathbf{A}_G[V \backslash v]$, and by (7.1), $D(\mathbf{A}_{G \backslash v}) = D(\mathbf{A}_G[V \backslash v]) = D(\mathbf{A}_G) \backslash v$. Thus if G is a fundamental graph of D, then $D * X = D(\mathbf{A}_G)$, for some $X \subseteq V$, and so $G \backslash v$ is a fundamental graph of $(D * X) \backslash v$, which is a minor of D. It follows that if H is a pivot-minor of G, then it is a fundamental graph of a minor of D.

Conversely, since any two fundamental graphs are related by pivots, it is enough to show that there are fundamental graphs of $D * v$ and $D \backslash v$ that are pivot-minors of a fundamental graph of D. For $D * v$, let X be a feasible set of D. Then $D * X$ is normal, and equals $(D * v) * (X \triangle v)$. Reading fundamental graphs from these gives that D and $D * v$ have a common fundamental graph, and hence, by Theorem 8.10, all their fundamental graphs are pivot-minors. For $D \backslash v$, suppose that there is some feasible set X of D that does not contain v (i.e., v is not a coloop). Then $D * X$ and $(D * X) \backslash v$ and are both normal. If G is the fundamental graph read from $D * X$, then $G \backslash v$ is the fundamental graph read from $(D * X) \backslash v$. But $(D * X) \backslash v = (D \backslash v) * X$, since $v \notin X$, so $G \backslash v$ is a fundamental graph of $D \backslash v$. On the other hand, if v is in every feasible set of D (i.e., v is a coloop), and X is a feasible set of D, then $D \backslash v$ and $D * v$ have identical feasible sets. Thus $D * X = (D * v) * (X \backslash v)$ and $(D \backslash v) * (X \backslash v)$ are normal delta-matroids with identical feasible sets and differ only in that v is in the ground set of one but not the other. Reading fundamental graphs from these delta-matroids gives that a fundamental graph of $D \backslash v$ can be obtained by deleting v from a fundamental graph of D. This completes the proof. $\qquad \square$

Theorems 8.10 and 8.11 gives that binary delta-matroids and their minors correspond to simple graphs and their pivot-minors:

$$\{\text{minors of even binary delta-matroids}\} \leftrightarrow \{\text{pivot-minors of simple graphs}\}. \quad (8.5)$$

Thus results about pivot-minors can be translated into results about delta-matroids.

8.3   Looped simple graphs and binary delta-matroids

Equation (8.4) identified even binary delta-matroids and simple graphs. What if the delta-matroid is not even? To answer this we need to consider looped simple graphs.

Recall that a *looped simple graph* is a graph obtained from a simple graph by adding a loop to some of its vertices. Each vertex has either exactly one loop or no loops.

The following definition provides versions of local complementation and pivots for looped simple graphs. For the definition it is convenient to think of a looped simple graph G as a graft. A *graft* is a pair, (H, T), consisting of a graph H together with a subset T of its vertices. (Grafts will be the topic of Section 10.) A looped simple graph G is then exactly a graft (G_s, T) where G_s is the simple graph obtained from G by deleting all of its loops, and T is the set of vertices of G with loops.

Definition 8.12 (Elementary pivots). Let G be a looped simple graph. Consider G as a graft (G_s, T). Then *local complementation at the looped vertex v* is defined as the operation

$$(G_s, T) \mapsto (G_s * v, T \triangle N_G(v)), \quad \text{where } v \in T.$$

(So form the local complement of the underlying simple graph, then 'toggle' the loops and non-loops on the neighbours of v.) We use $G * v$ to denote the looped simple graph resulting from local complementation of G at v,

Pivoting an edge between non-looped vertices is defined as the operation

$$(G_s, T) \mapsto (G_s \wedge uv, T), \quad \text{where } uv \in E(G_s), u, v \notin T \text{ and } u \neq v.$$

(So form the edge pivot on the underlying simple graph. Do not change the loops.) We use $G \wedge uv$ to denote the looped simple graph resulting from pivoting uv in G.

These two operations on looped simple graphs are collectively called *elementary pivots*.

It is worth emphasising that elementary pivots only act on looped vertices, and on edges incident to two loopless vertices.

Recall from Definition 2.10 that the *adjacency matrix* $\mathbf{A}_G$ of a looped simple graph G is the matrix over GF(2) whose (u, v)-entry is 1 if and only if uv is an edge of G. In particular it has diagonal entry 1 if and only if the corresponding vertex has a loop. Every symmetric matrix over GF(2) can be written as $\mathbf{A}_G$ for some looped simple graph G, giving

$$\{\text{symmetric matrices over GF(2)}\} \xleftrightarrow{\text{1-1}} \{\text{looped simple graphs}\}.$$

Versions of (8.1) and (8.2) hold for looped simple graphs. For a vertex v,

$$\mathbf{A}_G[V \setminus v] = \mathbf{A}_{G \setminus v}, \tag{8.6}$$

and if v has a loop,

$$\mathbf{A}_G * \{v\} = \mathbf{A}_{G * v}. \tag{8.7}$$

For an edge uv of G between two loopless vertices,

$$\mathbf{A}_G * \{u, v\} = \mathbf{A}_{G \wedge uv}. \tag{8.8}$$

Exercise 8.13. *Verify Equations (8.6)–(8.8).*

Passing to delta-matroids, and using (7.1), (7.2), and (8.6)–(8.8) gives:

$$D(\mathbf{A}_G)\backslash v = D(\mathbf{A}_{G\backslash v}); \tag{8.9}$$

when $\{v\}$ is feasible in $D(\mathbf{A}_G)$,

$$D(\mathbf{A}_G) * \{v\} = D(\mathbf{A}_{G*v}); \tag{8.10}$$

and when $\{u, v\}$, but not $\{u\}$ nor $\{v\}$, is feasible in $D(\mathbf{A}_G)$,

$$D(\mathbf{A}_G) * \{u, v\} = D(\mathbf{A}_{G\wedge uv}). \tag{8.11}$$

Example 8.14. Consider the looped simple graph G in Figure 24a. Its pivot $G \wedge 12$ is shown in Figure 24b. The adjacency matrices of these graphs are

$$\mathbf{A}_G = \begin{array}{c} \\ 1 \\ 2 \\ 3 \\ 4 \end{array}\begin{array}{cccc} 1 & 2 & 3 & 4 \\ \left[\begin{array}{cccc} 0 & 1 & 1 & 1 \\ 1 & 0 & 1 & 0 \\ 1 & 1 & 1 & 0 \\ 1 & 0 & 0 & 0 \end{array}\right] \end{array}, \quad \text{and so} \quad \mathbf{A}_{G\wedge 12} = \begin{array}{c} \\ 1 \\ 2 \\ 3 \\ 4 \end{array}\begin{array}{cccc} 1 & 2 & 3 & 4 \\ \left[\begin{array}{cccc} 0 & 1 & 1 & 0 \\ 1 & 0 & 1 & 1 \\ 1 & 1 & 1 & 1 \\ 0 & 1 & 1 & 0 \end{array}\right] \end{array}.$$

It was verified in Example 7.3 that $\mathbf{A}_{G\wedge 12} = \mathbf{A}_G * \{1, 2\}$. We have $D(\mathbf{A}_G) = (E, \mathcal{F})$ and $D(\mathbf{A}_{G\wedge 12}) = (E, \mathcal{F}')$ where $E = \{1, 2, 3, 4\}$,

$$\mathcal{F} = \{\emptyset, \{3\}, \{1, 2\}, \{1, 3\}, \{1, 4\}, \{2, 3\}, \{1, 2, 3\}, \{1, 3, 4\}, \{1, 2, 3, 4\}\},$$

and

$$\mathcal{F}' = \{\emptyset, \{3\}, \{1, 2\}, \{1, 3\}, \{2, 3\}, \{2, 4\}, \{3, 4\}, \{1, 2, 3\}, \{2, 3, 4\}\},$$

from which we can verify that $D(\mathbf{A}_G) * \{1, 2\} = D(\mathbf{A}_{G\wedge 12})$.

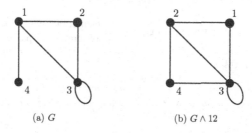

(a) G (b) $G \wedge 12$

Figure 24: A simple graph G and its pivot $G \wedge 12$

Lemma 8.15. *Let D and D' be normal binary delta-matroids on V. Then $D' = D * X$, for some $X \subseteq V$, if and only if there are looped simple graphs G and G' such that $D = D(\mathbf{A}_G)$, $D' = D(\mathbf{A}_{G'})$ and G' can be obtained from G by a sequence of elementary pivots.*

Proof. One direction follows from (8.10) and (8.11). For the other direction, suppose D and D' are normal binary delta-matroids and $D' = D * X$. Since D' is normal, $\emptyset, X \in \mathcal{F}(D)$. In any delta-matroid D'', if $\emptyset, Y \in \mathcal{F}(D'')$, the symmetric exchange axiom gives that, for, each $y \in Y$, either $\{y\} \in \mathcal{F}(D'')$ or $\{y, y'\} \in \mathcal{F}(D'')$, for some $y' \in Y$ (where $\{y'\}$ may or may not be in $\mathcal{F}(D'')$). This means that for any nonempty feasible set Y we can find some $\{y\} \subseteq Y$ that is feasible, or some $\{y, y'\} \subseteq Y$ that is feasible but where neither $\{y\}$ nor $\{y'\}$ is. Thus, since D is binary, for some looped simple graph G we can write

$$D' = D * X = D(\mathbf{A}_G) * X = D(\mathbf{A}_G) * \{x_1\} * \cdots * \{x_i\} * \{x_{i+1}, x'_{i+1}\} * \cdots * \{x_k, x'_k\},$$

where each $\{x_j\}$ is feasible in the relevant delta-matroid, or $\{x_j, x'_j\}$ is but neither $\{x_j\}$ nor $\{x'_j\}$ are. It follows from (8.10) and (8.11) that $D' = D(\mathbf{A}_{G'})$ where G' is obtained from G by a sequence of elementary pivots. $\quad\square$

Theorem 8.16. *Let $D = (E, \mathcal{F})$ be a binary delta-matroid, and G and H be looped simple graphs. Then $D * X = D(\mathbf{A}_G)$ and $D * Y = D(\mathbf{A}_H)$, for some $X, Y \subseteq E$, if and only if H can be obtained from G by a sequence of elementary pivots.*

Proof. If $D * X = D(\mathbf{A}_G)$ and $D * Y = D(\mathbf{A}_H)$, by Lemma 8.15, G and H are related by elementary pivots.

Conversely, if H can be obtained from G by elementary pivots, then, by Lemma 8.15, $D(\mathbf{A}_G) * Y = D(\mathbf{A}_H)$, for some Y. It follows that for some X, $D * (X \triangle Y) = D(\mathbf{A}_H)$. $\quad\square$

Thus we have shown a correspondence between binary delta-matroids and looped simple graphs:

$$\left\{ \begin{array}{c} \text{binary delta-matroids} \\ \text{up to twists} \end{array} \right\} \overset{\text{1-1}}{\longleftrightarrow} \left\{ \begin{array}{c} \text{looped simple graphs} \\ \text{up to elementary pivots} \end{array} \right\}.$$

Note that the correspondence in (8.4) between simple graphs and even binary delta-matroids can be deduced from this since G has a loop if and only if $D(\mathbf{A}_G)$ is an odd delta-matroid.

We say that a looped simple graph H is an *elementary pivot-minor* of G if it can be obtained from G through a sequence of elementary pivots and vertex deletions. By adapting the proof of Theorem 8.11, it can be shown that minors of binary delta-matroids correspond to elementary pivot-minors of looped simple graphs:

$$\{\text{minors of binary delta-matroids}\} \longleftrightarrow \left\{ \begin{array}{c} \text{elementary pivot-minors} \\ \text{of looped simple graphs} \end{array} \right\}. \tag{8.12}$$

Stated as a theorem, this is:

Theorem 8.17. *Let D and D' be a binary delta-matroids on E such that $D * X = D(\mathbf{A}_G)$ and $D' * Y = D(\mathbf{A}_H)$, for some $X, Y \subseteq E$. Then a graph H is an elementary pivot-minor of G if and only if D' is a minor of D.*

9 Circle graphs, and ribbon-graphic and Eulerian delta-matroids

We have just seen a connection between simple graphs and binary delta-matroids. The case when the graph is a circle graph turns out to be of particular interest in delta-matroid theory as it is related to ribbon-graphic delta-matroids. From Section 6 we know that Eulerian delta-matroids are the delta-matroids of ribbon graphs meaning that we can phrase the ribbon-graphic delta-matroid results in this section in terms of Eulerian delta-matroids.

A *chord diagram* consists of a circle in the plane and a number line segments, called *chords*, whose end-points lie on the circle. The end-points of chords should all be distinct.

The *intersection graph* of a chord diagram is the graph $G = (V, E)$ where V is the set of chords, and where $uv \in E$ if and only if the chords u and v intersect. A graph is a *circle graph* if it is the intersection graph of a chord diagram. A *looped circle graph* is a looped graph obtained by adding loops to a circle graph. Figure 25 shows a circle graph and a corresponding chord diagram.

Figure 25: A circle graph and a corresponding chord diagram

Circle graphs are closed under vertex deletion, local-complementation, and edge-pivots. Thus they are closed under taking vertex-minors and pivot-minors. The class of circle graphs has excluded-minor characterisations with respect to both types of minor. Bouchet, in [15], gave the following excluded-vertex-minor characterisation of circle graphs.

Theorem 9.1. *A graph is a circle graph if and only if it has no vertex-minor isomorphic to any of the graphs shown in Figure 26.*

Figure 26: Excluded vertex-minors for circle graphs

Building upon Bouchet's characterisation, Geelen and Oum, in [40], gave an excluded-pivot-minor characterisation of circle graphs.

Theorem 9.2. *A graph is a circle graph if and only if it has no pivot-minor isomorphic to any of the graphs shown in Figure 27.*

As pivot-minors of simple graphs correspond to minors of even binary delta-matroids, by (8.5), it is reasonable to expect this theorem to find an application to delta-matroids. This is what we find next.

There is a natural way to associate a chord diagram with an orientable 1-vertex ribbon graph $\mathbb{G}$: take the boundary of the vertex as the circle, and place a chord between the two ends of each edge of $\mathbb{G}$. By forming the intersection graph of this chord diagram, we have a natural way to associate a circle graph with a ribbon graph. Moreover, this circle graph is a fundamental graph of $D(\mathbb{G})$, and so by (8.5) we obtain an excluded-minor characterisation of even ribbon-graphic delta-matroids from [40].

Theorem 9.3. *A delta-matroid is even ribbon-graphic if and only if has no minor isomorphic to $D(\mathbf{A}_G)$ where G is one of the graphs shown in Figure 27, or to one of the excluded minors for binary delta-matroids given in Theorem 7.11.*

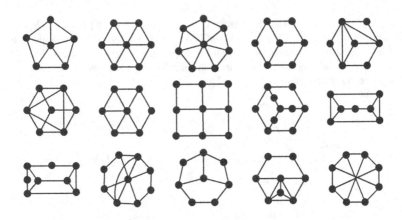

Figure 27: Excluded pivot-minors for circle graphs

More generally, by relating the pivot minors of the graph G to circle graphs, in the case where $D = D(\mathbf{A}_G)$ is ribbon-graphic, Geelen and Oum, in [40], were able to find a set of 171 excluded minors for the class of ribbon-graphic delta-matroids.

Theorem 9.4. *A delta-matroid is ribbon-graphic if and only if has no minor isomorphic to $D(\mathbf{A}_G)$ where G is one of the looped simple graphs shown in Figure 28, or to one of the excluded minors for binary delta-matroids given in Theorem 7.11.*

10   Grafts and graphic delta-matroids

A *graft* is a pair (G, T) consisting of a graph G together with a subset T of its vertices. Vertices in T are called T-*vertices*. A graft is shown in Figure 6a. Grafts, introduced by Seymour in [68], are useful in matroid theory. For example, they can be used to give a characterisation of graphic matroids [69]. We do not pursue this classical matroid direction here. Instead we consider a method due to Oum [59] for obtaining a delta-matroid (that need not be a matroid) from a graft, and consider the interaction between grafts, their delta-matroids, and rank-width.

Delta-matroids that arise from grafts are called *graphic* delta-matroids. While circle graphs and bipartite graphs are closed under pivot-minors, line graphs are not. Via their fundamental graphs, the closure of circle and bipartite graphs under pivot-minors corresponds to the closure of even Eulerian delta-matroids, and twists of matroids (see Corollary 11.7) under taking minors. That line graphs are not closed under pivots means that the class of delta-matroids whose fundamental graphs are line graphs is not minor-closed. Graphic delta-matroids were introduced to get around this. They are defined in such a way that pivot-minors of line graphs are exactly fundamental graphs of graphic delta-matroids.

Except where otherwise stated, all of the results in this section are due to Oum and from [59].

Definition 10.1 (T-spanning subgraph). Let (G, T) be a graft. A subgraph H of G is said to be T-*spanning* if $V(H) = V(G)$, and each component of the graft (H, T) has either:

1. an odd number of T-vertices, or

Figure 28: Non-binary excluded minors. Looped vertices are shown in squares. (Image from [40].)

2. spans a component of G that has no T-vertex.

Definition 10.2 (Graphic delta-matroid). Let (G,T) be a graft. Let $D(G,T)$ denote the set system $(E(G), \mathcal{F})$, where, for each $A \subseteq E(G)$

$$A \in \mathcal{F} \iff (V, A) \text{ a } T\text{-spanning forest of } G.$$

We call $D(G,T)$ the *delta-matroid of the graft* (G,T), and a delta-matroid D is said to be *graphic* if there exists a graft (G,T) such that D is a twist of $D(G,T)$.

Oum showed that $D(G,T)$ is indeed a delta-matroid.

Theorem 10.3. *Let (G,T) be a graft. Then the set system $D(G,T)$ defined in Definition 10.2 is a delta-matroid.*

Example 10.4. Example 2.15 shows the delta-matroid of a graft. The delta-matroid on $\{1,2,3,4,5\}$ with feasible sets

$$\mathcal{F} = \{\emptyset, \{3,4\}, \{1,2\}, \{1,2,3,4\}, \{1,4,\}, \{2,4\}\}$$

is graphic since it is $D * \{3,5\}$ where D is the delta-matroid from Example 2.15.

Exercise 10.5. *Consider a connected graph (G,T). Show that if $|T| \leq 1$ then $D(G,T)$ is the cycle matroid of G. Show that if $|T| = 2$ and G' is the graph obtained by identifying the two T-vertices of G, then $D(G,T) = D(G', \emptyset)$. Conclude that if $|T| \leq 2$ then $D(G,T)$ is a graphic matroid. (These results are from [58, 59].)*

Let (G,T) be a graft, e be one of its edges and v one of its vertices. Edge and vertex *deletion* for grafts is defined in the obvious way: $(G,T) \backslash e$ is defined to be the graft $(G \backslash e, T)$, and $(G,T) \backslash v$ is defined to be the graft $(G \backslash v, T \backslash v)$. *Contraction* is defined by setting $(G,T)/e$ as the graft $(G/e, T')$ where, if w is the vertex created by contracting an edge $e = uv$,

$$T' := \begin{cases} (T \backslash \{u,v\}) \cup \{w\} & \text{if exactly one of } u \text{ or } v \text{ is in } T, \\ T \backslash \{u,v\} & \text{otherwise.} \end{cases}$$

See Figure 29.

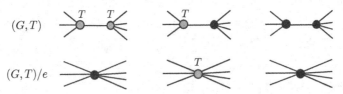

Figure 29: Contracting an edge in a graft

Exercise 10.6. *Let (G,T) be a graft, with a vertex u that is not a T-vertex. Let (G',T') be the graft obtained from (G,T) by adding a vertex v and an edge $e = uv$ to G, then making both u and v into T-vertices. Prove that $D(G,T) = D(G',T')/e$. Deduce that, for any graft (G,T), $D(G,T)$ can be obtained as a minor of $D(G', V(G'))$ for some graft $(G', V(G'))$. (Note this can also be deduced from Theorem 10.8 below.)*

While graphs have bridges and loops, grafts have T-bridges, T-tunnels, and loops. An edge e of a graft (G, T) is a T-*bridge* if $(G, T) \backslash e$ has more components without T-vertices than (G, T). See Figure 30a. An edge $e = uv$ is said to be a T-*tunnel* if u and v are the only T-vertices in the component of G containing them. See Figure 30b.

(a) A T-bridge, e (b) A T-tunnel, e

Figure 30: T-bridges and T-tunnels

Just as loops and coloops in cycle matroids correspond to loops and bridges in graphs, loops and coloops in delta-matroids correspond to their analogues in grafts.

Theorem 10.7. *Let (G, T) be a graft, and e be an edge of G. Then:*

1. *e is a loop in $D(G, T)$ if and only if e is a loop or a T-tunnel in (G, T),*

2. *e is a coloop in $D(G, T)$ if and only if e is a T-bridge in (G, T).*

Theorem 10.8. *Let (G, T) be a graft, and e be an edge of G. Then*

$$D((G, T) \backslash e) = D(G, T) \backslash e \qquad and \qquad D((G, T)/e) = D(G, T)/e.$$

Deletion and contraction in delta-matroids act differently on coloops and loops, respectively, compared to other types of elements. The proof of Theorem 10.8, proceeds by analysing what it means in terms of the graft when e is a loop or coloop in the delta-matroid, then tracks through how the change in the graft under deletion and contraction changes the delta-matroids.

With Theorem 10.8, it follows trivially that when v is a vertex of G, $D(G, T) \backslash v = D((G, T) \backslash v)$. Thus the theorem shows that minor theory for graphic delta-matroids and for grafts are compatible with one another:

$$\text{Graft minors} \quad \xleftrightarrow{\text{compatible}} \quad \text{delta-matroid strong-minors.}$$

A consequence of this is that the class of graphic delta-matroids is minor-closed.

Theorem 10.9. *A minor of a graphic delta-matroid is graphic.*

Graphic delta-matroids are another example of even binary delta-matroids, and so their properties are tied to simple graphs and pivoting.

Theorem 10.10. *Graphic delta-matroids are even binary.*

The idea behind the proof of this theorem is to show that (i) $D(G, V(G))$ is even binary, (ii) that minors of graphic delta-matroids are graphic (Theorem 10.9), and (iii) that any graft (G, T) can be obtained as a minor of some graft $(G', V(G'))$ (see Exercise 10.6). The proof of (i) depends upon line graphs, using a result of Kishi and Uetake [44] that the adjacency matrix (over GF(2)) of the line graph of a simple graph G is non-singular if and only every component of G is a tree with an odd number of vertices. (This also provides some insight into Definition 10.1.) An alternative approach is to check that the excluded minors for binary delta-matroids from Theorem 7.11 do not arise from grafts.

Oum's interest in graphic delta-matroids arose from his conjecture (also in [59]) that if H is a bipartite circle graph, then every graph G with sufficiently large rank-width must have a pivot-minor isomorphic to H. (Rank-width, is a tree-width-like graph parameter introduced by Oum and Seymour in [60] to investigate clique-width.) This conjecture implies Robertson and Seymour's Grid Theorem [65], as well as its version for binary matroids from [38]. Oum proved that the conjecture holds when G is a line graph. In order to do this he had to navigate the difficulty that line graphs are not closed under pivot-minors. This was done by introducing graphic delta-matroids. With this concept he obtained the following rank-width results in [59].

Theorem 10.11. *Let Γ be the fundamental graph of the delta-matroid $D(G,T)$ of a graft (G,T). If the branch-width of G is k, then the rank-width of Γ is k, $k-1$, or $k-2$.*

Theorem 10.12. *Let H be a bipartite circle graph. Then there is a constant $c(H)$ such that if the fundamental graph Γ of the delta-matroid $D(G,T)$ of a graft (G,T) has rank-width larger than $c(H)$, then Γ has a pivot-minor isomorphic to H.*

11   Matchings and delta-matroids

For a graph $G = (V,E)$, and a subset $U \subseteq V$, let $G[U]$ denote the *induced subgraph* on U (so $G[U]$ is obtained by deleting any vertices of G that are not in U). A *matching* on G is a set of its edges that do not share a vertex. A matching is *perfect* if every vertex is incident with an edge in the matching. A set $U \subseteq V$ is said to be *matchable* if $G[U]$ has a perfect matching.

Definition 11.1 (Matching delta-matroid). Let $G = (V,E)$ be a simple graph. Let $\mathcal{F}$ be the collection of its matchable sets:

$$\mathcal{F} := \{X \subseteq V : G[X] \text{ has a perfect matching}\}.$$

We call $(V, \mathcal{F})$ the *matching delta-matroid of G*.

Example 2.16 gives the matching delta-matroid of a simple graph.

Bouchet, in [14], proved that matching delta-matroids are indeed delta-matroids.

Theorem 11.2. *The matching delta-matroid of a simple graph is a delta-matroid.*

Sketch. Let $X, X' \in \mathcal{F}$, and let M and M' be perfect matchings of $G[X]$ and $G[X']$, respectively. Any $x \in X \triangle X'$ is incident to an edge in exactly one of the matchings. Let H be the subgraph of G on the edge set $M \triangle M'$, then the component of H that contains x is a chain C with one end equal to x. Let y be the other end C. Then $M \triangle C$ is a perfect matching of $G[X \triangle \{x,y\}]$, and so $X \triangle \{x,y\} \in \mathcal{F}$. $\square$

Exercise 11.3. *Prove that a matching delta-matroid is always even. Realise the delta-matroid of Item 5 of Theorem 7.11 as a matching delta-matroid, and hence show that matching delta-matroids need not be binary. Give an example to show that a matching delta-matroid may be binary.*

In Section 8.2 we met the fundamental graph $G_{D,X}$ of an even binary delta-matroid $D = (E, \mathcal{F})$, where X was a feasible set. For the construction we do not actually need that D is binary, and so we can construct $G_{D,X}$ for any delta-matroid D. The *fundamental graph* $G_{D,X}$ is then the graph with vertex set V, and with an edge uv if and only if $\{u, v\}$ is a feasible set of $D * X$ (or equivalently, if and only if $\{u, v\} \triangle X$ is a feasible set of D).

Exercise 11.4. *Prove that if D is a matroid with a basis X, then $G_{D,X}$ is bipartite. (Hint, suppose that $G_{D,X}$ has an odd cycle.)*

The delta-matroid structure of a matching matroid can be used to gain insight into matchable sets, as in the following theorem from [14].

Theorem 11.5. *Let $D = (V, \mathcal{F})$ be an even delta-matroid and $X, X' \in \mathcal{F}$ be two feasible sets. Then $X \triangle X'$ is matchable in $G_{D,X}$.*

Remark 8. For those familiar with matroids, it is worth noting that Theorem 11.5 is a generalisation of a theorem of Brualdi's (Theorem 1 of [23]) which states that given two bases F and F' of a matroid, there is a bijection $\sigma : F \backslash F' \to F' \backslash F$ such that $(F \backslash e) \cup \sigma(e)$ is a base for all $e \in F \backslash F'$.

At the end of Section 6 we met the partition problem. A special case (taking $D = D'$) of this asks if the ground set of a delta-matroid can be partitioned into two of its feasible sets. This is related to perfect matchings as follows.

Suppose that F_1 and F_2 are two complementary feasible sets of a delta-matroid D, and that $G = G_{D,X} = G_{D*X,\emptyset}$ is a fundamental graph of D. Then, for $i = 1, 2$, applying Theorem 11.5 to the feasible sets $F_i \triangle X$ and $\emptyset$ of $D * X$, gives perfect matchings M_i in $G[F_i \triangle X]$. Since $F_i \triangle X$ and $F_i \triangle X$ are complementary, $M_1 \cup M_2$ is a perfect matching for G. Thus we have the following result of Bouchet [16].

Corollary 11.6. *If an even delta-matroid admits two complementary feasible sets then each of its fundamental graphs admits a perfect matching.*

In Exercise 11.4 we saw that the fundamental graphs of matroids were necessarily bipartite. Theorem 11.5 can be used to show the converse.

Corollary 11.7. *A delta-matroid is a twist of a matroid if and only if its fundamental graphs are bipartite. That is, if D is a normal delta-matroid and X is a feasible set, then $D * X$ is a matroid if and only if $G_{D,\emptyset}$ is bipartite.*

The result is from [14], where its (short) proof can be found. The result was extended by Duchamp in [33].

Exercise 11.8. *Let $\mathbb{G} = (V, E)$ be a 1-vertex ribbon graph, and $A \subseteq E$. Use Corollary 11.7 to prove that $\mathbb{G}^A$ is a plane ribbon graph if and only if $\mathbb{G} \backslash A$ and $\mathbb{G} \backslash (E \backslash A)$ are both plane ribbon graphs. (This is a special case of the rough structure theorem for partial duals of plane graph from [53]. A delta-matroid analogue of the rough structure theorem was given in [29].)*

Exercise 11.9. *Use Corollary 11.7 and the results of Section 9 to find a characterisation of the class of Eulerian delta-matroids that are twists of matroids.*

Another family of delta-matroids that is intimately connected with matchings is linking delta-matroids. A *red-blue graph* is a simple graph $G = (V, E)$ in which each edge is coloured either red or blue. A *red-blue path* is a path in it whose edges alternate in colour.

Definition 11.10 (Linking delta-matroid). Let $G = (V, E)$ be a red-blue graph. Let $\mathcal{F}$ be the collection of subsets of V given by

$$\mathcal{F} := \{X : X \text{ is the end vertex set of a collection of}$$
$$\text{pairwise vertex disjoint red-blue paths}\}$$

Then the pair (V, F) is called the *linking delta-matroid* of G

Linking delta-matroids were shown to be delta-matroids in [16, 21].

Example 11.11. Figure 31a shows a red-blue graph, with the two colour classes indicated as black or grey edges. Then its linking delta-matroid has ground set $V = \{1, 2, 3, 4, 5\}$ and feasible sets

$$\mathcal{F} = \{\emptyset, \{1,2\}, \{1,3\}, \{1,4\}, \{1,5\}, \{2,3\}, \{2,4\}, \{2,5\}, \{3,4\}, \{3,5\}, \{4,5\},$$
$$\{1,2,3,4\}, \{1,2,3,5\}, \{1,2,4,5\}, \{1,3,4,5\}, \{2,3,4,5\}\}.$$

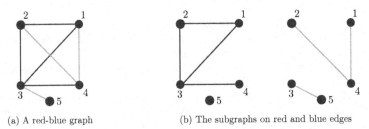

(a) A red-blue graph   (b) The subgraphs on red and blue edges

Figure 31: A red-blue (or black-grey) graph

To see how this relates to matchings we need the notion of the *delta-sum*. The *delta-sum*, $D \bigtriangleup D'$, of two delta-matroids $D = (V, \mathcal{F})$ and $D' = (V, \mathcal{F}')$ is defined to be $D \bigtriangleup D' := (V, \mathcal{F} \bigtriangleup \mathcal{F}')$, where $\mathcal{F} \bigtriangleup \mathcal{F}' := \{F \bigtriangleup F' : F \in \mathcal{F}, F' \in \mathcal{F}'\}$. It was introduced by Duchamp, and while it is cited as 'in preparation' in early delta-matroid papers, he does not appear to have ever published the work. A proof that it does result in a delta-matroid can be found in [21].

Bouchet and Schwärzler, in [21], used the delta-sum to express linking delta-matroids is terms of matching delta-matroids:

Theorem 11.12. *Let G be a red-blue graph. Let D_ℓ be the linking delta-matroid of G, and D_r and D_b be the matching delta-matroids of the subgraph induced by the red and blue edges, respectively. Then*
$$D_\ell = D_r \bigtriangleup D_b.$$

Example 11.13. Consider again Example 11.11 which gave the linking delta-matroid of the red-blue graph of Figure 31a. Figure 31b shows the subgraphs induced by the red and blue edges. One has matching delta-matroid $D_r = (V, \mathcal{F}_r)$ where

$$\mathcal{F}_r = \{\emptyset, \{1,2\}, \{1,3\}, \{2,3\}, \{3,4\}, \{1,2,3,4\}\}.$$

The other has matching delta-matroid $D_b = (V, \mathcal{F}_b)$ where

$$\mathcal{F}_b = \{\emptyset, \{1,4\}, \{2,4\}, \{3,5\}, \{1,3,4,5\}, \{2,3,4,5\}\}.$$

Then the feasible sets of $D_r \triangle D_b$ are

$$\mathcal{F}_r \triangle \mathcal{F}_b = \{\emptyset, \{1,2\}, \{1,3\}, \{1,4\}, \{1,5\}, \{2,3\}, \{2,4\}, \{2,5\}, \{3,4\}, \{3,5\}, \{4,5\},$$
$$\{1,2,3,4\}, \{1,2,3,5\}, \{1,2,4,5\}, \{1,3,4,5\}, \{2,3,4,5\}\},$$

and we see $D_r \triangle D_b$ agrees with the linking delta-matroid from Example 11.11.

Bouchet and Schwärzler found a formula for the polyhedral rank function of a linking delta-matroid. We will not discuss this formula here, although we will point out of one nice graph theoretic corollary from of their delta-matroid work: the recovery of the following result of Gallai [37].

Theorem 11.14. *The maximum number of vertex disjoint paths in a graph $G = (V, E)$ having both end vertices in $U \subseteq V$ is*

$$\min_{S \subseteq V} \left(|S| + \sum_C \lfloor |C \cap U|/2 \rfloor \right),$$

where the sum ranges over all components C of $G \backslash S$ with $|C \cap U|$ odd.

Notation

$\mathbf{A}$	A matrix.
$\mathbf{A}[X]$	The principal submatrix of $\mathbf{A}$ on the rows/columns X.
$\mathbf{A}_G$	Adjacency matrix of a graph G.
$\mathbf{A} * X$	Pivot of $\mathbf{A}$ w.r.t. a set of rows/columns X.
$C(G)$	Cycle matroid of a graph G.
D	A delta-matroid.
$D * A$	The twist of delta-matroid D w.r.t. A.
$D(\mathbf{A})$	The delta-matroid of a matrix $\mathbf{A}$.
$D(\mathbb{G})$	The delta-matroid of a ribbon graph $\mathbb{G}$.
$D(\vec{G}, T_P)$	The delta-matroid of an Eulerian 4-regular digraph $\vec{G}$ w.r.t T_P.
$D(G, T)$	The delta-matroid of a graft (G, T).
$D(G, T_F, T_P)$	The Eulerian delta-matroid of 4-regular graph G w.r.t T_F and T_P.
G, H	Graphs.
$G * v$	Local complementation of graph G a graph w.r.t. a vertex v.
$G \wedge uv$	Pivot of graph G a graph w.r.t. an edge uv.
(G, T)	A graft.
$\mathbb{G}, \mathbb{H}$	Ribbon graphs.
$\mathbb{G}^A$	The partial dual of a ribbon graph $\mathbb{G}$ w.r.t. A.
$\Bbbk$	A field.

References

[1] M. Aigner, *The Penrose polynomial of a plane graph*, Math. Ann. **307** (1997), no. 2, 173–189.

[2] M. Aigner, *The Penrose polynomial of graphs and matroids*, Surveys in combinatorics, 2001 (Sussex), London Math. Soc. Lecture Note Ser., vol. 288, Cambridge Univ. Press, Cambridge, 2001, pp. 11–46.

[3] R. Arratia, B. Bollobás, and G. B. Sorkin, *The interlace polynomial: a new graph polynomial*, Proceedings of the Eleventh Annual ACM-SIAM Symposium on Discrete Algorithms (San Francisco, CA, 2000), ACM, New York, 2000, pp. 237–245.

[4] R. Arratia, B. Bollobás, and G. B. Sorkin, *The interlace polynomial of a graph*, J. Combin. Theory Ser. B **92** (2004), no. 2, 199–233.

[5] R. Arratia, B. Bollobás, and G. B. Sorkin, *A two-variable interlace polynomial*, Combinatorica **24** (2004), no. 4, 567–584.

[6] B. Bollobás and O. Riordan, *A polynomial invariant of graphs on orientable surfaces*, Proc. London Math. Soc. (3) **83** (2001), no. 3, 513–531.

[7] B. Bollobás and O. Riordan, *A polynomial of graphs on surfaces*, Math. Ann. **323** (2002), no. 1, 81–96.

[8] A. V. Borovik, *Matroids and Coxeter groups*, Surveys in combinatorics, 2003 (Bangor), London Math. Soc. Lecture Note Ser., vol. 307, Cambridge Univ. Press, Cambridge, 2003, pp. 79–114.

[9] A. Bouchet, *Greedy algorithm and symmetric matroids*, Math. Programming **38** (1987), no. 2, 147–159.

[10] A. Bouchet, *Isotropic systems*, European J. Combin. **8** (1987), no. 3, 231–244.

[11] A. Bouchet, *Graphic presentations of isotropic systems*, J. Combin. Theory Ser. B **45** (1988), no. 1, 58–76.

[12] A. Bouchet, *Representability of △-matroids*, Combinatorics (Eger, 1987), Colloq. Math. Soc. János Bolyai, vol. 52, North-Holland, Amsterdam, 1988, pp. 167–182.

[13] A. Bouchet, *Maps and △-matroids*, Discrete Math. **78** (1989), no. 1-2, 59–71.

[14] A. Bouchet, *Matchings and △-matroids*, Discrete Appl. Math. **24** (1989), no. 1-3, 55–62, First Montreal Conference on Combinatorics and Computer Science, 1987.

[15] A. Bouchet, *Circle graph obstructions*, J. Combin. Theory Ser. B **60** (1994), no. 1, 107–144.

[16] A. Bouchet, *Coverings and delta-coverings*, Integer programming and combinatorial optimization (Copenhagen, 1995), Lecture Notes in Comput. Sci., vol. 920, Springer, Berlin, 1995, pp. 228–243.

[17] A. Bouchet, *Multimatroids. I. Coverings by independent sets*, SIAM J. Discrete Math. **10** (1997), no. 4, 626–646.

[18] A. Bouchet, *Multimatroids. III. Tightness and fundamental graphs*, European J. Combin. **22** (2001), no. 5, 657–677, Combinatorial geometries (Luminy, 1999).

[19] A. Bouchet and W. H. Cunningham, *Delta-matroids, jump systems, and bisubmodular polyhedra*, SIAM J. Discrete Math. **8** (1995), no. 1, 17–32.

[20] A. Bouchet and A. Duchamp, *Representability of △-matroids over* GF(2), Linear Algebra Appl. **146** (1991), 67–78.

[21] A. Bouchet and W. Schwärzler, *The delta-sum of matching delta-matroids*, Discrete Math. **181** (1998), no. 1-3, 53–63.

[22] R. Brijder and H. J. Hoogeboom, *Interlace polynomials for multimatroids and delta-matroids*, European J. Combin. **40** (2014), 142–167.

[23] R. A. Brualdi, *Comments on bases in dependence structures.*, Bull. Aust. Math. Soc. **1** (1969), 161–167 (English).

[24] T. H. Brylawski, *A decomposition for combinatorial geometries*, Trans. Amer. Math. Soc. **171** (1972), 235–282.

[25] R. Chandrasekaran and S. N. Kabadi, *Pseudomatroids*, Discrete Math. **71** (1988), no. 3, 205–217.

[26] S. Chmutov, *Generalized duality for graphs on surfaces and the signed Bollobás-Riordan polynomial*, J. Combin. Theory Ser. B **99** (2009), no. 3, 617–638.

[27] C. Chun, D. Chun, and S. D. Noble, *Inductive tools for connected delta-matroids and multimatroids*, European J. Combin. **63** (2017), 59–69.

[28] C. Chun, R. Hall, C. Merino, I. Moffatt, and S. D. Noble, *The structure of delta-matroids with width one twists*, Electron. J. Combin. **25** (2018), no. 1, Paper 1.9, 12.

[29] C. Chun, I. Moffatt, S. D. Noble, and R. Rueckriemen, *Matroids, delta-matroids and embedded graphs*, preprint (2014).

[30] C. Chun, I. Moffatt, S. D. Noble, and R. Rueckriemen, *On the interplay between embedded graphs and delta-matroids*, Proc. London Math. Soc. (in press).

[31] O. T. Dasbach, D. Futer, E. Kalfagianni, X.-S. Lin, and N. W. Stoltzfus, *The Jones polynomial and graphs on surfaces*, J. Combin. Theory Ser. B **98** (2008), no. 2, 384–399.

[32] A. Dress and T. F. Havel, *Some combinatorial properties of discriminants in metric vector spaces*, Adv. in Math. **62** (1986), no. 3, 285–312.

[33] A. Duchamp, *Delta matroids whose fundamental graphs are bipartite*, Linear Algebra Appl. **160** (1992), 99–112.

[34] J. A. Ellis-Monaghan and I. Moffatt, *Twisted duality for embedded graphs*, Trans. Amer. Math. Soc. **364** (2012), no. 3, 1529–1569.

[35] J. A. Ellis-Monaghan and I. Moffatt, *Graphs on surfaces: Dualities, polynomials, and knots*, Springer, New York, 2013.

[36] J. A. Ellis-Monaghan and I. Moffatt, *A Penrose polynomial for embedded graphs*, European J. Combin. **34** (2013), no. 2, 424–445.

[37] T. Gallai, *Maximum-minimum Sätze und verallgemeinerte Faktoren von Graphen*, Acta Math. Acad. Sci. Hungar. **12** (1961), 131–173.

[38] J. Geelen, B. Gerards, and G. Whittle, *Excluding a planar graph from GF(q)-representable matroids*, J. Combin. Theory Ser. B **97** (2007), no. 6, 971–998.

[39] J. Geelen, B. Gerards, and G. Whittle, *Structure in minor-closed classes of matroids*, Surveys in combinatorics 2013, London Math. Soc. Lecture Note Ser., vol. 409, Cambridge Univ. Press, Cambridge, 2013, pp. 327–362.

[40] J. Geelen and S. Oum, *Circle graph obstructions under pivoting*, J. Graph Theory **61** (2009), no. 1, 1–11.

[41] J. F. Geelen, *A generalization of Tutte's characterization of totally unimodular matrices*, J. Combin. Theory Ser. B **70** (1997), no. 1, 101–117.

[42] J. F. Geelen, S. Iwata, and K. Murota, *The linear delta-matroid parity problem*, J. Combin. Theory Ser. B **88** (2003), no. 2, 377–398.

[43] F. Jaeger, *On transition polynomials of 4-regular graphs*, Cycles and rays (Montreal, PQ, 1987), NATO Adv. Sci. Inst. Ser. C Math. Phys. Sci., vol. 301, Kluwer Acad. Publ., Dordrecht, 1990, pp. 123–150.

[44] G. Kishi and Y. Uetake, *Rank of edge incidence matrix*, IEEE Trans. Circuit Theory **CT-16** (1969), 230–232.

[45] A. Kotzig, *Eulerian lines in finite 4-valent graphs and their transformations*, Theory of Graphs (Proc. Colloq., Tihany, 1966), Academic Press, New York, 1968, pp. 219–230.

[46] T. Krajewski, I. Moffatt, and A. Tanasa, *Hopf algebras and Tutte polynomials*, Adv. in Appl. Math. **95** (2018), 271–330.

[47] J. P. S. Kung, *Bimatroids and invariants*, Adv. in Math. **30** (1978), no. 3, 238–249.

[48] L. Lovász, *Matroid matching and some applications*, J. Combin. Theory Ser. B **28** (1980), no. 2, 208–236.

[49] L. Lovász, *Selecting independent lines from a family of lines in a space*, Acta Sci. Math. (Szeged) **42** (1980), no. 1-2, 121–131.

[50] L. Lovász, *The matroid matching problem*, Algebraic methods in graph theory, Vol. I, II (Szeged, 1978), Colloq. Math. Soc. János Bolyai, vol. 25, North-Holland, Amsterdam-New York, 1981, pp. 495–517.

[51] F. Mazoit and S. Thomassé, *Branchwidth of graphic matroids*, Surveys in combinatorics 2007, London Math. Soc. Lecture Note Ser., vol. 346, Cambridge Univ. Press, Cambridge, 2007, pp. 275–286.

[52] I. Moffatt, *A characterization of partially dual graphs*, J. Graph Theory **67** (2011), no. 3, 198–217.

[53] I. Moffatt, *Partial duals of plane graphs, separability and the graphs of knots*, Algebr. Geom. Topol. **12** (2012), no. 2, 1099–1136.

[54] I. Moffatt, *Ribbon graph minors and low-genus partial duals*, Ann. Comb. **20** (2016), no. 2, 373–378.

[55] I. Moffatt and E. Mphako-Banda, *Handle slides for delta-matroids*, European J. Combin. **59** (2017), 23–33.

[56] I. Moffatt and B. Smith, *Matroidal frameworks for topological tutte polynomials*, J. Combin. Theory Ser. B **133** (2018), 1–31.

[57] A. Morse, *Interlacement and activities in delta-matroids*, preprint (2017).

[58] S. Oum, *Graphic delta-matroids*, slides from Kyoto RIMS Workshop on Combinatorial Optimization and Discrete Algorithms. Avaliable at mathsci.kaist.ac.kr/~sangil/pdf/2008kyoto.pdf, 2008.

[59] S. Oum, *Excluding a bipartite circle graph from line graphs*, J. Graph Theory **60** (2009), no. 3, 183–203.

[60] S. Oum and P. Seymour, *Approximating clique-width and branch-width*, J. Combin. Theory Ser. B **96** (2006), no. 4, 514–528.

[61] J. Oxley, *Matroid theory*, Oxford Science Publications, The Clarendon Press, Oxford University Press, New York, 1992.

[62] J. Oxley, *On the interplay between graphs and matroids*, Surveys in combinatorics, 2001 (Sussex), London Math. Soc. Lecture Note Ser., vol. 288, Cambridge Univ. Press, Cambridge, 2001, pp. 199–239.

[63] R. Penrose, *Applications of negative dimensional tensors*, Combinatorial Mathematics and its Applications (Proc. Conf., Oxford, 1969), Academic Press, London, 1971, pp. 221–244.

[64] L. Q. Qi, *Directed submodularity, ditroids and directed submodular flows*, Math. Programming **42** (1988), no. 3, (Ser. B), 579–599, Submodular optimization.

[65] N. Robertson and P. D. Seymour, *Graph minors. X. Obstructions to tree-decomposition*, J. Combin. Theory Ser. B **52** (1991), no. 2, 153–190.

[66] G. F. Royle, *Recent results on chromatic and flow roots of graphs and matroids*, Surveys in combinatorics 2009, London Math. Soc. Lecture Note Ser., vol. 365, Cambridge Univ. Press, Cambridge, 2009, pp. 289–327.

[67] P. D. Seymour, *A note on the production of matroid minors*, J. Combinatorial Theory Ser. B **22** (1977), no. 3, 289–295.

[68] P. D. Seymour, *Decomposition of regular matroids*, J. Combin. Theory Ser. B **28** (1980), no. 3, 305–359.

[69] P. D. Seymour, *On Tutte's characterization of graphic matroids*, Ann. Discrete Math. **8** (1980), 83–90, Combinatorics 79 (Proc. Colloq., Univ. Montréal, Montreal, Que., 1979), Part I.

[70] A. D. Sokal, *The multivariate Tutte polynomial (alias Potts model) for graphs and matroids*, Surveys in combinatorics 2005, London Math. Soc. Lecture Note Ser., vol. 327, Cambridge Univ. Press, Cambridge, 2005, pp. 173–226.

[71] É. Tardos, *Generalized matroids and supermodular colourings*, Matroid theory (Szeged, 1982), Colloq. Math. Soc. János Bolyai, vol. 40, North-Holland, Amsterdam, 1985, pp. 359–382.

[72] L. Traldi, *The transition matroid of a 4-regular graph: an introduction*, European J. Combin. **50** (2015), 180–207.

[73] W. T. Tutte, *A ring in graph theory*, Proc. Cambridge Philos. Soc. **43** (1947), 26–40.

[74] W. T. Tutte, *Lectures on matroids*, J. Res. Nat. Bur. Standards Sect. B **69B** (1965), 1–47.

[75] W. T. Tutte, *The coming of the matroids*, Surveys in combinatorics, 1999 (Canterbury), London Math. Soc. Lecture Note Ser., vol. 267, Cambridge Univ. Press, Cambridge, 1999, pp. 3–14.

[76] D. Welsh, *Matroid theory*, Academic Press [Harcourt Brace Jovanovich, Publishers], London-New York, 1976, L. M. S. Monographs, No. 8.

[77] D. Welsh, *Colouring problems and matroids*, Surveys in combinatorics (Proc. Seventh British Combinatorial Conf., Cambridge, 1979), London Math. Soc. Lecture Note Ser., vol. 38, Cambridge Univ. Press, Cambridge-New York, 1979, pp. 229–257.

[78] H. Whitney, *On the abstract properties of linear dependence.*, Am. J. Math. **57** (1935), 509–533 (English).

Department of Mathematics
Royal Holloway, University of London
iain.moffatt@rhul.ac.uk

Extremal theory of vertex or edge ordered graphs[1]

Gábor Tardos[2]

Abstract

We enrich the structure of finite simple graphs with a linear order on either the vertices or the edges. Extending the standard question of Turán-type extremal graph theory we ask for the maximal number of edges in such a vertex or edge ordered graph on n vertices that does not contain a given pattern (or several patterns) as a subgraph. The forbidden subgraph itself is also a vertex or edge ordered graph, so we forbid a certain subgraph with a specified ordering, but we allow the same underlying subgraph with a different (vertex or edge) order. This allows us to study a large number of extremal problems that are not expressible in the classical theory. In this survey we report on ongoing research. For easier access we include sketches of proofs of selected results.

1   Definitions

A *vertex ordered graph* or simply an *ordered graph* is a simple graph with a linear order on the vertices. Formally, an ordered graph is a triple $(V, E, <)$, where V is the vertex set, $E \subseteq \binom{V}{2}$ is the edge set (so that (V, E) is a simple graph) and $<$ is a linear order relation on V. Similarly, an *edge ordered graph* is a simple graph with a linear order on its edges, that is $(V, E, <)$, where (V, E) is a simple graph and $<$ is a linear order relation on E. In this survey we assume that V is finite. We say that (V, E) is the simple graph *underlying* the vertex or edge ordered graph $(V, E, <)$ and $(V, E, <)$ is a *vertex or edge ordering* of the simple graph (V, E). The notions of isomorphism and subgraph naturally extend to these graphs: An isomorphism between vertex or edge ordered graphs is an isomorphism between the underlying simple graphs that preserves the ordering. A subgraph of a vertex or edge ordered graph is a subgraph of the underlying simple graph with the inherited vertex or edge order.

Armed with these definitions we can extend some classic areas of graph theory to ordered graphs. Here we do this for Turán-type extremal graph theory and for most of this survey we consider vertex orderings. (See the last section for some preliminary results regarding ongoing research on edge ordered graphs.) Extremal graph theory asks for the maximal number of edges in a simple graph of given size that *avoids* (i.e., does not contain as a subgraph) a specified pattern or any member of a given family of patterns. In particular, we are interested in the maximal number, $\mathrm{ex}(\mathcal{P}, n)$, of edges in an n-vertex simple graph that has no subgraph isomorphic to any member of the family $\mathcal{P}$. Note that we must require that $\mathcal{P}$ does not contain empty graphs (i.e., each member has at least one edge) in order for this definition to make sense. In the case where the forbidden family consists of a single pattern we write $\mathrm{ex}(P, n)$ to denote $\mathrm{ex}(\{P\}, n)$. We call $\mathrm{ex}(\mathcal{P}, n)$ the *extremal function* of the family $\mathcal{P}$ and will concentrate on its asymptotic behavior. Accordingly, all the asymptotic notations like $O(\cdot)$ and $o(\cdot)$ should be interpreted for a fixed family $\mathcal{P}$ and, in particular, the implied constants in $O(\cdot)$ may depend on this family.

For a natural extension of this theory to ordered graphs, we consider a family $\mathcal{P}$ of ordered graphs and we look for the largest number $\mathrm{ex}_<(\mathcal{P}, n)$ of edges in an n-vertex ordered

[1] This survey is an extended version of my talk at ICM 2008, see [29].

[2] Supported by the Cryptography "Lendület" project of the Hungarian Academy of Sciences and by the National Research, Development and Innovation Office, NKFIH projects K-116769 and SNN-117879.

graph with no ordered subgraph isomorphic to any member of $\mathcal{P}$. As before, we require that each member of $\mathcal{P}$ has at least one edge and simplify the notation for singleton families by writing $\mathrm{ex}_<(P, n)$ to denote $\mathrm{ex}_<(\{P\}, n)$. Our remark on the asymptotic notation also applies here.

Let us first observe that the extremal theory of ordered graphs is strictly richer than classical extremal graph theory in the sense that the classical questions can be equivalently asked in this setting, but we can also ask new questions. In particular, for any family $\mathcal{P}$ of simple graphs one can form the family $\mathcal{P}_<$ consisting all orderings of the patterns in $\mathcal{P}$ and then we trivially have:

$$\mathrm{ex}(\mathcal{P}, n) = \mathrm{ex}_<(\mathcal{P}_<, n).$$

On the other hand, if we forbid, say, a single ordered graph P, the corresponding extremal function $\mathrm{ex}_<(P, n)$ has no direct analogue in the classical theory. We naturally have $\mathrm{ex}_<(P, n) \geq \mathrm{ex}(\overline{P}, n)$, where $\overline{P}$ is the simple graph underlying P, but this lower bound is typically very weak, since avoiding $\overline{P}$ in a particular order is often much easier than avoiding it in all possible orders.

In the paper [5] Braß, Károlyi and Valtr establish a very similar theory. Instead of a linear order on the vertices, they consider a *cyclic order*. They have very nice results on the extremal function of certain cyclically ordered graphs. These results have natural translations in the extremal theory of ordered graphs. Let us also give credit to Füredi and Hajnal, who in the last paragraph of their paper [12] explicitly ask for developing both the extremal theory of vertex ordered graphs surveyed here and for that of the cyclically ordered graphs as done later by Braß, Károlyi and Valtr.

Extensions of Ramsey theory to ordered graphs are also studied extensively, see [1,6].

Most of this survey is about the extremal theory of vertex ordered graphs. In Section 2 the classical Erdős–Stone–Simonovits theorem and its generalization to vertex ordered graphs is presented. This result determines the asymptotics of the extremal function of ordered graphs with interval chromatic number three or higher. We are satisfied with asymptotical results, so we concentrate exclusively on *ordered bipartite* graphs (ordered graphs with interval chromatic number two) in later sections. In Section 3 the close connection between the extremal theory of ordered bipartite graphs and the somewhat older extremal theory of 0-1 matrices is explained. Classical extremal graph theory always gives us a lower bound on the corresponding ordered questions. In Section 4 we explore how far this lower bound can be from the ordered extremal function. The case of forests is treated separately in Section 5. Here the unordered theory gives a linear lower bound, and a prominent open question is to decide if an almost linear (say, $n^{1+o(1)}$) bound also holds for all ordered bipartite forests. This is known for a wide class of such forests, but not for all. In Section 6 we present two simple results on the class of ordered graphs with linear extremal functions. Finally, in Section 7 we present results about extremal functions arising from simultaneously forbidding two (or more) ordered graphs. While it is not known in the classical extremal theory of graphs whether forbidding several graphs can result in an extremal function of lower order of magnitude than the ones obtained from just one forbidden graph, we have many such examples in the ordered setting.

In Section 8, the last section of this survey, we summarize recent research on the extremal theory of edge ordered graphs.

2 Basic results

Any survey about extremal graph theory should start with the following classical theorem of Turán from 1941, [30], of which the $r = 2$ special case (the maximal number of edges in a triangle-free graph) was proved by Mantel in 1907, [19]. The result gives the exact extremal function when the forbidden graph is a complete graph. Further, for the $(r + 1)$-vertex complete graph K_{r+1} the theorem states that the unique (up to isomorphism) n-vertex graph with the maximum number of edges avoiding K_{r+1} is the *Turán graph* $T(n, r)$ formed by partitioning the vertices into r almost equal parts and letting a pair of vertices form an edge if and only if they are from distinct parts. Note that the number of edges of the Turán graph $T(n, r)$ is $(1 - 1/r)n^2/2 - O(1)$, where the $O(1)$ error term comes from unequal parts and can go as high as $r/8$. As a consequence, we have:

Theorem 2.1 (Turán [30]). *For every $r \geq 1$ we have*

$$\text{ex}(K_{r+1}, n) = (1 - 1/r)\frac{n^2}{2} - O(1).$$

A trivial generalization of this result to ordered graphs involves the ordered clique, the unique ordering of the complete graph. Let $K_{r+1,<}$ stand for the $(r + 1)$-vertex ordered clique and we trivially have $\text{ex}_<(K_{r+1,<}, n) = \text{ex}(K_{r+1}, n)$. A more revealing generalization concerns the ordered path $P_{r+1,<}$ obtained from the $(r + 1)$-vertex path P_{r+1} with the natural order on the vertices where edges connect neighboring vertices in the order. We have $\text{ex}_<(P_{r+1,<}, n) = \text{ex}(K_{r+1}, n)$. Here the direction $\leq$ follows from the fact that $P_{r+1,<}$ is an ordered subgraph of $K_{r+1,<}$ and $\geq$ follows from the fact that if we order the vertices of $T(n, r)$ in a way that the r parts become intervals in the ordering, then the resulting ordered graph does not contain $P_{r+1,<}$ as an ordered subgraph. Note, however, that in the case r does not divide n, this process yields several non-isomorphic extremal ordered graphs. Note also that the path P_{r+1} has several non-isomorphic orderings for $r > 1$, and by Theorem 2.3 below, all other orderings have substantially smaller extremal functions.

The most general result in Turán-type extremal graph theory is the following consequence of the Erdős–Stone theorem, [7]. It basically states that the extremal function of *any* simple graph is close to the extremal function of the complete graph with the same *chromatic number*.

Theorem 2.2 (Erdős–Stone–Simonovits [7,9]). *Let $\mathcal{P}$ be a family of simple graphs and $r + 1 = \min_{P \in \mathcal{P}} \chi(P)$ be the smallest chromatic number of a member of this family. We have*

$$\text{ex}(\mathcal{P}, n) = (1 - 1/r)\frac{n^2}{2} + o(n^2).$$

Pach and Tardos, [24] gave a generalization of this result for ordered graphs. It is based on finding the "correct" version of the chromatic number for ordered graph.

An *interval coloring* of an ordered graph is a proper coloring of the underlying simple graph in which each color class is an interval of the linear order. The *interval chromatic number* of an ordered graph P is the smallest number of colors in an interval coloring of P. We write $\chi_<(P)$ to denote the interval chromatic number of P.

Note that the interval chromatic number is much simpler to compute than the chromatic number because a greedy strategy suffices. Indeed, we can form the first color class by taking longest initial segment of the vertices that form an independent set and proceed similarly for subsequent color classes. The process yields an interval coloring with the fewest possible colors. This is because the first color class of any interval coloring is a subset of the first color class found above, so greedily choosing the longest possible interval cannot hurt us later. Using this definition, the generalization of the Erdős–Stone–Simonovits theorem is rather straightforward:

Theorem 2.3 (Erdős–Stone–Simonovits theorem for ordered graphs [24]). *Let $\mathcal{P}$ be a family of ordered graphs and $r + 1 = \min_{P \in \mathcal{P}} \chi_<(P)$ be the smallest interval chromatic number of a member of this family. We have*

$$\mathrm{ex}_<(\mathcal{P}, n) = (1 - 1/r)\frac{n^2}{2} + o(n^2).$$

Proof. Order the vertices of the Turán graph with r-classes such that each class becomes an interval of the ordering. This way we obtain an ordered graph with interval chromatic number r, so it avoids all ordered graphs with higher interval chromatic number, including all members of $\mathcal{P}$. This provides the lower bound for the extremal function $\mathrm{ex}_<(\mathcal{P}, n)$.

Let $P \in \mathcal{P}$ be an ordered graph with interval chromatic number $r + 1$. The upper bound comes from the classical Erdős–Stone theorem. Let m be the number of vertices of P. By the theorem, a simple graph on n vertices with at least $(1 - \frac{1}{r} + \varepsilon)n^2$ edges contains the Turán graph T with $r + 1$ classes, each containing m vertices if n is large as a function of r, m and ε. We show that any vertex ordering of $T_<$ of T contains P by induction on r. This statement holds trivially for $r = 0$, so we start the induction here (despite the fact that the theorem itself requires $r > 0$ as $\mathcal{P}$ cannot contain an empty graph). For $r > 0$ we explicitly find a monotonic homomorphism from P to $T_<$. For this, identify the first interval I in an optimal interval coloring of P. We map the vertices in I to the first $k = |I|$ vertices of $T_<$ in a single class of the underlying Turán graph. Let x be the last vertex used, so it is the kth vertex of the chosen class in $T_<$. We choose the class so as to make x appear as early in the vertex order as possible. Let us obtain the induced subgraph T' of $T_<$ by deleting all vertices in the class of x and the first k vertices in all other classes. Note that T' is an ordering of the Turán graph with r classes, each containing $m - k$ vertices, so by the inductive hypothesis it contains $P - I$, an ordered graph of interval chromatic number r on $m - k$ vertices. This gives a mapping from $P - I$ to T' (and so also to $T_<$) and together with our mapping of I provides the required monotonic homomorphism. Indeed, the mapping is monotonic as the image of $P - I$ is inside T', therefore strictly after x, the last vertex in the image of I. It is a homomorphism as $T_<$ contains a complete bipartite graph connecting the image of I (and the whole class containing it) with T'. □

Just as the classic version of this theorem, it gives exact asymptotics for the extremal function of ordered graphs unless the ordered graph is *ordered bipartite* (i.e., has interval chromatic number 2). We will therefore concentrate on ordered bipartite graphs. Containment between ordered bipartite graphs can also be visualized using the language of containment in 0-1 matrices. This connection is explored in the next section.

3   Connection to 0-1 matrices

A 0-1 matrix is simply a matrix with all entries being 0 or 1. The *weight* of such a matrix is the number of its 1-entries. The 0-1 matrix A is said to *dominate* the 0-1 matrix B of the same size if $A_{ij} \geq B_{ij}$ for all i and j, that is, if $B = A$ or B is obtained from A by replacing some 1-entries by 0-entries. A 0-1 matrix A is said to *contain* another 0-1 matrix P, if P is dominated by a submatrix of A. Note that permuting rows or columns is not allowed. If A

does not contain P, we say it *avoids* P. The extremal problem for 0-1 matrix containment can be formulated as computing (or estimating) the following extremal function for families $\mathcal{P}$ of 0-1 matrices: $\mathrm{Ex}(\mathcal{P}, n)$ is the maximal weight of an n-by-n 0-1 matrix that avoids all matrices in $\mathcal{P}$. We require that all matrices in $\mathcal{P}$ have positive weights. We write $\mathrm{Ex}(P, n)$ to denote $\mathrm{Ex}(\{P\}, n)$.

For a 0-1 matrix P, let G_P stand for the ordered bipartite graph whose vertices correspond to the rows and columns of P, the order of the vertices agrees with the order of rows and columns in P with all row-vertices preceding all column vertices, and with an edge between a row-vertex and a column-vertex if and only if the corresponding entry in P is 1. This makes P the bipartite adjacency matrix of G_P and turns the weight of P into the number of edges in G_P. The close connection between the extremal theory of ordered bipartite graphs and 0-1 matrices follows from the trivial observation that if a 0-1 matrix A contains another 0-1 matrix P, then the ordered graph G_A also contains G_P. The converse is also true if the homomorphism of G_P to G_A maps row-vertices to row-vertices and column-vertices to column-vertices. This extra condition is automatically satisfied if both the last row and first column of P contain at least one 1-entry, so in this case we have $\mathrm{Ex}(P, n) \leq \mathrm{ex}_<(G_P, 2n)$. There is no equality in general, because $\mathrm{ex}_<(G_P, 2n)$ is the maximum number of edges among all ordered graphs on $2n$ vertices avoiding G_P and the extremal ones may not be ordered bipartite. Still, the two extremal functions are really close to each other as shown by the following observation:

Theorem 3.1 ([24]). *For a 0-1 matrix P and the corresponding ordered bipartite graph G_P we have*

$$\mathrm{Ex}(P, n) \leq \mathrm{ex}_<(G_P, 2n) = O(\mathrm{Ex}(P, n) \log n).$$

The logarithmic term in the bound above is needed even for some small matrices, e.g., for the matrix

$$P = \begin{pmatrix} 1 & 1 \\ 0 & 1 \end{pmatrix}.$$

For this matrix, we have $\mathrm{Ex}(P, n) = 2n - 1$, but for the corresponding ordered graph G_P one has $\mathrm{ex}_<(G_P, n) = n \log n + O(n)$, where log stands for the binary logarithm. A construction showing the lower bound for this latter estimate is an ordered graph whose vertices are adjacent if and only if their distance in the ordering is a power of 2. To see that $\mathrm{Ex}(P, n) \leq 2n - 1$ notice that by removing the first 1 entry in every row and the last 1 entry in every column in an n-by-n 0-1 matrix one removes at most $2n - 1$ 1 entries (any 1 entry in the first column is also the first 1 entry in its row) and if the remaining matrix still contains a 1 entry, then the original matrix contains P. To see the reverse inequality $\mathrm{Ex}(n, P) \geq 2n - 1$ simply consider the n-by-n matrix with ones in the last row and first column and zeros elsewhere and notice that it does not contain P.

The extremal theory of 0-1 matrices predates the related theory of ordered graphs. Zoltán Füredi [11] established the extremal function for a specific 2-by-3 0-1 matrix and used this result for a problem in combinatorial geometry: he bounded the number of diagonals of equal length in a convex n-gon. Independently, Bienstock and Győri [3] found the extremal function of a few small 0-1 matrices. Later Füredi and Hajnal [12] started a systematic study of the extremal theory of 0-1 matrices. This latter paper not only contained many nice results, but was also rich in conjectures and had a significant effect on future research.

4   Connections between ordered and unordered extremal functions

Füredi and Hajnal [12] made the general conjecture that $\text{Ex}(P, n) = O(\text{ex}(\overline{G}_P, n) \log n)$, where P is any 0-1 matrix with positive weight and $\overline{G}_P$ is the simple graph underlying the ordered graph G_P. We will see that his conjecture is way too strong and fails in general, but it was still influential in subsequent research. It is not hard to see that a reverse of the conjectured inequality, namely $\text{ex}(\overline{G}_P, n) \le \text{Ex}(P, n)$ always holds, so the conjecture states that the two quantities $\text{Ex}(P, n)$ and $\text{ex}(\overline{G}_P, n)$ are close. We saw in Theorem 3.1 that $\text{Ex}(P, n)$ and $\text{ex}_<(G_P, n)$ are close, so here we focus on comparing $\text{ex}_<(G_P, n)$ and $\text{ex}(\overline{G}_P, n)$. In other words, we ask how much more edges can we have in an ordered graph of a given size if we only forbid one particular bipartite ordering of a simple graph G as a subgraph compared to forbidding all orderings of G. Note that it is important to insist that the forbidden ordering of a bipartite graph should have interval chromatic number two, otherwise Theorem 2.3 provides the easy answer: the two quantities can be very far apart. Indeed, for an ordered tree P of interval chromatic number larger than 2 one has $\text{ex}_<(P, n) = \Theta(n^2)$ but $\text{ex}(\overline{P}, n) = \Theta(n)$.

These considerations lead us to formulate the following question:

Question 1. *How high can the ratio $\frac{\text{ex}_<(P,n)}{\text{ex}(\overline{P},n)}$ be for an ordered bipartite graph P with more than one edge, where $\overline{P}$ stands for its underlying simple graph?*

The paper [24] gives the example with the largest known ratio. Let Q_k be the following ordered graph on $2k$ vertices $v_1, v_2, \ldots, v_{2k}$ (in this order) and the following $2k$ edges: $v_i v_{k+i}$ for $1 \le i \le k$, $v_{i+1} v_{k+i}$ for $1 \le i \le k-1$ and $v_1 v_{2k}$. Note that the interval chromatic number of Q_k is two and its underlying simple graph is the cycle C_{2k}.

Theorem 4.1 ([24]). *There exist ordered bipartite graphs on n vertices with $\Theta(n^{4/3})$ edges that contain none of the ordered graphs Q_k.*

Proof. Consider an arrangement of $n/2$ points and $n/2$ lines in the Euclidean plane with m *incidences*, i.e., point-line pairs with the point on the line. The celebrated Szemerédi–Trotter theorem [27] states that $m = O(n^{4/3})$. This theorem is tight, so we can select the points and the lines in such a way that $m = \Theta(n^{4/3})$. (For example, points of an appropriate square grid and the lines containing the most of these points will work.)

We turn this arrangement into a graph whose vertices are the points and lines in the arrangement and the edges correspond to pairs forming an incident. This graph has n vertices and $m = \Theta(n^{4/3})$ edges. We order the vertices of the constructed graph to turn it into an ordered graph. In our ordering point-vertices preceed line-vertices, this makes the constructed ordered graph ordered bipartite. We fix a coordinate system such that no line in the arrangement is vertical and order the point-vertices according their x-coordinates and we order the line-vertices according their slope (we break ties arbitrarily in both cases).

To finish the proof of the theorem we need to show that the constructed ordered graph does not contain Q_k for any k. Assume for a contradiction that it does, so the vertices v_i of Q_k correspond to points and lines in the arrangement. Clearly, a vertex v_i with $i \le k$ must correspond to a point p_i, while v_i with $i > k$ corresponds to a line l_i. Using the incidences represented by the edges $v_i v_{k+i}$ and $v_{i+1} v_{k+i}$ we conclude that the line segments $p_i p_{i+1}$ (belonging to the line l_{k+i}) form a convex chain for $i = 1, \ldots, k-1$. The contradiction comes from the observation that l_{2k} is the line $p_1 p_k$ connecting the two end points of this chain, thus its slope cannot exceed the slopes of all the segments in the chain (as it should since v_{2k} is the last in the vertex order). $\qquad\square$

Theorem 4.1 implies that $ex_<(Q_k, n) = \Omega(n^{4/3})$ for every k. On the other hand the simple graph underlying Q_k is the cycle C_{2k} and by the Bondy–Simonovits theorem [4] we have $ex(C_{2k}, n) = O(n^{1+1/k})$. This gives a lower bound of $\Omega(n^{1/3-1/k})$ for the ratio in Question 1 for the ordered graph Q_k. This also shows that conjecture mentioned in the beginning of this section fails for the bipartite adjacency matrix of Q_k whenever $k > 3$. We do not know if any pattern achieves a ratio of $\Omega(n^{1/3})$ in Question 1. For an upper bound for the same ratio we trivially have $O(n)$, as both the enumerator and the denominator are functions between n and n^2. In fact, they are $O(n^{2-\varepsilon})$ for some $\varepsilon > 0$ depending on the size of P by the Kővári–Sós–Turán theorem [18], so the ratio is always $O(n^{1-\varepsilon})$, but no better upper bound is known in general.

Question 1 asks how far the extremal function of a forbidden ordered bipartite can be from the extremal function of the family of all orderings of the same underlying graph. In a similar vein one can ask how far the extremal functions of two distinct bipartite orderings of the same underlying graph might be from each other. The author does not know of any results in this very interesting direction, but results of Győri, Korándi, Methuku, Tomon, Tompkins and Vizer on the extremal function of various bipartite orderings of even cycles, [15], might later prove useful to establish such gaps. This question is also related to the problem discussed in Section 7. The gap established in Theorem 4.1 will also show up between the extremal functions of two different bipartite orderings of the same even cycle or it is the case that forbidding all orderings of an even cycle of length at least 8 results in a substantially smaller extremal function than forbidding just one ordering. The latter would answer a variant of Question 2 in Section 7 (note that for simplicity we ask Question 2 about forbidding a pair of ordered graphs and here we need to consider larger families).

5 Forests

The Füredi–Hajnal paper [12] formulated the special case of their conjecture mentioned in the previous section separately for cycle-free patterns. Here we call a 0-1 matrix P cycle-free if the corresponding simple graph $\overline{G}_P$ is cycle-free, that is a forest. In this case, $ex(\overline{G}_P, n)$ (the extremal function of an unordered forest) is trivially linear, so their conjecture boils down to stating $Ex(P, n) = O(n \log n)$ for any cycle-free 0-1 matrix P. The log factor in the conjecture probably came from the the first matrix considered in this context [3, 37]:

$$T = \begin{pmatrix} 1 & 0 & 1 \\ 1 & 1 & 0 \end{pmatrix}$$

that happen to be cycle-free and its extremal function is $\Theta(n \log n)$.

Here we formulate a closely related but somewhat weaker conjecture:

Conjecture 1 *For an ordered bipartite forest P and any $c > 1$, we have*

$$ex_<(P, n) = o(n^c).$$

Note first that if this conjecture is true, then it characterizes the ordered graphs with almost linear extremal functions. Indeed, if P is not ordered bipartite, then $ex_<(P, n) = \Theta(n^2)$ by Theorem 2.3, while if the underlying graph $\overline{P}$ contains a cycle, then $ex_<(P, n) \geq ex(\overline{P}, n) = \Omega(n^c)$ for some $c > 1$. The latter statement follows from a simple application of the probabilistic method.

Note that stronger conjectures could be formulated by replacing $o(n^c)$ with a bound $O(n \log^c n)$ for a constant $c = c_P$ depending on P, or even with an $O(n \log n)$ bound. Conjecture 1 and the conjecture with the $O(n \log^c n)$ bound are still open and by Theorem 3.1 are equivalent to the similar conjectures about $\text{Ex}(M, n)$ for cycle-free 0-1 matrices M. The strongest form of the conjecture (an $O(n \log n)$ bound) was also considered for a while and was supported by the fact that it was easy to find an extremal function of order $\Theta(n \log n)$, but there was no known example of an ordered bipartite forest whose extremal function grows faster. If true, it would imply the Füredi–Hajnal conjecture for cycle-free patterns mentioned above. But Seth Pettie, [26], found a cycle-free 0-1 matrix with extremal function slightly higher than $n \log n$: for this matrix M one has $\text{Ex}(M, n) = \Omega(n \log n \log \log n)$. By this, he also disproved the strengthening of Conjecture 1 with the $O(n \log n)$ upper bound, but the conjecture may still hold with the bound $O(n \log^2 n)$. Pettie's result was slightly improved and the current best lower bound is due Park and Shi [25]. They found cycle-free 0-1 matrices M_m with $\text{Ex}(M_m, n) = \Omega(n \log n \log \log n \cdots \log^{(m)} n)$, where $\log^{(m)}$ denotes the m-times-iterated logarithm function.

On the positive side, $\text{ex}_<(P, n) = O(n \log^c n)$ was established in [24] for all ordered bipartite forests with at most 6 vertices. The exponent c in this result can be chosen to be three less than the number of vertices in P. For most of the small ordered bipartite graphs the bound follows from this simple observation.

Lemma 5.1 ([24]). *Let P be a 0-1 matrix. Suppose that the last column of P contains a single 1 entry and let us obtain P' from P by deleting this last column. We have*

$$\text{Ex}(P, n) = O(\text{Ex}(P', n) \log n).$$

As the example of the matrix T above shows, the extra log factor is sometimes necessary in Lemma 5.1. It is reasonable to conjecture the following stronger form of this lemma also holds. If so, it easily implies Conjecture 1, even with the stronger $O(n \log^c n)$ bound.

Conjecture 2. *Let P be 0-1 matrix and let us obtain P' from P by deleting a column that contains a single 1 entry. We have*

$$\text{Ex}(P, n) = O(\text{Ex}(P', n) \log n).$$

The most general positive result toward Conjecture 1 appears in the paper [17] by Korándi, Tardos, Tomon and Weidert. They call a split of a 0-1 matrix P into two matrices P' and P'' a *legal horizontal split* if P is obtained by placing P' atop P'' (so in particular all three matrices have the same number of columns) and at most one of the columns have a one entry in both P' and P''. A 0-1 matrix P is *vertically degenerate* if it can be partitioned into single line matrices by a sequence of legal horizontal splits. Note that all vertically degenerate 0-1 matrices are cycle-free. All cycle-free 0-1 matrices with at most three rows are vertically degenerate, but there are 4-row cycle-free 0-1 matrices that are not vertically degenerate (see below). Using a density increment argument, they prove the following theorem.

Theorem 5.2 ([17]). *Let P be a vertically degenerate 0-1 matrix with l rows. We have*

$$\text{Ex}(P, n) = n2^{O(\log^{1-1/l} n)}.$$

This result implies that Conjecture 1 holds for all ordered graphs G_P, where P is a vertically degenerate 0-1 matrix. By symmetry, Conjecture 1 is also true for all G_P, where P is *horizontally degenerate*, that is, the transpose of P is vertically degenerate, but it has not been proved for any larger class of ordered bipartite forests. The smallest open case is an ordered path on 8 vertices, specifically G_M for the matrix

$$M = \begin{pmatrix} 1 & 0 & 1 & 0 \\ 0 & 1 & 0 & 1 \\ 0 & 0 & 1 & 0 \\ 1 & 0 & 0 & 1 \end{pmatrix}.$$

Note that M has no legal horizontal or vertical split (where a *legal vertical split* is the legal horizontal split of the transpose). The following matrix N can be split into trivial (one-by-one) matrices using a sequence of vertical and horizontal splits, but still it is neither vertically nor horizontally degenerate because the splits alternate in direction. Verifying Conjecture 1 for such matrices is probably simpler than for matrices like M above and may be the next logical step in verifying Conjecture 1.

$$N = \begin{pmatrix} 1 & 0 & 1 & 0 & 1 \\ 0 & 1 & 0 & 1 & 0 \\ 1 & 0 & 0 & 0 & 0 \\ 0 & 1 & 0 & 0 & 1 \end{pmatrix}$$

In the rest of this section we give a very rough sketch of the proof of Theorem 5.2 because it may give some insights as to the limitation of this particular technique. For the technical details (for example the exact settings of the parameters) we refer to the paper [17]. Let us introduce some notation. Let A be an m-by-kn 0-1 matrix. We consider A to be the union of k vertical blocks, each consisting of n consecutive columns. We say that the 0-1 matrix Q with k columns has a *block respecting embedding* in A if A contains Q in such a way that the submatrix of A dominating Q has all its columns coming from distinct vertical blocks (that is, the column corresponding to the ith column of Q must come from the ith vertical block of A for every i). The advantage of block respecting embeddings is that if Q has a legal horizontal split into matrices Q' and Q'', then it is much easier to combine the block respecting embeddings of Q' and Q'' to form a block respecting embedding of Q than it is without the extra condition. Indeed, one only has to check that Q' uses rows of A higher than the ones used by Q'' and the single column that has a 1 entry in both Q' and Q'' use the same column of A. The disadvantage is that simply requiring that A has a lot of 1 entries is not enough to force the existence of a block respecting embedding. Instead we will insist that the 1 entries are evenly distributed with the following definition: we say that A is (k, u)-*complete* if among the n entries in the intersection of any row and any vertical block, one always finds at least u 1-entries.

We say that a 0-1 matrix Q with k columns *easily embeds* if for a certain range of the parameters n, m and u and for every m-by-kn (k, u)-complete 0-1 matrix A either Q has a block respecting embedding in A or one can find a submatrix of A which is significantly denser than A itself. We do not give the precise values of m and u required here, but the reader may think of $u = n^\varepsilon$ for some $\varepsilon > 0$ and "significantly denser" may mean an s-by-s submatrix with weight $s^{1+\varepsilon'}$, where $\varepsilon' > \varepsilon$ depends on ε but not on n. The proof of Theorem 5.2 is based on two lemmas. The first states that if Q has a legal horizontal split to Q' and Q'' and both Q' and Q'' easily embed, then so does Q (with a slight deterioration in the parameters). As single line matrices easily embed, this lemma implies the same for all vertically degenerate matrices. The second lemma takes care of the extra uniformity condition. It states that any 0-1 matrix A has a (k, u)-complete submatrix with comparable size and density to A or a submatrix of significantly larger density.

Consider the 0-1 matrix

$$Q = \begin{pmatrix} 1 & 1 & 0 \\ 0 & 0 & 1 \\ 1 & 0 & 0 \\ 0 & 1 & 1 \end{pmatrix}.$$

Q is horizontally degenerate, so G_Q satisfies the statement of Conjecture 1, but it is not vertically degenerate, and so we do not know if it easily embeds. It is easy to see that permuting the columns of a 0-1 matrix does not ruin the property that it easily embeds, neither does adding extra columns with a single 1 entry. In this way showing that Q easily embeds would imply the same for both matrices M and N above and would establish the statement of Conjecture 1 for G_M and G_N.

6   Linear extremal functions

Füredi and Hajnal [12] conjectured and later Marcus and Tardos [20] proved that $Ex(P, n) = O(n)$ for permutation matrices P. It is not hard to see that this result can be restated in the following equivalent form (although Theorem 3.1 does not directly imply this equivalence).

Theorem 6.1 ([20]). *The extremal function of any ordered bipartite matching P is linear. That is,*

$$ex_<(P, n) = O(n).$$

Conjecture 1, if true, characterizes all ordered graphs with almost linear extremal functions. It would be nice to find a characterization of ordered graphs or 0-1 matrices with linear extremal functions. One possibility is finding all *minimally nonlinear matrices*. We call a 0-1 matrix P minimally nonlinear, if its extremal function $Ex(P, n)$ is nonlinear, but $Ex(P', n) = O(n)$ for all 0-1 matrices $P' \neq P$ contained in P. It might be possible to find such a characterization, but the following theorem indicates that this is a difficult task:

Theorem 6.2 (Geneson and Keszegh [13, 16]). *There are infinitely many minimally nonlinear matrices.*

Note that "minimally nonlinear" simple graphs (for the classic extremal graph theory) are well understood despite the fact that there are infinitely many of them: they are the cycles. Keszegh, [16], found a sequence of 0-1 matrices $H_0, H_1, \ldots$ (shown below with the zeros omitted for clarity) that show some repetitive behavior. He did not prove that they are minimally nonlinear, instead he showed that they are nonlinear, specifically $Ex(H_i, n) = \Omega(n \log n)$ for all i, and thus each contains a minimally nonlinear matrix. Then Geneson, [13], showed that no two of them can contain the same nonlinear matrix.

$$H_0 = \begin{pmatrix} & 1 & 1 & \\ & & & 1 \\ & & & 1 \\ 1 & & & \end{pmatrix}$$

$$H_1 = \begin{pmatrix} & 1 & 1 & & & & & \\ & & & & & 1 & & \\ & & & & 1 & & & \\ 1 & & & & & & & \\ & & & & & & 1 & \\ & & & & & & 1 & \\ & & & 1 & & & & \end{pmatrix}$$

$$H_2 = \begin{pmatrix} & 1 & 1 & & & & & & \\ & & & & & 1 & & & \\ & & & & 1 & & & & \\ 1 & & & & & & & & \\ & & & & & & & 1 & \\ & & & & & & 1 & & \\ & & & 1 & & & & & \\ & & & & & & & & 1 \\ & & & & & & & & 1 \\ & & & & 1 & & & & \end{pmatrix}$$

7 Interaction between ordered graphs

In this section we compare the extremal functions of families of several forbidden patterns with the extremal function of just a single member of that family. Let us start with the classical extremal theory of graphs. Clearly, we have

$$\text{ex}(\{G, H\}, n) \leq \min(\text{ex}(G, n), \text{ex}(H, n)). \qquad (*)$$

By Theorem 2.2, the two sides are asymptotically the same for non-bipartite graphs G and H. It is easy to see that they differ by a factor of less than 2 if only one of the graphs is bipartite. Indeed, if G is bipartite and H is not, one can make any simple graph avoiding G itself bipartite by deleting less than half of its edges. For bipartite graphs, the situation is more complicated. We say that G and H *interact* if the two sides of $(*)$ differ more than by a constant factor. It is not known if there exists any interacting pair of graphs. Erdős and Simonovits, [8], conjecture that there exists no interaction between graphs, but more recently Faudree and Simonovits, [10], conjecture the opposite. Specifically, they conjecture that the cycle C_4 and the subdivision of the complete graph K_4, in which each edge is subdivided with a single new vertex, do interact.

Let us emphasize that here we do not care for constant factors in the extremal functions. Finding *weakly interacting* pairs, that is, where $\text{ex}(\{G, H\})$ is a constant factor less than $\min(\text{ex}(G, n), \text{ex}(H, n))$ is considerably simpler. The Erdős–Stone–Simonovits theorem prevents even weak interactions between non-bipartite graphs, but a bipartite and a non-bipartite graph can interact weakly. Specifically, Erdős and Simonovits [8] prove that C_4 and C_5 do interact weakly. Similar questions were also studied in the context of uniform hypergraphs, where answering a question of Mubayi and Rödl, [23], Mubayi and Pikhurko, [22] find weakly interacting pairs of r-uniform hypergraphs with extremal function $\Theta(n^r)$ for all $r > 2$. The weakly interacting pairs (or families) with such high extremal functions are called *non-principal* families. Earlier József Balogh, [2], found non-principal families of 3-uniform hypergraphs of larger finite size. Note that for graphs (i.e., $r = 2$) non-principal families do not exist.

In contrast to graphs, it is not hard to find a lot of interactions in the extremal theory of ordered graphs and 0-1 matrices. Consider the 3-by-2 matrix $T = \begin{pmatrix} 1 & 0 & 1 \\ 1 & 1 & 0 \end{pmatrix}$. Füredi [11] and Bienstock and Győri [3] proved that $\mathrm{Ex}(T_1) = \Theta(n \log n)$. By symmetry, the extremal functions of the matrices $T_2 = \begin{pmatrix} 1 & 1 & 0 \\ 1 & 0 & 1 \end{pmatrix}$, $T_3 = \begin{pmatrix} 1 & 0 & 1 \\ 0 & 1 & 1 \end{pmatrix}$ and $T_4 = \begin{pmatrix} 0 & 1 & 1 \\ 1 & 0 & 1 \end{pmatrix}$ are the same. The following theorem implies that each of T_2, T_3 and T_4 interacts with T:

Theorem 7.1 ([28]).
$$\mathrm{Ex}(\{T, T_2\}, n) = \Theta(n)$$
$$\mathrm{Ex}(\{T, T_3\}, n) = \Theta(n \log n / \log \log n)$$
$$\mathrm{Ex}(\{T, T_4\}, n) = \Theta(n \log \log n)$$

The close connection between the extremal functions of 0-1 matrices and ordered graphs makes it easy to turn these interactions into interactions between ordered graphs.

These results represent the first step toward exploring interactions between different patterns. It would be interesting to find stronger interactions, where the ratio between the right and left sides of $(*)$ is larger than logarithmic, ideally a power on n. We remark that the conjectured interaction between bipartite graphs in [10] is of this stronger nature.

Question 2 *Are there ordered graphs G and H such that*
$$\mathrm{ex}_<(\{G, H\}, n) = O(\min(\mathrm{ex}_<(G, n), \mathrm{ex}_<(H, n))/n^\epsilon)$$
for some $\epsilon > 0$?

8 Edge ordered graphs

In this final section we survey some preliminary results from ongoing research of the author with Dániel Gerbner, Abhishek Methuku, Dániel T. Nagy, Dömötör Pálvölgyi and Máté Vizer [14] on the extremal theory of edge ordered graphs. Recall that we defined edge ordered graphs in Section 1 analogously to (vertex) ordered graphs, but now the linear order is on the edges of a simple graph. The extremal function for a family $\mathcal{P}$ of edge ordered graphs is also defined analogously: we are looking for the largest number $\mathrm{ex}'_<(\mathcal{P}, n)$ of edges in an n-vertex edge ordered graph with no edge ordered subgraph isomorphic to any member of $\mathcal{P}$. We require that $\mathcal{P}$ does not contain empty graphs and we write $\mathrm{ex}'_<(\mathcal{P}, n)$ to denote $\mathrm{ex}'_<(\mathcal{P}, n)$ when $\mathcal{P} = \{P\}$ is a singleton.

As a natural first step we generalize the Erdős–Stone–Simonovits theorem for edge ordered graphs. As for the corresponding theorem for vertex ordered graphs, namely Theorem 2.3, the main thing here is to find the "correct" notion of the chromatic number. Then the result follows easily from the original Erdős–Stone theorem [9]. We say that a simple graph *strongly contains* the edge ordered graph G if every edge ordering of H contains G as an edge ordered subgraph. We define the *order chromatic number* of an edge ordered graph G to be the smallest chromatic number of a simple graph strongly containing G. In case no graph strongly contains G we say that the order chromatic number of G is infinity.

Theorem 8.1 (Erdős–Stone–Simonovits Theorem for a single edge ordered forbidden graph). *If the order chromatic number of the edge ordered graph G is infinity we have*
$$\mathrm{ex}'_<(G, n) = \binom{n}{2}.$$

If the order chromatic number of G is $r+1$ we have

$$\mathrm{ex}'_<(G,n) = (1 - 1/r)\frac{n^2}{2} + o(n^2).$$

This theorem determines the extremal function $\mathrm{ex}'_<(G,n)$ exactly if the order chromatic number of G is infinity; it determines the extremal function asymptotically, if the order chromatic number is larger than 2, but tells fairly little in case the order chromatic number is 2.

As an example consider the cycle C_4 and its three different edge orderings: C_4^{1234}, C_4^{1324} and C_4^{1243}. Here we number the edges according the ordering and the upper index represents the numbers along the cycle, so in C_4^{1243} the first and last edges are opposite in the cycle. Any simple graph can be edge ordered by first imposing a linear order on the vertices, then ordering the edges ab with $a < b$ by the lexicographic order on the pairs (a,b). We call this the *lexicographic edge order*. It is easy to see that no graph with a lexicographic edge order contains either C_4^{1234} or C_4^{1324}. This means that these two edge ordered graphs have order chromatic number infinity. The order chromatic number of C_4^{1243} is 2. This can be seen directly or follows from the following result. For the statement of the result we need to define the edge ordered graph $K_{n,n}^{\mathrm{lex}}$. This is the lexicographic edge ordering of the complete bipartite graph $K_{n,n}$ obtained from a vertex order in which one vertex class precedes the other.

Theorem 8.2. *The non-empty edge ordered graph G on n vertices has order chromatic number 2 if and only if it is contained in $K_{n,n}^{\mathrm{lex}}$.*

Theorem 8.1 says little about the extremal function of C_4^{1243}. Naturally, the edge ordering can only increase the extremal function, so we have

$$\mathrm{ex}'_<(C_4^{1243}, n) \geq \mathrm{ex}(C_4, n) = \Theta(n^{3/2}).$$

Applying techniques of the paper [21] we prove a nearly matching upper bound:

Theorem 8.3. $\mathrm{ex}'_<(C_4^{1243}, n) = O(n^{3/2} \log n)$.

We do not know if the logarithmic term is needed in this estimate.

We list here results on a selection of other specific forbidden edge ordered graphs of order chromatic number two. We start with edge ordered graphs whose connected components are (edge ordered) stars. We call them edge ordered star forests. Their extremal functions are obtained through known estimates in generalized Davenport–Schinzel theory.

Theorem 8.4. *For any edge ordered star forest F we have $\mathrm{ex}'_<(F,n) \leq n 2^{\alpha(n)^c}$ for some exponent c depending on F. Here α denotes the inverse of the Ackermann function. For the edge ordered star forest F_0 consisting of two components and five edges with one component consisting of the second and fourth edge we have $\mathrm{ex}'_<(F_0, n) = \Omega(n\alpha(n))$.*

We specify edge orderings of a path P_{k+1} by an upper index listing the ranks of the k edges along the path, so for example P_5^{1342} stands for the edge ordered path where the edges along the path follow as first-third-fourth-second. This is the shortest edge ordered path where we could not find the exact order of magnitude of the extremal function.

Theorem 8.5. *For an edge ordered path P on three edges we have $\mathrm{ex}'_<(P,n) = \Theta(n)$. For an edge ordered path P on four edges we have either $\mathrm{ex}'_<(P,n) = \Theta(n)$ or $\mathrm{ex}'_<(P,n) = \Theta(n\log n)$ or $\mathrm{ex}'_<(P,n) = \binom{n}{2}$ or P is isomorphic to P_5^{1342} or the equivalent P_5^{4213}. In this last case we have $\mathrm{ex}'_<(P,n) = \Omega(n\log n)$ and $\mathrm{ex}'_<(P,n) = O(n\log^2 n)$.*

In the rest of this section we consider forbidden families of edge ordered graphs and possible weak interaction (see Section 7) between the members of such a family. Note that we formulated Theorem 8.1 for a single forbidden edge ordered graph. This is because some families of forbidden edge ordered graphs behave differently than any of their members alone, see Theorem 8.7(ii) below for an example. This contrasts with the situation for simple graphs and vertex ordered graphs, where the Erdős–Stone–Simonovits theorem (Theorem 2.2) and its variant for vertex ordered graphs (Theorem 2.3) prevent any such weak interaction. To be more specific, we will see that such weak interaction does not happen between two edge ordered graphs of order chromatic number three but it does happen between certain edge ordered graphs of order chromatic number four and above. To be able to generalize the Erdős–Stone–Simonovits theorem to families of forbidden edge ordered graphs we need to extend the definition of order chromatic number from edge ordered graphs to families of such graphs. Let the *order chromatic number of a family* $\mathcal{P}$ be the smallest chromatic number of a simple graph H such every edge ordering of H contains a member of $\mathcal{P}$ as an edge ordered subgraph. Again, if no such H exists, the order chromatic number is infinite. With this definition we have the following generalization of Theorem 8.1.

Theorem 8.6 (Erdős–Stone–Simonovits Theorem for a family of forbidden edge ordered graphs). *If the order chromatic number of the family $\mathcal{P}$ of edge ordered graphs is infinity we have*

$$\mathrm{ex}'_<(\mathcal{P}, n) = \binom{n}{2}.$$

If the order chromatic number of $\mathcal{P}$ is $r+1$ we have

$$\mathrm{ex}'_<(\mathcal{P}, n) = (1 - 1/r)\frac{n^2}{2} + o(n^2).$$

The following result gives a specific example when a pair of edge ordered graphs behaves differently than either members.

Theorem 8.7.

(i) *The order chromatic number of a family of edge ordered graphs is 2 if and only if the family has a member with order chromatic number 2.*

(ii) *The order chromatic number of both edge orderings P_5^{1423} and P_5^{2314} of the path P_5 is infinity, but the order chromatic number of the family $\{P_5^{1423}, P_5^{2314}\}$ is 3.*

Part (i) of this theorem follows easily from Theorem 8.2. For part (ii) note that no graph with a lexicographic edge ordering contains P_5^{2314}, so the order chromatic number of P_5^{2314} is infinity. The same statement about P_5^{1423} follows from symmetry. The statement on the pair $\{P_5^{1423}, P_5^{2314}\}$ can be derived directly or follows from an analogue of Theorem 8.2 for families of edge ordered graphs of order chromatic number 3. While these analogous results for order chromatic number three and above can be easily deduced from Ramsey's theorem, one has to deal with exceedingly many different *homogeneous* edge orderings. Here we state the result for order chromatic number infinity only, where we have to consider only four homogeneous edge orderings of the complete graph K_n as follows: Let $K_n^{(1)}$ be the lexicographic edge ordering of K_n. For $K_n^{(2)}$ consider the vertex set of K_n to be $\{1, 2, \ldots, n\}$ and order the edges ab with $a < b$ according to the lexicographic order on the pairs $(a, -b)$. Let us obtain $K_n^{(3)}$ and $K_n^{(4)}$ by reversing the edge order in $K_n^{(1)}$ and $K_n^{(2)}$, respectively.

Theorem 8.8. *The order chromatic number of a family $\mathcal{P}$ of edge ordered graphs is infinity if and only if there exists $1 \leq i \leq 4$ such that the graphs $K_n^{(i)}$ contain no member of $\mathcal{P}$ for any n.*

Acknowledgment

The author is indebted to the anonymous referee for careful reading, important corrections and useful suggestions.

References

[1] M. Balko, J. Cibulka, K. Král, and J. Kynčl, Ramsey numbers of ordered graphs, *Electronic Notes in Discrete Mathematics* **49** (2015), 419–424.

[2] J. Balogh, The Turán density of triple systems is not principal, *Journal of Combinatorial Theory, Series A* **100** (2002), 176–180.

[3] D. Bienstock and E. Győri, An extremal problem on sparse 0-1 matrices, *SIAM Journal on Discrete Mathematics* **4** (1991), 17–27.

[4] J.A. Bondy and M. Simonovits, Cycles of even length in graphs, *Journal of Combinatorial Theory, Series B* **16** (1974), 97–105.

[5] P. Braß, Gy. Károlyi, and P. Valtr, A Turán-type extremal theory of convex geometric graphs, in: *Discrete and Computational Geometry–The Goodman-Pollack Festschrift*, editors B. Aronov et al., Springer, Berlin, 2003, pp. 275–300.

[6] D. Conlon, J. Fox, C. Lee, and B. Sudakov, Ordered Ramsey numbers, *Journal of Combinatorial Theory, Series B* **122** (2017), 353–383.

[7] P. Erdős and M. Simonovits, A limit theorem in graph theory, *Studia Scientiarum Mathematicarum Hungarica* **1** (1966), 51–57.

[8] P. Erdős and M. Simonovits, Compactness results in extremal graph theory, *Combinatorica* **2** (1982), 275–288.

[9] P. Erdős and A.H. Stone, On the structure of linear graphs, *Bulletin of the American Mathematical Society* **52** (1946), 1087–1091.

[10] R.J. Faudree and M. Simonovits, On a class of degenerate extremal graph problems II, manuscript.

[11] Z. Füredi, The maximal number of unit distances in a convex n-gon, *Journal of Combinatorial Theory, Series A* **55** (1990), 316–320.

[12] Z. Füredi and P. Hajnal, Davenport-Schinzel theory of matrices, *Discrete Mathematics* **103** (1992), 233–251.

[13] J.T. Geneson, Extremal functions of forbidden double permutation matrices, *Journal of Combinatorial Theory, Series A* **116** (2009), 1235–1244.

[14] D. Gerbner, A. Methuku, D.T. Nagy, D. Pálvölgyi, G. Tardos, and M. Vizer, Edge-ordered Turán problems, in preparation.

[15] E. Győri, D. Korándi, A. Methuku, I. Tomon, C. Tompkins, and M. Vizer, On the Turán number of some ordered even cycles, *European Journal of Combinatorics* **73** (2018), 81–88.

[16] B. Keszegh, On linear forbidden submatrices, *Journal of Combinatorial Theory, Series A* **116** (2009), 232–241.

[17] D. Korándi, G. Tardos, I. Tomon, and C. Weidert, On the Turán number of ordered forests, manuscript, see arXiv:1711.07723.

[18] T. Kővári, V.T. Sós, and P. Turán, On a problem of K. Zarankiewicz, *Colloquium Mathematicae* **3** (1954), 50–57.

[19] W. Mantel, Problem 28 (Solution by H. Gouwentak, W. Mantel, J. Teixeira de Mattes, F. Schuh and W. A. Wythoff), *Wiskundige Opgaven* **10** (1907), 60–61.

[20] A. Marcus and G. Tardos, Excluded permutation matrices and the Stanley-Wilf conjecture, *Journal of Combinatorial Theory, Series A* **107** (2004), 153–160.

[21] A. Marcus and G. Tardos, Intersection reverse sequences and geometric applications, *Journal of Combinatorial Theory, Series A* **113** (2006), 675–691.

[22] D. Mubayi and O. Pikhurko, Constructions of non-principal families in extremal hypergraph theory, *Discrete Mathematics* **308** (2008), 4430–4434.

[23] D. Mubayi and V. Rödl, On the Turán number of triple systems, *Journal of Combinatorial Theory, Series A* **100** (2002), 135–152.

[24] J. Pach and G. Tardos, Forbidden paths and cycles in ordered graphs and matrices, *Israel Journal of Mathematics* **155** (2006), 359–380.

[25] S.G. Park and Q. Shi, New bounds on extremal numbers in acyclic ordered graphs, manuscript.[3]

[26] S. Pettie, Degrees of nonlinearity in forbidden 0-1 matrix problems, *Discrete Mathematics* **311** (2011), 2396–2410.

[27] E. Szemerédi and W.T. Trotter, A combinatorial distinction between the Euclidean and projective planes, *European Journal of Combinatorics* **4** (1983), 385–394.

[28] G. Tardos, On 0-1 matrices and small excluded submatrices, *Journal of Combinatorial Theory, Series A* **111** (2005), 266–288.

[29] G. Tardos, Extremal theory of ordered graphs, in: *Proceedings of the International Congress of Mathematics 2018 (ICM 2018)*, editors B. Sirakov et al., World Scientific, 2019, Volume 3, pp. 3219–3228.

[30] P. Turán, On an extremal problem in graph theory, *Matematikai és Fizikai Lapok* **48** (1941), 436–452.

Rényi Institute of Mathematics
Budapest, Hungary
tardosrenyi.hu

---

[3]See https://math.mit.edu/research/undergraduate/spur/documents/2013ParkShi.pdf

Some combinatorial and geometric constructions of spherical buildings

Hendrik Van Maldeghem and Magali Victoor

Abstract

We survey some known and new combinatorial and geometric constructions of the Lie incidence geometries related to some spherical buildings, in particular of some exceptional types.

1 Introduction

The current paper grew out of a geometric approach to the Freudenthal–Tits Magic Square. This Square is most popular in the area of Lie algebras and algebraic geometry. However, Tits [6] introduced an incidence geometric version of the square in his habilitation thesis, ten years before he gave a general formula for the Lie algebras appearing in the Square. These incidence geometries are now generally known as instances of Lie incidence geometries.

Our geometric approach focuses on characterizations of the point-line geometries of the Square and their embeddings into projective space. During this study a lot of natural geometric and combinatorial connections between the cells of the Square became apparent. This paper reports on these connections as far as constructions are concerned. However, also other types of Lie incidence geometries are involved, and certain generalizations lead to constructions of classes of Lie incidence geometries that have no direct connection anymore with the Square.

We present four types of constructions, many of them disposing links between Lie incidence geometries. Each time we mention which cells of the Square are involved, but we also give many examples outside the Square. A systematic treatment from this viewpoint of the Square is beyond the scope of this paper, and would lead us too far. Instead, we will present our constructions as examples of geometries related to the square.

In the final section we present some constructions of Coxeter complexes related to the exceptional types. This implies alternative constructions of some well-known graphs like some Johnson graphs and the Gosset graph. These constructions are mainly an application of the section about equator and trace geometries.

2 Definitions—Spherical Buildings and Lie Incidence Geometries

2.1 Coxeter systems and Coxeter complexes

Definition 2.1. Let W be a group generated by a finite nonempty set $S = \{s_1, \ldots, s_n\}$ of involutions and let, for each pair $(s_i, s_j) \in S \times S$, the number m_{ij} be the order of the product $s_i s_j$ (setting $m_{ij} = \infty$ if s_i, s_j generated an infinite group). Then (W, S) is a Coxeter system if W can be presented as $W = \langle S : (s_i s_j)^{m_{ij}} = 1, \forall i, j \in \{1, 2, \ldots, n\}\rangle$. The natural number n is called the *rank* of the system.

The symmetric matrix m_{ij} is called the *Coxeter matrix* belonging to (W, S). The *Coxeter diagram* is the edge labelled graph $\Gamma(W, S)$ with vertex set S and no edge between s_i and s_j if $m_{ij} = 2$; otherwise an edge with label (m_{ij}) between s_i and s_j, for all $i, j \in \{1, 2, \ldots, n\}$. The labels of edges with label (3) are usually omitted, those with label (4) are usually drawn as a a double edge, and those with label (6) are sometimes drawn as a triple edge.

Let (W, S) be a Coxeter system. If $S = S_1 \cup S_2$, with $W = \langle S_1 \rangle \times \langle S_2 \rangle$ (then automatically $S_1 \cap S_2 = \emptyset$), then we say that (W, S) is *reduced*. If (W, S) is not reduced, then it is called *irreducible*.

In this paper we will only be concerned with finite irreducible Coxeter groups, that moreover arise as automorphism group of a crystallographic root system. We will not need the precise definition; it will suffice to know that crystallographic root systems are classified by Dynkin diagrams, which are Coxeter diagrams where every edge labelled (ℓ), with $\ell \geq 4$, gets an orientation.

Here is the list of Dynkin diagrams of irreducible crystallographic root systems (i.e., the corresponding Coxeter group is irreducible). The nodes of the diagram are labelled according to the standard conventions introduced by Bourbaki [1] (we can think of the node with label i as the one corresponding with $s_i \in S$). Thinking away the arrow gives the corresponding Coxeter group.

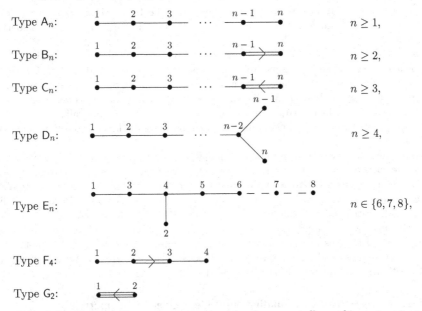

The irreducible finite Coxeter groups not arising from crystallographic root systems are the automorphism groups of the dodecahedron and the 600-cell in real Euclidean 4-space, and the finite dihedral groups D_{2n} (of which D_4 is reducible, D_6 is of type A_2, D_8 of types B_2 and C_2 and D_{12} is of type G_2).

The irreducible finite Coxeter groups and systems arising from the crystallographic root system of type X_n, with $X \in \{A, \ldots, G\}$ and n appropriate, is called of type X_n itself. This is unambiguous except for types B and C, which coincide for Coxeter groups and systems.

Definition 2.2. Let (W, S) be a Coxeter system. A *standard parabolic subgroup* is a subgroup of W generated by a proper subset of S. A *parabolic subgroup* is a conjugate of a standard parabolic subgroup. A *maximal* standard parabolic subgroup is one not contained in another one, i.e., generated by all but one elements of S.

Definition 2.3. Recall that a *simplicial complex* consists of a set X, whose elements are usually called *vertices*, and a family Ω of subsets of X, called *simplices*, with the only condition that every subset of a member of Ω is also contained in Ω. It is called a *chamber complex* if the maximal simplices, called *chambers*, are finite and all have the same cardinality, which is then called the *rank* of the complex. Two chambers C and C' with $|C \setminus C'| = 1$ are called *adjacent*, and in this case $C \cap C'$ is called a *panel*.

Definition 2.4. Let (W, S) be a Coxeter system. Then we define a simplicial complex $\Delta(W, S) = (X, \Omega)$ as follows. The set X consists of the cosets of the maximal standard parabolic subgroups. A subset of X belongs to Ω if it is the set of cosets of all maximal standard parabolic subgroups containing a coset of a not necessarily maximal standard parabolic subgroup.

It follows from the definition that $\Delta(W, S)$ is a chamber complex, that the chambers of $\Delta(W, S)$ are in one-to-one correspondence with the elements of W and that chambers have size $|S|$. Moreover, every panel is contained in exactly two chambers, since the standard parabolic subgroup generated by a single element of S has size 2.

2.2 Buildings and spherical buildings

We are now ready to define the concept of a building.

Definition 2.5. Let (W, S) be a Coxeter system. A building of type (W, S) is a simplicial chamber complex (X, Ω) endowed with a family $\mathcal{A}$ of subcomplexes called *apartments* all isomorphic to $\Delta(W, S)$ such that

(B1) Every pair of simplices of (X, Ω) is contained in a member of $\mathcal{A}$;

(B2) If two simplices A, A' are contained in two apartments Σ, Σ', then there exists an isomorphism $\Sigma \to \Sigma'$ fixing $A \cup A'$ pointwise.

The Coxeter group W is sometimes also called the *Weyl group* of the building. If (W, S) is of type X_n, with $\mathsf{X} \in \{\mathsf{A}, \ldots, \mathsf{G}\}$ and n appropriate, then the building is also said to be of type X_n itself.

The definition of a building is rather abstract and does not immediately link up with incidence geometry. However, every building gives rise to many point-line incidence geometries by the following procedure.

Let $\Delta = (X, \Omega)$ be a building of type (W, S). First we note that Δ is a *numbered complex*, i.e., one can assign types to the elements of X such that every chamber contains exactly one vertex of each type. These types are in one-to-one correspondence with the elements of S. Indeed, if C is a chamber and $x \in C$, then in any apartment containing C, the panel $C \setminus \{x\}$ corresponds to a coset of a standard parabolic subgroup of size 2, hence generated by a single element $s_i \in S$. Then we assign to x the type i. The type T of a simplex is then the union of the types of its elements, and the cotype is the complement $I \setminus T$. Set $I = \{1, 2, \ldots, n\}$. Then we choose a subset $J \subseteq I$ and define the following point-line geometry $\Delta_J = (\mathcal{P}_J, \mathcal{L}_J)$. The set $\mathcal{P}_J$ of points is just the set of all simplices of type J. A generic line is the set of simplices of type J for which the union with a fixed panel of cotype $j \in J$ is a chamber. We call Δ_J the J-Grassmannian geometry of Δ.

In the present paper we are interested mainly in J-Grassmannian geometries of spherical buildings which arise from simple algebraic groups (a *spherical* building is a building with finite Weyl group). Given an isotropic algebraic group G of relative rank n, there is a precise procedure that produces a spherical building of rank n, $n \geq 2$. In fact, this procedure is exactly the same as above for Coxeter groups, if we choose a fixed Borel in G and call the subgroups containing this Borel standard parabolic subgroups. When G is defined over an algebraically closed field $\overline{k}$, then G is completely determined by $\overline{k}$ and an irreducible crystallographic root system, or, equivalently, a connected Dynkin diagram X_n. We will denote the corresponding building by $\mathsf{X}_n(\overline{k})$. Note that the corresponding Coxeter complex is also a building, and we denote it by $\mathsf{X}_n(1)$, alluding to the point of view that these Coxeter groups are algebraic groups over the field of order 1. If k is finite, we also write $|k|$ instead of k.

Now, for an arbitrary field k, we can consider a simple algebraic group $G(\overline{k})$ over the algebraic closure $\overline{k}$ of k, and take the k-rational points (Galois descent) in such a way that the relative type (the corresponding Weyl group) is the same as the original one. In this case we denote the corresponding building by $\mathsf{X}_n(k)$. All such buildings are called *split buildings*.

For $J \subset I$, the J-Grassmannian geometry of $\mathsf{X}_n(k)$ is denoted by $\mathsf{X}_{n,J}(k)$ (where braces around the unique element of J are omitted if $|J| = 1$). These geometries are called *(split) Lie incidence geometries*. They are called *simple* when $|J| = 1$, and they are called *long root geometries* if J corresponds to the set of fundamental roots adjacent to the longest root. We have not defined the notions in the previous sentence, but it will suffice for us to list these subsets J:

type	J		type	J
A_n	$\{1,n\}$	$n \geq 2$	E_6	$\{2\}$
B_n	$\{2\}$	$n \geq 2$	E_7	$\{1\}$
C_n	$\{1\}$	$n \geq 3$	E_8	$\{8\}$
D_n	$\{2\}$	$n \geq 4$	F_4	$\{1\}$
			G_2	$\{2\}$

Now given a finite Coxeter system (W, S), there are other buildings of type (W, S) than the ones listed above. Tits [7] proves that, if the rank is at least 3, then every spherical building arises from an algebraic group in the broad sense (including classical groups and groups of mixed type). We will not introduce notations for all cases, but we content ourselves with mentioning the following terminology. We use the labelling of the diagrams introduced above.

1. If Δ is a building of type A_n, then the 1-Grassmannian is the point-line truncation (i.e., the restriction to the points and the lines) of a projective space. If the projective space arises from a vector space over a commutative field, then this 1-Grassmannian is the Lie incidence geometry $\mathsf{A}_{n,1}(k)$. If the projective space arises from a vector space over a non-commutative skew field ℓ, then we extend the previous notation substituting k with ℓ. If $n = 2$, there are other cases which we do not need.

2. If Δ is a building of type B_n (or equivalently, of type C_n if $n \geq 3$; for $n = 2$ the Bourbaki labelling does not agree), then the 1-Grassmannian is a polar space.

   (a) If this polar space arises from a parabolic quadric in $\mathrm{PG}(2n, k)$, for some field k and $n \geq 2$, i.e., from a non-degenerate symmetric bilinear form of Witt index n in a $(2n + 1)$-dimensional vector space, then Δ coincides with $\mathsf{B}_n(k)$.

(b) If this polar space arises from a linear line complex in $PG(2n - 1, k)$, for some field k and $n \geq 3$, i.e., from a non-degenerate alternating bilinear form in a $2n$-dimensional vector space, then Δ coincides with $C_n(k)$.

(c) If this polar space arises from a hyperbolic quadric in $PG(2n - 1, k)$, for some field k and $n \geq 4$, i.e., from a non-degenerate symmetric bilinear form of Witt index n in a $2n$-dimensional vector space, then Δ coincides with $D_n(k)$.

For the specific case $n = 2$, see below. Notice that also type D_n gives rise to polar spaces and to buildings of type B_n. Indeed, the so-called *flag complex* of a hyperbolic quadric in $PG(2n - 1, k)$ is a building of type B_n, but the so-called *oriflamme complex* is a building of type D_n. The difference between the two is that the latter is a *thick* building, meaning that every panel is contained in at least three chambers, while the former is not. There is a procedure to produce a thick building from every non-thick spherical one, except in the *thin* case, i.e., when every panel is contained in exactly two chambers, and then we have a Coxeter complex. So usually, one is only interested in thick spherical buildings.

3. All thick buildings of types D_n, $n \geq 4$ and E_m, $m \in \{6, 7, 8\}$, are isomorphic to $D_n(k)$, $E_m(k)$, respectively, for some commutative field k.

4. The 1- or 4-Grassmannian of any thick building of type F_4 is called a *metasymplectic space*.

In general, the J-Grassmannian geometry Δ_J, for a spherical building Δ, will be called a *Lie incidence geometry*, and it is called *simple* if $|J| = 1$.

Let Δ be a building of type (W, S). Then the Coxeter diagram of (W, S) can be used to deduce the structure of the links. More precisely, let F be a simplex of Δ with $|F| \leq |S| - 2$, then we can consider the set Δ_F of all simplices F' of Δ disjoint from F such that $F \cup F'$ is a simplex. This forms a simplicial chamber complex again, called the *residue in Δ of* F, and it is a building of type (W', S'), where S' corresponds to the cotype of F, and W' is generated by S'. Hence, in order to have a geometric feeling for buildings, we can do it inductively and start with rank 2, i.e., $|S| = 2$. In this case, the simplicial complex consists of singletons and pairs; the singletons are numbered 1 or 2 and every pair contains a vertex of either type. Hence we obtain a bipartite graph. In the non-spherical case, a bipartite graph corresponds to a building if and only if the graph is a (necessarily infinite) tree without vertices of valency 1. In the spherical case, the axioms (B1) and (B2) translate into the following definition.

Let $\Gamma = (V, E)$ be a bipartite graph. Then Γ is called a *generalized n-gon* if the following conditions hold.

(GP1) The diameter of Γ is n;

(GP2) The girth of Γ is $2n$.

If we restrict to thick buildings, then we restrict to bipartite graphs with bivalency (s, t), where $s, t \geq 3$. If n is odd, however, then $s = t$. This has the following consequence for a general thick building. Let Δ be a thick building of type (W, S). Delete in the Coxeter graph all edges labelled (∞) and delete the isolated vertices of the graph obtained. Then delete all edges with even label. Then every connected component Γ^* of the resulting graph can be assigned a cardinal number t such that, if $s_i \in S$ corresponds with any vertex of Γ^*, then the number of chambers containing any panel of cotype s_i is equal to $t + 1$. In the irreducible spherical case, this means that with every building Δ of type A_n, D_n and E_n is associated a unique cardinal number $t \geq 2$ such that every panel is contained in exactly $t + 1$ chambers. We say that t is the *order* of Δ. For the other types, there are two such cardinal numbers s, t, and we say that $\{s, t\}$ is the *order*. In the case where the rank is at least 3 and the building is finite, the orders are always powers of a prime number.

Example 2.6. Let Γ be a thick generalized 3-gon. Then Γ is the incidence graph of a projective plane. Hence the notions of a thick building of type A_2 and a projective plane are equivalent.

Example 2.7. Let Γ be a generalized 4-gon. Then Γ is the incidence graph of a *generalized quadrangle*, i.e., a point-line incidence structure such that two points determine at most one line, no point is contained in all lines, no line contains all points, and for each point p and each line L not incident with p, there exists a unique point q incident with L and a unique line M incident with p such that q is incident with M.

We now present two specific examples of generalized quadrangles. The first one is the building $\mathsf{B}_2(2)$ and the second one is not related to a Dynkin type.

Example 2.8. In general, the Lie incidence geometry $\mathsf{B}_{2,2}(k)$ is the generalized quadrangle, also denoted by $W(k)$, arising from a symplectic polarity or linear line system in $\mathrm{PG}(3, k)$ (or a non-degenerate alternating form in a 4-dimensional vector space over k). If k is perfect and has characteristic 2, then the Lie incidence geometries $\mathsf{B}_{2,2}(k)$ and $\mathsf{B}_{2,1}(k)$ are isomorphic. This happens for instance for $|k| = 2$. In this case, there is a combinatorial description of this geometry $W(2)$, or of the corresponding building, or of the corresponding bipartite graph (the incidence graph). Indeed, one considers a 6-set $N = \{1, 2, 3, 4, 5, 6\}$. The vertices of one bipartition class are the pairs of N, and the vertices of the other bipartition class are the partitions of N into 2-sets. Adjacency is natural (based on inclusion). We will use this representation below. Note that the order of $W(2)$, as a building, is $\{2, 2\}$.

We now introduce the unique generalized quadrangle $Q(2, 4)$ with lines of size 3 and such that every point is incident with exactly 5 lines (hence it has order $\{2, 4\}$).

Example 2.9. Let $N = \{1, 2, 3, 4, 5, 6\}$ and $N' = \{1', 2', 3', 4', 5', 6'\}$. Then we define the point set of $Q(2, 4)$ as $N \cup N' \cup \binom{N}{2}$, where $\binom{N}{2}$ is the set of all 2-subsets of N. The lines are given by the partitions of N into 2-sets (as above for $W(2)$), and also by the triples $\{i, \{i, j\}, j'\}$, for every ordered pair $(i, j) \in N \times N$ with $i \neq j$.

2.3 Graphs, embeddings and more conventions and notation

2.3.1 Graphs Given a spherical building Δ, there are a lot of graphs that one can associate with it. We will be mostly interested in *incidence graphs* and *collinearity graphs*. The full incidence graph of a building is just the 1-skeleton of Δ as a simplicial complex, i.e., only considering the vertices and the simplices of size 2. For buildings of rank 2, this is common to use. One can also restrict to certain types, but more common is to use the incidence graph of Lie incidence geometries related to Δ, i.e., the vertices of the graph are the points and lines of a certain Lie incidence geometry, and adjacency in the graph is incidence in the geometry.

For Lie incidence geometries, another commonly used graph is the *collinearity graph* whose vertex set is just the points set of the geometry, two distinct vertices being adjacent if the corresponding points are collinear (are contained in a common line).

2.3.2   Embeddings An embedding of a Lie incidence geometry $(\mathcal{P}, \mathcal{L})$ into a projective space $\mathsf{PG}(d, k)$ consists of an injective mapping of $\mathcal{P}$ into the point set of $\mathsf{PG}(d, k)$ such that all points of any line $L \in \mathcal{L}$ are mapped onto all points of a line of $\mathsf{PG}(d, k)$. The theory of representations of algebraic groups yields embeddings of many Lie incidence geometries. We will construct some of those in this paper in an alternative way.

To accomplish this we will have to use some notions typical for projective geometry.

Let $\mathsf{PG}(d, k)$ be the d-dimensional projective space over the commutative field k. Then we can assign to a quadruple $(p_1, p_2; q_1, q_2)$ of collinear points a unique scalar $r \in k$, called the *cross-ratio*. It is a number invariant under any linear permutation and base-change.

Dually, the cross-ratio can also be defined for a quadruple of *concurrent* hyperplanes. It can be defined directly by considering a line disjoint from the intersection of the hyperplanes and then identifying the hyperplanes with their intersections with this line. Likewise, one can define in a completely similar way the cross ratio of every quadruple of *concurrent* i-subspaces, i.e., four i-spaces having an $(i-1)$-space in common and being contained in a common $(i+1)$-space.

An abstract Segre geometry $\mathcal{S}_{n,m}(k)$ is a Lie incidence geometry $\mathsf{A}_{n,1}(k) \times \mathsf{A}_{m,1}(k)$. It has a canonical embedding in the projective space $\mathsf{PG}(nm+n+m, k)$. Indeed, we can map the point $((x_i)_{0 \le i \le n}, (y_j)_{0 \le j \le m})$ to the point $(x_i y_j)_{0 \le i \le n; 0 \le j \le m}$. This canonical embedding is called a *Segre variety*. There are many geometric constructions; we mention one below in Section 5.

When $n = 1$, then fix a maximal 1-space L. Every m-space of $\mathcal{S}_{1,m}(k)$ (if $m = 1$, then we refer to a 1-space of the other system) intersects L in a unique point, and hence this induces a cross-ratio for the m-spaces. It is independent of the choice of L (but it is dependent on the embedding, hence it is only defined for Segre varieties, and not for arbitrary Segre geometries).

Finally, for an arbitrary point of an arbitrary point-line geometry, we denote collinearity always by $\perp$; in particular $x^{\perp}$ always means the set of points collinear to the point x.

3   Intersections of Quadrics

In this section we will describe a method to construct the smallest dimensional representations of the buildings of type E_6 and E_7 over a field k, namely those in respective projective dimensions 26 and 55, as an intersection of (degenerate) quadrics. For type E_6, such a set of quadrics was given by Cohen [2], but we present a combinatorial logical way to write these down, as opposed to just list them.

3.1   A baby example

It is well know that there is a very explicit way to write any Grassmannian of any projective space as a subset of quadrics, see Hirschfeld & Thas [4]. We will present the example of the line Grassmannian $\mathsf{A}_{5,2}(k)$ of a projective 5-space over any field k (which corresponds to the third cell in the second row of the Freudenthal-Tits Magic Square).

A generic point of $\mathsf{A}_{5,2}(k)$ has coordinates of the form

$$(p_{ij})_{1 \le i < j \le 6}, \quad \text{where } p_{ij} = \begin{vmatrix} x_i & x_j \\ y_i & y_j \end{vmatrix},$$

for two independent vectors $(x_i)_{1 \le i \le 6}$ and $(y_i)_{1 \le i \le 6}$. If we set $p_{ji} = -p_{ij}$, then $\mathsf{A}_{5,2}(k)$ is contained in the quadric with equation $p_{ij}p_{m\ell} + p_{i\ell}p_{jm} + p_{im}p_{\ell j} = 0$, with $i, j, m, \ell \in$

$\{1, 2, \ldots, 6\}$ all distinct. Varying over all possibilities, the intersection of all such quadrics yields exactly $A_{5,2}(k)$. In fact, it is easy to see that we can restrict to all i, j, m, ℓ with $1 \leq i < j < m < \ell \leq 6$. In this case, we even do not need to define p_{ji} for $j > i$, since we can write the above equation as $p_{ij}p_{m\ell} + p_{i\ell}p_{jm} = p_{im}p_{j\ell}$. So $A_{5,2}(k)$ is the intersection of 15 quadrics. Note that it lives in 14-dimensional space, so the vector dimension of the embedding is equal to the number of quadrics. A similar thing will happen for the Lie incidence geometry $E_{6,1}(k)$ of exceptional type.

3.2   Buildings of type E_6

The combinatorial data underlying the subscripts of the coordinates of our baby example were the 2-subsets of a 6-set (the 2-subsets can be taken as unordered by insisting that p_{ij} is only defined for $i < j$). It is well known that the 2-subsets of a 6-set can be identified with the points of the smallest nontrivial generalized quadrangle denoted by $W(2)$, where the lines are the partitions of the 6-set into three 2-subsets. The subscripts of the coordinates appearing in the equation of a single quadric then correspond with the points collinear to a fixed point of $W(2)$ (and there are 15 points in $W(2)$, so all points appear as this fixed point). Within the equation of a fixed quadric, the subscripts of the coordinates that are multiplied together, correspond to collinear points, while those that are in distinct terms are always non-collinear. However, there is no homogeneous description of the subscripts of the coordinates of the term that has to go into the right hand side.

We generalize the above construction as follows. Let $Q(2, 4)$ be the unique generalized quadrangle of order $(2, 4)$ with points set $\mathcal{P}$ and line set $\mathcal{L}$. As already mentioned before, we have $|\mathcal{P}| = 27$ and $|\mathcal{L}| = 45$. Every line is incident with three points and every point is collinear with 10 other points. Let a basis of $\mathsf{PG}(26, k)$ be indexed by the points of $Q(2, 4)$. Hence an arbitrary point has coordinates of the form $(x_i)_{i \in \mathcal{P}}$, $x_i \in k$ for all $i \in \mathcal{P}$. Given a point $i \in \mathcal{P}$ and a line $L \in \mathcal{L}$ with $i \in L$, we define the quadric $Q_{i,L}$ with equation

$$x_{j_1}x_{j_2} + x_{j_3}x_{j_4} + x_{j_5}x_{j_6} + x_{j_7}x_{j_8} = x_{j_9}x_{j_{10}},$$

where $i^\perp = \{j_k : k = 1, 2, \ldots, 10\}$ and $L = \{i, j_9, j_{10}\}$. If we want a set of 27 quadrics, like in Cohen, or similarly as in the baby example above, then we should make a unique choice for L given i. This is accomplished by introducing a spread. A *spread* of $Q(2, 4)$ is a partition of $\mathcal{P}$ into nine lines. It turns out that, up to isomorphism, $Q(2, 4)$ contains exactly two spreads. Only one is a so-called *regular* spread $\mathcal{S}$, i.e., given any pair of lines $L_1, L_2 \in \mathcal{S}$, the unique line L_3 composed of the three points outside $L_1 \cup L_2$ that are collinear with collinear points of $L_1 \cup L_2$ also belongs to $\mathcal{S}$. The lines L_1, L_2, L_3 form a *regulus*, i.e., a set of three lines such that every line of $Q(2, 4)$ intersecting two of them also intersects the third. The reguli of $\mathcal{S}$ define an affine plane of order 3 on $\mathcal{S}$.

Now we have the following result.

Theorem 3.1. *The Lie incidence geometry $E_{6,1}(k)$ is the intersection of the twenty seven quadrics $Q_{i,L}$ in $\mathsf{PG}(26, k)$, with i ranging over $\mathcal{P}$ and L the unique member of $\mathcal{S}$ incident with i.*

Preparing for the construction of the Lie incidence geometry $E_{7,7}(k)$ in the next subsection, we make a few additional observations regarding the case $E_{6,1}(k)$.

Consider the complement Γ^{e_6} of the collinearity graph of $Q(2, 4)$. It is well known that this is isomorphic to the thin Lie incidence geometry $E_{6,1}(1)$. We identify each vertex of Γ^{e_6} with the corresponding basis point of $\mathsf{PG}(26, k)$. Then every vertex of Γ^{e_6} belongs to $Q_{i,L}$, for all $i \in \mathcal{P}$ and $L \ni i$ in $\mathcal{S}$. Moreover, two basis points i, j are non-adjacent as vertices of Γ^{e6} if and only if they are opposite points of precisely one of those quadrics, namely $Q_{k,M}$, with $\{i, j, k\}$ a line of $Q(2, 4)$ and $k \in M \in \mathcal{S}$. In this case, i and j are incident with a unique element Σ of type 6 of $\mathsf{E}_{6,1}(1)$, which is isomorphic to the Lie incidence geometry $\mathsf{D}_{5,1}(1)$. The points of Σ are precisely the subscripts appearing in the equation of $Q_{k,M}$.

3.3 Buildings of type E_7

We now construct the Lie incidence geometry $\mathsf{E}_{7,7}(k)$ as an intersection of quadrics. The Lie incidence geometries of the last three subsections, $\mathsf{A}_{5,2}(k), \mathsf{E}_{6,1}(k), \mathsf{E}_{7,7}(k)$ are situated as follows in the Freudenthal-Tits Magic Square (and we only write the types):

		$\mathsf{A}_{5,2}$	$\mathsf{E}_{6,1}$
			$\mathsf{E}_{7,7}$

Hence just as $\mathsf{A}_{5,2}(k)$ served as a baby example for $\mathsf{E}_{6,1}(k)$ because it is at the left of it, $\mathsf{E}_{6,1}(k)$ serves as a warming-up for $\mathsf{E}_{7,7}(k)$ because it is just above it.

It is well known that the Gosset graph is a model for the thin Lie incidence geometry $\mathsf{E}_{7,7}(1)$. Let us therefore denote the Gosset graph by Γ^{e7}. Vertices of type 1 of $\mathsf{E}_7(1)$ correspond to copies of $\mathsf{D}_{6,1}$. There are 126 such copies inside Γ^{e7}. Also, Γ^{e7} has 56 vertices. We again identify the vertices of Γ^{e7} with the points of a basis of $\mathrm{PG}(55, k)$. It turns out, similarly as above for $\mathsf{E}_{6,1}(k)$, that the representation of the Lie incidence geometry $\mathsf{E}_{7,7}(k)$ in $\mathrm{PG}(55, k)$ is contained in the intersection of 126 quadrics Q_Σ with equations of the form (signs being indetermined for now, hence the $\pm$ notation)

$$\sum_{a=1}^{6} \pm x_{i_{2a-1}} x_{i_{2a}} = 0,$$

where $\{i_1, i_2, \ldots, i_{12}\}$ is the set of points of a copy Σ of $\mathsf{D}_{6,1}$ inside Γ^{e7}. We first determine the signs in these equations. This is not possible by one homogeneous rule, i.e., one cannot choose the signs in such a way that all equations have the same number of plus and minus signs. Instead there are two different possiblities. In order to explain this, we need the following construction of the Gosset graph.

Consider two copies $\Gamma_1 = (V_1, E_1)$ and $\Gamma_2 = (V_2, E_2)$ of Γ^{e6}, and let θ be an isomorphism between them (we shall use θ in both directions, i.e., we view θ as an involutory permutation of $V_1 \cup V_2$ interchanging Γ_1 and Γ_2). Let ∞_1, ∞_2 be two symbols not belonging to $V_1 \cup V_2$. Then $\Gamma^{e7} = (V, E)$ can be described as follows. The vertex set V is equal to $\{\infty_1\} \cup V_1 \cup V_2 \cup \{\infty_2\}$. Adjacency inside V_1 and V_2 is the one of Γ_1 and Γ_2, respectively. The vertex ∞_a is adjacent to all vertices in V_a, $a = 1, 2$. Finally, a vertex $v_1 \in V_1$ is adjacent to a vertex $v_2 \in V_2$ if v_1^θ is not adjacent to v_2 in Γ_2 (which is equivalent to v_1 not being adjacent to v_2^θ in Γ_1).

Note that, in Γ^{e6}, the graph induced on the set of vertices not adjacent to a given vertex is precisely a copy of $\mathsf{D}_{5,1}(1)$. Hence every vertex in V_1 or V_2 is adjacent to a set of vertices of V_2 or V_1, respectively, for which the induced subgraph is $\mathsf{D}_{5,1}(1)$. Let us briefly call a copy of $D_{5,1}(1)$ and $D_{6,1}(1)$ in Γ^{e_6} and Γ^{e_7}, respectively, a D_5 and a D_6, respectively. Let $\{a, b\} = \{1, 2\}$. Pick $v_a \in V_a$ arbitrary. Then the set W_a of points in V_b collinear to v_a is a D_5 in Γ_b; hence it is easy to see that $W_a \cup \{v_a, \infty_b\}$ is a D_6 in Γ^{e_7}. This way, we recover $2 \times 27 = 54$ subgraphs D_6 of the 126 in total in Γ^{e_7}. We call these *of type* 1.

Now let P be a vertex of type 2 in the building $E_6(1)$. In the Lie incidence geometry $E_{6,1}(1)$, P corresponds to the subgraph isomorphic to the Lie incidence geometry $A_{5,1}(1)$. This, in turn, corresponds to a 6-clique in the graph Γ^{e_6}. So let P be a 6-clique in, say, Γ_1. There is a unique 6-clique P' in Γ_1 (the opposite one) with the property that P' is incident with each D_5 opposite some element of P. This can best be seen in $Q(2, 4)$ as follows. The set P is a set of six points no two of which are collinear. Let i_1 be one of them. Then there exists a point j_1 of $Q(2, 4)$ opposite i_1 such that P is equal to $\{i_1\}$ union the set of points of $Q(2, 4)$ collinear with j_1 but not collinear with i_1. Then P' is the union of $\{j_1\}$ with the set of points of $Q(2, 4)$ collinear with j_1 but not collinear with i_1. Clearly, each point of P is not collinear to a unique point of P' and vice versa. This means that, in Γ_1, each element of P is adjacent to a unique element of P'. Hence each element of P' is not adjacent to a unique element of P^θ. Consequently $P' \cup P^\theta$ is a D_6 in Γ^{e_7}, called *of type* 2. Since there are $72 = 126 - 54$ vertices of type 2 in $E_6(1)$, this takes care of the other subgraphs D_6 in Γ^{e_7}.

We can now define the signs in the equation of Q_Σ, for each D_6. So let Σ be any D_6. We fix a regular spread S_1 in the generalized quadrangle corresponding to Γ_1 and we let S_2 be its image under θ. First let Σ be of type 1, and suppose, to fix the ideas, that Σ contains ∞_1. Then exactly two non-collinear vertices i and j of $\Sigma \cap V_1$ belong to a common member of S_1 and we define all the signs in the equation of Q_Σ to be positive, except for the sign of the term $x_i x_j$, which is defined to be negative.

Now let Σ be of type 2 and suppose $\Sigma = P_1 \cup P_2$, with P_a a 6-clique in Γ_a, $a = 1, 2$. We define the following set Π of thin projective planes in Γ_1. The set Π will have the property that each line of Γ_1 is incident with a unique member of Π. Let L be a thin line of Γ_1, i.e., an edge $\{i, j\}$. Then i and j can be seen as two non-collinear points of $Q(2, 4)$. Let L, M be the members of S containing i, j, respectively. Let $\{L, M, N\}$ be the regulus of $Q(2, 4)$ containing L and M. Then $N \in S$ and there is a unique point ℓ on N collinear in $Q(2, 4)$ to neither i nor j. Hence $\{i, j, \ell\}$ is a thin projective plane in Γ_1. We would have obtained the same plane starting from the edges $\{i, \ell\}$ and $\{j, \ell\}$. Hence each line of Γ_1 is incident with a unique member of Π. A double count now reveals that $|\Pi| = 72$.

Below we show that each 6-clique of Γ_1 contains exactly two members of Π, which moreover are disjoint. Let π and π' be the two members of Π contained in P_1. Then we give the terms of the equation of Q_Σ containing subscripts in different planes π and π' different signs. This completes the description of 126 quadrics which all contain $E_{7,7}(k)$.

We still need three quadrics to completely determine $E_{7,7}(k)$. Indeed, opposite points in Γ^{e_7} do not appear as subscripts in a common equation yet, hence the corresponding points are collinear in the intersection of the 126 quadrics, whereas these points are not collinear in $E_{7,7}(k)$ (neither in its embedding in $PG(55, k)$).

We construct a set of 63 quadrics $Q_{\Sigma, \Sigma'}$ with equation

$$\sum_{a=1}^{12} \pm x_{i_a} x_{i_{a+12}} = 0,$$

where $\{i_1, i_2, \ldots, i_{12}\}$ is the set of points of a copy Σ of $D_{6,1}(1)$ in Γ^{e_7}, and $\{i_{13}, i_{14}, \ldots, i_{24}\}$ is the unique other copy Σ' of $D_{6,1}(1)$ opposite Σ, i.e., every point i_j of Σ is opposite (at distance 3 of) a unique point i_{j+12} of Σ' (so Σ' just consists of the opposites of Σ), $1 \le j \le 12$. There are two possibilities.

1. Suppose first that $\infty_1 \in \Sigma$. Then $\infty_2 \in \Sigma'$. Let i_1 and i_2 be the vertices of Σ and Σ', respectively, opposite ∞_1 and ∞_2, respectively, in Σ and Σ', respectively. Then the two terms $X_{\infty_1} X_{\infty_2}$ and $X_{i_1} X_{i_2}$ get the same sign, and all the others the opposite sign. Note there are 27 such quadrics.

2. If $\infty_1 \notin \Sigma \cup \Sigma'$, then we may choose the indices such that $\{i_1, \ldots, i_6\}$ and $\{i_7, \ldots, i_{12}\}$ are 6-cliques of Γ^{e_7} contained in Γ_1. Then $\{i_{13}, \ldots, i_{18}\}$ and $\{i_{19}, \ldots, i_{24}\}$ are two 6-cliques in Γ_2. We then choose the signs of $X_{i_1} X_{i_{13}}, \ldots, X_{i_6} X_{i_{18}}$ all equal, and those of $X_{i_7} X_{i_{19}}, \ldots, X_{i_{12}} X_{i_{24}}$ get the opposite sign. Note that there are 36 such quadrics.

We now have the following theorem.

Theorem 3.2.   (i) *The intersection of the 126 quadrics Q_Σ, with Σ a subgraph of Γ^{e_7} isomorphic to $\mathsf{D}_{6,1}(1)$, and the 63 quadrics $Q_{\Sigma,\Sigma'}$, with Σ and Σ' opposite copies in Γ^{e_7} isomorphic to $\mathsf{D}_{6,1}(1)$, is an embedding of the Lie incidence geometry $\mathsf{E}_{7,7}(k)$ in $\mathsf{PG}(55, k)$.*

(ii) *Let i_1, i_2, i_3 be three vertices of Γ_2 corresponding to the points on a line of the generalized quadrangle $Q(2, 4)$ underlying Γ_2 (said differently, $\{i_1, i_2, i_3\}$ is a maximal coclique of Γ_2). Then the intersection of the 126 quadrics Q_Σ, with Σ a subgraph of Γ^{e_7} isomorphic to $\mathsf{D}_{6,1}(1)$, and the 3 quadrics $Q_{\Sigma,\Sigma'}$, with Σ and Σ' opposite copies in Γ^{e_7} isomorphic to $\mathsf{D}_{6,1}(1)$, with $\infty_1 \in \Sigma$ and $\{i_1, i_2, i_3\} \cap \Sigma \ne \emptyset$, is an embedding of the Lie incidence geometry $\mathsf{E}_{7,7}(k)$ in $\mathsf{PG}(55, k)$.*

We finally prove the result announced and used above.

Proposition 3.3. *Let $\mathcal{S}$ be a regular spread of $Q(2, 4)$ and let Π be a set of 72 thin projective planes constructed as above such that every thin line of $\mathsf{E}_{6,1}(1)$ is contained in precisely one member of Π. Then every thin 5-space of $\mathsf{E}_{6,1}(1)$ contains exactly two members of Π, which moreover are disjoint.*

Proof. Let P be a 5-space of $\mathsf{E}_{6,1}(1)$, hence a 6-clique in Γ^{e_6}. For each point $i_1 \in P$, we now construct a member of Π containing i_1 and contained in P. Indeed, as above, there exists a point j_1 of $Q(2, 4)$ opposite i_1 such that P is equal to $\{i_1\}$ union the set $\{i_2, i_3, i_4, i_5, i_6\}$ of points of $Q(2, 4)$ collinear with j_1 but not collinear with i_1. Let L_1 be the unique member of $\mathcal{S}$ containing i_1. We can choose the indices so that the line of $Q(2, 4)$ joining j_1 with i_2 meets L. Let M be the unique member of $\mathcal{S}$ containing j_1. Then we can choose indices such that $i_3 \in M$. Then clearly $\{i_1, i_2, i_3\} \in \Pi$.

Considering i_4, we can construct a second member of Π in P. Hence every thin 5-space contains at least two members of Π. A double count of the pairs (P', π), with P' a 5-space of $\mathsf{E}_{6,1}(1)$ and $\pi \in \Pi$ with $\pi \subseteq P'$, reveals that this must be exactly two. Since every point of P must be contained in a member of Π contained in P, the two members of Π in P must cover all points and hence are disjoint. $\square$

4   Equator and Trace Geometries

In this section, we construct Lie incidence geometries of lower rank inside Lie incidence geometries of higher rank. This might seem like a trivial exercise from the point of view of buildings, thinking about residues, but it is actually an interesting general open question which Lie incidence geometries are contained in a given one. The constructions that we will present give ordinary and predictable residue geometries, but also a geometric interpretation of inclusions of quadratic composition algebras. Hence it happens typically for the geometries of the Freudenthal–Tits Magic Square, since the construction of the Lie algebra in the cell (i, j) uses quadratic alternative composition algebras of dimensions 2^{i-1} and 2^{j-1} over the base field.

Also, we will use equator geometries in the next two sections, where they will prove a useful tool.

All in all, we will treat the following cells in the Freudenthal-Tits Magic Square.

		$C_{3,2}$	$F_{4,4}$
		$A_{5,2}$	$E_{6,1}$
	$A_{5,3}$	$D_{6,6}$	$E_{7,7}$
$E_{6,2}$	$E_{7,1}$	$E_{8,8}$	

4.1   The general principle and some easy examples

There are many possible definitions for what in general an equator geometry should be. We shall here give a practical definition that is easy to apply in different situations.

The starting point is the fact that in many Lie incidence geometries Δ the residue of a point is not only a quotient geometry, but it is also a subgeometry of Δ, which can be seen as follows.

Proposition 4.1. *Let p be a point of a Lie incidence geometry Δ of rank at least 3 containing planes. Let X be an object opposite p. Then on each line L through p there is a unique point p_L not opposite X. The set of such p_L, with lines inherited from the planes through p, is a Lie incidence geometry of the residue of p in the building associated with Δ.*

For a simple Lie incidence geometry, there is only one case where it is of rank at least 3 and does not contain planes and this corresponds to the dual polar spaces. In this case, the residue can still be recovered as a subgeometry by considering the same point set as above, but by defining the line set as the set of so-called *hyperbolic lines* contained in it.

We generalise the above situation and define in general a trace geometry.

Definition 4.2. Let Δ be a Lie incidence geometry related to a building $\mathfrak{B}$, say of rank n and with type set I; more exactly let Δ be the K-Grassmannian of $\mathfrak{B}$, $K \subseteq I$. Let F_1 and F_2 be two opposite flags of $\mathfrak{B}$, say of common type J. Let F be a flag of type, say, J' incident with F_1 and suppose that the projection of F_2 onto F contains a point p_F. Then the set $T(F_1, F_2; J')$ of points p_F, for F ranging over the set of all flags of type J' incident with F_1 is called a (J, J')-*trace* in Δ. Let J^* be the union of J with the type sets of the connected components of $I \setminus J$ disjoint from J'. Then $T(F_1, F_2; J')$ is obviously in natural bijective correspondence with the point set of the Lie incidence geometry Δ' of the J^*-residue $\mathfrak{B}'$ of $\mathfrak{B}$ corresponding to the J'-Grassmannian. Endowing $T(F_1, F_2; J')$ with the lines inherited from Δ' turns $T(F_1, F_2; J')$ into a Lie incidence geometry, called the $(J, J')_K$-*trace geometry*. If we want a nontrivial set of lines, we must require $|J^*| < n - 1$.

In many cases, the lines inherited from Δ' are just the lines of Δ contained (as point sets) in $T(F_1, F_2; J')$. We will see a prominent counter example below.

So, trace geometries in Lie incidence geometries are in fact either subgeometries, or point sets endowed with a set of abstract lines turning them into Lie incidence geometries. The most satisfying (and also most interesting) situation occurs when F_1 and F_2 play the same role, i.e., when $T(F_1, F_2; J') = T(F_2, F_1; J')$.

Definition 4.3. A trace geometry with point set $T(F_1, F_2; J')$ is called an *equator geometry* if $T(F_1, F_2; J') = T(F_2, F_1; J')$. We talk about the $(J, J')_K$-equator geometry. If $J = K$, then we omit the subscript K and write (J, J')-equator geometry.

In fact, one can be slightly more general and not require that F_1 and F_2 have the same type. But we do not insist on that since in all our examples they have the same type.

A convenient situation is where F_1 and F_2 are just points of Δ, i.e., $J = K$. We now present some examples related to polar spaces. We start with the standard example.

Example 4.4. Let Δ be a polar space of rank n, hence a Lie incidence geometry of relative type $\mathsf{B}_{n,1}(k)$ or absolute type $\mathsf{D}_{n,1}(k)$. Let $J = \{1\}$ and $J' = \{2\}$. Then $J^* = J$. Let p_1 and p_2 be two opposite points. Then $T(p_1, p_2; J')$ is the set of points obtained by intersecting a line through p_1 with a line through p_2, hence $T(p_1, p_2; J') = p_1^\perp \cap p_2^\perp = T(p_2, p_1; J')$. The corresponding $(1, 2)$-equator geometry is a polar space of rank $n - 1$ and its lines are precisely the lines of Δ contained in $p_1^\perp \cap p_2^\perp$.

Example 4.5. Let Δ again be a polar space of rank n, over the type set I. Let $j \in I$ with $j \leq n - 2$. Put $J = \{i\}$ and $J' = \{i+1\}$. Then $J^* = \{1, 2, \ldots, i\}$. Then the corresponding trace geometry is again an equator geometry coinciding with the subspace with point set $U_1^\perp \cap U_2^\perp$ for singular subspaces U_1 and U_2 of projective dimension $i - 1$.

Example 4.6. Now let Δ be the Lie incidence geometry $\mathsf{B}_{n,i}(k)$, $i \geq 1$, $n \geq 2i$, or any Lie incidence geometry obtained from the i-Grassmannian of a polar space of rank n at least $2i$. Set $J = \{i\}$ and $J' = \{2i\}$. Then the (J, J')-trace geometry is an equator geometry and isomorphic to $\mathsf{B}_{n-i,i}(k)$, or the Lie incidence geometry obtained from the i-Grassmannian of a polar space of rank $n - i$, respectively.

The case $i = 2$ in the previous example is a long root geometry. More generally, we have the following proposition.

Proposition 4.7. *Let Δ be a long root geometry $\mathsf{X}_{n,J}(k)$, with $n \geq 4$. Let Y_m be the connected diagram of longest length m when removing the vertices of types J from the Dynkin diagram X_n. Let J' be such that $\mathsf{Y}_{m,J'}(k)$ is a long root geometry. Then there is a (J, J')-equator geometry and it is isomorphic to $\mathsf{Y}_{m,J'}(k)$.*

The proposition is easily proved noting that long root geometries are so-called hexagonal geometries; we do not go into detail here.

We now concentrate on some exceptional types. First we note that Proposition 4.7 gives us the following examples: The $(8, 1)$-equator geometry of $\mathsf{E}_{8,8}(k)$ is $\mathsf{E}_{7,1}(k)$; The $(1, 6)$-equator geometry of $\mathsf{E}_{7,1}(k)$ is $\mathsf{D}_{6,2}(k)$; The $(2, \{1, 6\})$-equator geometry of $\mathsf{E}_{6,2}(k)$ is $\mathsf{A}_{5,\{1,5\}}(k)$; The $(1, 4)$-equator geometry of $\mathsf{F}_{4,1}(k)$ is $\mathsf{C}_{3,1}(k)$. Note that only in the latter case, the lines of the equator geometry are not induced by the lines of the larger Lie incidence geometry.

4.2   Two examples related to F_4

Since the diagram of type F_4 is symmetric, the Lie incidence geometry $F_{4,4}(k)$ is also a hexagonal geometry, and so we also expect a $(4,1)$-equator geometry here. In fact, this equator geometry, which is isomorphic to $B_{3,1}(k)$ (and note it is not a long root geometry) is extensively studied by De Schepper, Sastry &Van Maldeghem [3]. We will use it in the next section for a combinatorial construction of $E_{6,1}(k)$. We content ourselves here with mentioning that the lines of this $(4,1)$-equator geometry are hyperbolic lines in the symplecta, which are symplectic polar spaces of rank 3. Such hyperbolic lines are determined by any pair of its points, unlike the situation in the $(1,4)$-equator geometry.

A second example is the $(1,3)_4$-equator geometry in $F_{4,4}(k)$. It is isomorphic to $C_{3,2}(k)$. Note that the same can be done with any metasymplectic space.

4.3   Examples related to E_6

First we consider the Lie incidence geometry $E_{6,1}(k)$. Let $J = \{2\}$, $J' = \{5\}$. The corresponding $(2,5)_1$-equator geometry is $A_{5,2}(k)$. This is exactly the Lie incidence geometry appearing in the cell next to $E_{6,1}(k)$. It is well-known that the latter can be viewed as a "projective plane" over the split octonions over k and the former as a "projective plane" over the split quaternions over k. The embedding of $A_{5,2}(k)$ into $E_{6,1}(k)$ as an equator geometry is the geometric evidence for the algebraic inclusion of the split quaternions in the split octonions.

We can go one step down and consider the Lie incidence geometry $A_{5,2}(k)$, set $J = \{3\}$ and $J' = \{1,4\}$. Then $J^* = J$ and the corresponding $(3,\{1,4\})_2$-equator geometry is $A_{2,1}(k) \times A_{2,1}(k)$, the Segre geometry $\mathcal{S}_{2,2}(k)$. This witnesses the inclusion of the ring $k \times k$ (with componentwise addition and multiplication) into the split quaternion algebra over k.

In the two cases considered in this subsection, we can characterize the equator geometries in a seemingly different way as follows (and the proofs are exercises in parapolar spaces).

Proposition 4.8.

(i) *The $(2,5)_1$-equator geometry of $E_{6,1}(k)$ with respect to the opposite 5-spaces W and W' of $E_{6,1}(k)$ consists of the set of points of $E_{6,1}(k)$ collinear with all points of a 3-space of W and also with all points of a 3-space of W'.*

(ii) *The $(3,\{1,4\})_2$-equator geometry of $A_{5,2}(k)$ with respect to the opposite planes U and U' of $E_{5,2}(k)$ consists of the set of points of $A_{5,2}(k)$ collinear with all points of a line of U and also with all points of a line of U'.*

Next we fix $J = \{2\}$, and we list some examples $E_{6,K}(k)$ for which there exists J' such that the $(J,J')_K$-equator geometry $A_{5,J''}(k)$ exists.

$E_{6,K}$	$(J,J')_K$	$A_{5,J''}$
$E_{6,1}$	$(2,3)_1$	$A_{5,2}$
$E_{6,2}$	$(2,\{1,6\})_2$	$A_{5,\{1,5\}}$
$E_{6,3}$	$(2,1)_3$	$A_{5,1}$
	$(2,\{1,4\})_3$	$A_{5,\{1,3\}}$
$E_{6,4}$	$(2,4)_4$	$A_{5,3}$
	$(2,\{1,5\})_4$	$A_{5,\{1,4\}}$
$E_{6,\{1,6\}}$	$(2,\{3,5\})_{\{1,6\}}$	$A_{5,\{2,4\}}$

4.4 Examples related to E_7

There is no $J' \subseteq \{1, 2, \ldots, 7\}$ such that there is a $(7, J')$-equator geometry. However, in order to see $D_{6,6}(k)$ as a subgeometry of $E_{7,7}(k)$, we set $J = \{1\}$ and $J' = \{2\}$. Then the $(1, 2)_7$-equator geometry exists and is actually isomorphic to $D_{6,6}(k)$. Again, this is geometric evidence for the algebraic inclusion of the split quaternions in the split octonions. To see the geometric evidence of the inclusion of the product of the field k with itself (split quadratic extension) in the split quaternion algebra over k, we need to look at the $(5, 3)_6$-equator geometry of $D_{6,6}(k)$, which is isomorphic to $A_{5,3}(k)$.

Finally, we see the inclusion of the Lie incidence and long root geometry $E_{6,2}(k)$ in $E_{7,1}(k)$ through the $(7, 2)_1$-equator geometry of the latter.

4.5 Examples related to E_8

There are equator geometries in $E_{8,8}(k)$ for $J = \{i\}$ any endpoint of the diagram, i.e., for all $i \in \{1, 2, 8\}$. We already mentioned the $(8, 1)$-equator geometry, giving the geometric evidence $E_{7,1}(k) \subseteq E_{8,8}(k)$ of the fourth row of the Freudenthal–Tits Magic Square for the inclusion of the split quaternions in the split octonions over k. For $i = 2$, we have the $(2, \{1, 8\})_8$-equator geometry, isomorphic to $A_{7,\{1,7\}}(k)$. Finally, for $i = 1$, we have the $(1, 3)_8$-equator geometry giving rise to $D_{7,2}(k)$. Note that all these equator geometries are actually long root geometries.

5 Projective Constructions

In this section, we aim at constructing the $E_{6,1}(k)$-variety in $PG(26, k)$, for an arbitrary field k, in a purely geometric way.

To that aim, we first present an obviously but deliberately complicated construction of a Segre variety, proceed with a warming-up example and then explain the case of $E_{6,1}(k)$. These three Lie incidence geometries are related to following three cells on the second row of the Freudenthal-Tits Magic Square.

	$A_2 \times A_2$	$A_{5,2}$	$E_{6,1}$
		$D_{6,6}$	

We also construct the half spin Lie incidence geometries in an inductive manner. This corresponds to the third cell in the third row above.

We begin with the latter.

5.1 Hyperbolic polar spaces and half spin geometries

In this section we will consider many isomorphisms between two structures; we will always assume that an isomorphism φ acts on both structures and is involutive.

The following is a well known construction of the Lie incidence geometry $D_{2,2}(k)$, otherwise known under the name "ruled quadric in $PG(3, k)$".

Example 5.1. Consider two skew lines L_1 and L_2 in $\mathsf{PG}(3, k)$. Let φ be a projectivity between them, i.e., φ preserves the cross ratio of quadruples of points on L_i, $i = 1, 2$. Then the union of the lines pp^φ with p ranging over L_1 is a hyperbolic quadric.

This is a special case of the following straight forward construction.

Proposition 5.2. *Let U_1 and U_2 be two disjoint subspaces of dimension $n - 1$ in the projective space $\mathsf{PG}(2n - 1, k)$, and let φ be a linear duality between U_1 and U_2, i.e., φ maps points of U_i onto $(n - 2)$-spaces in $U_{i'}$ such that collinear points are mapped onto concurrent subspaces in a bijective way preserving the cross ratio. Then the union of all lines p_1p_2, with $p_1 \in U_1$ and $p_2 \in p_1^\varphi$ is the point set of a hyperbolic quadric, i.e., the Lie incidence geometry $\mathsf{D}_{n,1}(k)$.*

Proof. Let Q be any hyperbolic quadric in $\mathsf{PG}(2n - 1, k)$, and let U_1 and U_2 be two disjoint maximal singular subspaces, i.e., U_1 and U_2 are two disjoint $(n - 1)$-space completely contained in Q. Let $p \in Q \setminus U_2$ be arbitrary and let $U \ni p$ be the unique maximal singular subspace of Q intersecting U_2 in an $(n - 2)$-space. Then, since the parity of the sum of the dimensions of the intersection of a maximal singular subspace with two given maximal singular subspaces is always constant, U intersects U_1 in a point p_1. Hence p is contained in the line p_1p_2, with $p_2 \in p_1^\perp \cap U_2$. Now the mapping $\theta : U_1 \to U_2^* : p_1 \mapsto p_1^\perp \cap U_2$, with U_2^* the dual space of U_2, is a linear duality. Indeed, we can take for Q the quadric with equation $X_{-1}X_1 + X_{-2}X_2 + \cdots + X_{-n}X_n = 0$, for U_1 the space with trivial positively indexed coordinates, and U_2 the space with trivial negatively indexed coordinates. Then θ maps the point $(x_{-1}, x_{-2}, \ldots, x_{-n}, 0, 0, \ldots, 0)$ to the subspace with equations

$$X_{-1} = X_{-2} = \cdots = X_{-n} = x_{-1}X_1 + x_{-2}X_2 + \cdots + x_{-n}X_n = 0.$$

This defines φ completely and uniquely and shows that it is a linear duality. $\square$

Interestingly, also the half spin geometries can be constructed in a similar fashion, but this time starting with two half spin geometries of lower rank. Let us first demonstrate this in some low ranks.

The Lie incidence geometry $\mathsf{D}_{3,3}(k)$ is just the projective space $\mathsf{PG}(3, k)$ viewed as point-line geometry. Application of the construction in Proposition 5.2 produces $\mathsf{D}_{4,1}(k)$, which is isomorphic to the half spin geometry $\mathsf{D}_{4,4}(k)$ via triality. Now we consider the Lie incidence geometry $\mathsf{D}_{5,5}(k)$, corresponding to a hyperbolic quadric Q in $\mathsf{PG}(9, k)$. Each point p of Q defines via its residue a unique subspace Q_p of $\mathsf{D}_{5,5}(k)$ isomorphic to $\mathsf{D}_{4,1}(k)$. If p and q are non-collinear points, then there is no maximal singular subspace containing both, hence Q_p and Q_q are disjoint in this case. So let p, q be non-collinear points of Q and consider an arbitrary point of $\mathsf{D}_{5,5}(k)$, i.e., a maximal singular subspace U of Q of certain prescribed type, say Type I and call the other type Type II. Assume that p and q do not belong to U. Then $W := p^\perp \cap q^\perp \cap U$ is a singular subspace of dimension $n - 3$ and hence defines a unique line L of $\mathsf{D}_{5,5}(k)$ containing U, $\langle p, W \rangle$ and $\langle q, W \rangle$. Hence every point of $\mathsf{D}_{5,5}(k)$ not contained in $q_p \cap Q_q$ lies on a unique line that intersects both Q_p and Q_q nontrivially. Hence, if we embed Q_p and Q_q in disjoint 7-dimensional subspaces of $\mathsf{PG}(15, k)$, then $\mathsf{D}_{5,5}(k)$ is contained in $\mathsf{PG}(15, k)$. Now suppose U belongs to Q_p. Then all points of Q_q that are incident with $\langle q, q^\perp \cap U \rangle$ (the latter viewed as a—singular—subspace of Q_q) are collinear with U in $\mathsf{D}_{5,5}(k)$. Hence $\mathsf{D}_{5,5}(k)$ is the union of all lines p^*q^*, where $p \in Q_p$ and $q^* \in Q_q$ such that $q^* \in (p^*)^\varphi$ for some linear duality φ between Q_p and Q_q.

A similar argument proves the following construction result. First note that, by taking the residue of an element of type $n - 2$ in a building of type D_n, we see that the structure of the set of maximal subspaces of $D_{n,n}(k)$, $n \geq 4$ (maximal subspaces have dimension $n-1$—those corresponding to the elements of type $n-1$ of the corresponding building— and 3—corresponding to the elements of type $n-3$ of the corresponding building), containing a fixed line of $D_{n,n}(k)$ is a Segre variety $\mathcal{S}_{n-3,1}(k)$. Hence it makes sense to talk about the cross ratio of a quadruple of disjoint maximal subspaces of maximal dimension.

Proposition 5.3. *Let Q_1 and Q_2 be two half spin geometries isomorphic to $D_{n,n}(k)$, $n \geq 4$, embedded in disjoint subspaces of dimension $2^{n-1}-1$ of $PG(2^n-1,k)$. Let φ be any linear duality between Q_1 and Q_2 (i.e., φ maps (collinear) points to maximal singular subspace intersecting in a line and preserves the cross ratio). Then the union of all lines p_1p_2, with $p_1 \in Q_1$ and $p_2 \in p^{\varphi}$ is an embedding of the Lie incidence geometry $D_{n+1,n+1}(k)$.*

Using an obvious induction argument, this construction easily implies that we are dealing here with the universal embedding.

5.2   Two constructions of the Segre variety $\mathcal{S}_{2,2}(k)$

Let $\pi_1, \pi_2, \pi_3, \pi_4$ be four planes in $PG(8,k)$ such that no three of them are contained in a hyperplane. Then every point x of π_1 is contained in a unique plane π_x of $PG(8,k)$ intersecting all of π_2, π_3, π_4 in points. The union of all such planes (for x ranging over π_1) is the Segre variety $\mathcal{S}_{2,2}(k)$.

For the second construction we remind the reader of the following property of $\mathcal{S}_{2,2}(k)$, proved by Schillewaert & Van Maldeghem in [5] .

Proposition 5.4.

(i) *Every pair of points of $\mathcal{S}_{2,2}(k)$ not contained in a common plane of $\mathcal{S}_{2,2}(k)$ is contained in a unique subgeometry isomorphic to $\mathcal{S}_{1,1}(k)$.*

(ii) *Every pair subgeometries isomorphic to $\mathcal{S}_{1,1}(k)$ of $\mathcal{S}_{2,2}(k)$ intersect in either a point or a line.*

Let p be a point of $PG(8,k)$, and let π_1, π_2 be two planes of $PG(8,k)$ intersecting in $\{p\}$. Let Q be some hyperbolic quadric in a 3-space disjoint from $\langle \pi_1, \pi_2 \rangle$. If the configuration $\pi_1 \cup \pi_2 \cup Q$ is to be part of a Segre variety $\mathcal{S}_{2,2}(k)$, then each line of Q is contained in a plane of $\mathcal{S}_{2,2}(k)$ that intersects $\pi_1 \cup \pi_2$ in a unique point. Moreover, by the existence of sub-Segre varieties isomorphic to $\mathcal{S}_{1,1}(k)$ (which are just hyperbolic quadrics in 3-spaces), we see that these points constitute two lines, one line L_1 in π_1 and a line L_2 in π_2. Let $\mathcal{R}_1$ and $\mathcal{R}_2$ be the two reguli of Q. Each of these has the natural structure of a projective line over k. Then indices can be chosen such that $\mathcal{S}_{2,2}(k)$ defines projectivities $\theta_i : \mathcal{R}_i \to L_i$ with the property that the planes spanned by $R \in \mathcal{R}_i$ and R^{θ_i} belong to $\mathcal{S}_{2,2}(k)$, $i = 1, 2$.

Adding these planes to our data π_1, π_2, Q does not yet determine $\mathcal{S}_{2,2}(k)$ uniquely. Note, however, that all these data are independent and can be chosen freely.

Now, by Proposition 5.4 we know that every point of $\mathcal{S}_{2,2}(k)$ outside $\pi_1 \cup \pi_2$ is contained in a unique hyperbolic quadric Q_q containing p and intersecting Q in a unique point q. Let M_i, $i = 1, 2$ be defined as $q \in M_i \in \mathcal{R}_i$. Set $x_i = M_i^{\theta_i}$, $i = 1, 2$. Then the lines $\langle p, x_1 \rangle$, $\langle p, x_2 \rangle$, $\langle q, x_1 \rangle$ and $\langle q, x_2 \rangle$ are four lines of Q_q. We claim that fixing the quadric Q_q for one point q determines $\mathcal{S}_{2,2}(k)$ uniquely.

Indeed, fix $q \in Q$. Then, following Example 5.1, and with the notation of the previous paragraph, Q_q is determined by a projectivity $\varphi : \langle p, x_1 \rangle \to \langle q, x_2 \rangle$. Now let q' be any other point of Q. By connectivity of Q, we may assume that $\langle q, q' \rangle$ is a generator of Q, and without loss we can assume it belongs to $\mathcal{R}_1$. Let $M_2' \in \mathcal{R}_2$ be such that $q' \in M_2'$ and set $x_2' = M_2'^{\theta_2}$. Then the planes $\langle q, q', x_1 \rangle$ and π_2 belong to $\mathcal{S}_{2,2}(k)$ and the three lines $\langle p, x_1 \rangle$, $\langle q, x_2 \rangle$ and $\langle q', x_2' \rangle$ also belong to $\mathcal{S}_{2,2}(k)$ and each meet each of these planes. It follows that, for an arbitrary point $x \in \langle p, x_1 \rangle$, the plane π_x of $\mathcal{S}_{2,2}(k)$ containing the line $\langle x, x^\varphi \rangle$ also intersects $\langle q', x_2' \rangle$. It follows that π_x contains the line K distinct from $\langle q, x_2 \rangle$ belonging to the quadric of $\mathcal{S}_{2,2}(k)$ determined by L_2 and M_1 and θ_2. In this quadric, K is determined by a projectivity $\phi : \langle q, x_2 \rangle \to \langle q', x_2' \rangle$. We conclude that Q_x is determined by the projectivity $\varphi' : \langle p, x_1 \rangle \to \langle q', x_2' \rangle : x \mapsto x^{\varphi\phi}$.

Hence $\mathcal{S}_{2,2}(k)$ is completely determined by $\pi_1, \pi_2, L_1, L_2, Q, \theta_1, \theta_2$ and φ.

5.3 A line Grassmannian

We now construct $\mathsf{A}_{5,2}(k)$ in a similar fashion. This is a warming up for the next subsection where we will construct $\mathsf{E}_{6,1}(k)$.

We again start with a point p in $\mathrm{PG}(14, k)$. Instead of the lines L_1, L_2, we now take the local structure of $\mathsf{A}_{5,2}(k)$, which is a Segre variety $\mathcal{S}_{1,3}(k)$. So we consider a 7-space U_1 in $\mathrm{PG}(14, k)$ not containing p and three mutually disjoint 3-spaces in U_1. Let $\mathfrak{S}$ be the Segre variety consisting of all lines intersecting all of these three 3-spaces in points. We call these lines the *line-generators* of $\mathfrak{S}$. The family $\mathcal{F}$ of all line-generators naturally has the structure of a projective 3-space (by intersection with an arbitrary 3-space intersecting all line-generators), which we denote by Π. Let $\mathfrak{C}$ be the cone with vertex p and base $\mathfrak{S}$. Let U_2 be a 5-space disjoint from $\langle p, U_1 \rangle$ and consider a hyperbolic quadric (a Klein quadric) Q in U_2. Let $\mathcal{R}_i$, $i = 1, 2$, be the two systems of generators of Q.

Since Q is Lie incidence geometry $\mathsf{D}_{3,1}(k) = \mathsf{A}_{3,2}(k)$, the family $\mathcal{R}_1$ is the point set of a 3-space, and hence there is a natural isomorphism $\theta : \mathcal{F} \to \mathcal{R}_1$. We can choose θ linear, i.e., such that it preserves the cross ratio. For each line $L \in \mathcal{F}$, we add the 4-space $\langle L, L^\theta \rangle$ to our data.

As in the previous subsection, it now remains to add, for each point $q \in Q$, a unique Klein quadric $\mathcal{K}_q$ containing p and q. Fix $q \in Q$. The inverse image of q under θ is a line of Π, hence a hyperbolic quadric $\mathcal{H}$ in $\mathfrak{S}$. Fix two arbitrary points a, b of $\mathcal{H}$, not on a common generator. Let L and M be disjoint generators of $\mathcal{H}$ containing a and b, respectively. Then the planes $\alpha := \langle p, L \rangle$ and $\beta := \langle q, M \rangle$ are opposite planes in $\mathcal{K}_q$ and so, in view of Proposition 5.2, $\mathcal{K}_q$ is completely determined by a linear duality δ between α and β. For each point x on L, there is a unique point $y \in M$ such that $\langle x, y \rangle$ is a generator of $\mathcal{H}$. Hence $x^\delta = \langle q, y \rangle$. Moreover, $p^\delta = M$. Hence it suffices to know the image of one more point in $\alpha \setminus (L \cup \{p\})$. We claim that this is enough to complete the construction of $\mathsf{A}_{5,2}(k)$. Indeed, it suffices to prove that any point $q' \in Q$ can be included in a Segre geometry isomorphic to $\mathcal{S}_{2,2}(k)$ containing p and q, because then we can apply the construction in Subsection 5.2.

Clearly there exist opposite points x_1 and x_2 in Q collinear to both q and q'. Let L_i be a line of $\mathfrak{S}$ all of whose points are collinear to x_i, $i = 1, 2$, with L_1 and L_2 contained in distinct 3-spaces of $\mathfrak{S}$. Since q and q' are collinear to x_1 and x_2, both q and q' are collinear with unique points of both L_1 and L_2. It follows that q and q' are collinear with all points of unique lines in the planes $\langle x_1, L_1 \rangle$ and $\langle x_2, L_2 \rangle$. Since also p is collinear with the points of unique lines in these planes (namely, L_1 and L_2), we deduce from Proposition 4.8(ii) that p, q, q' are contained in a Segre geometry isomorphic to $\mathcal{S}_{2,2}(k)$.

5.4   The Cartan variety

A construction of $E_{6,1}(k)$ similarly to the previous subsection exists by considering a cone with vertex some point p and base a half spin geometry $\mathfrak{D}$ isomorphic to $D_{5,5}(k)$ as constructed in Proposition 5.3 (this cone spans a 16-dimensional subspace U_1 of $PG(26, k)$) and a hyperbolic quadric Q in a complementary 9-space U_2. We can identify each point of $\mathfrak{D}$ with a certain 4-space of Q, and we can do so in a linear way, i.e., respecting the cross ratio inherited from U_1 and U_2. If we identify a point x with the 4-space W, then we add the 5-space $\langle x, W \rangle$ to our data. Then a point q of Q is collinear to the points of a hyperbolic quadric Q_q on $\mathfrak{D}$ isomorphic to $D_{4,1}(k)$. So we have two cones $\mathfrak{C}_p$ and $\mathfrak{C}_q$ with vertex p and q, respectively, and with base Q_q. Again, p, q and Q_q define a unique hyperbolic quadric isomorphic to $D_{5,1}(k)$ if we know for one point $a \in \mathfrak{C}_p \setminus (Q_q \cup \{p\})$ a collinear point $b \in \mathfrak{C}_q \setminus (Q_q \cup \{q\})$. Fixing such a collinear pair determines all others, just as in the previous subsection (now using Proposition 4.8(i) to construct an equator geometry isomorphic to $A_{5,2}(k)$).

6   Combinatorial Constructions

In this section, we construct bigger Lie incidence geometries out of smaller ones. Our main goal is here to report on the construction of $E_{6,1}(k)$ out of $F_{4,4}(k)$. In view of the Freudenthal–Tits Magic Square, there is an analogue for $A_{5,2}(k)$ out of $C_{3,2}(k)$. We will also explain this baby example. This means that we cover the following cells of the square.

		$C_{3,2}$	$F_{4,4}$
		$A_{5,2}$	$E_{6,1}$

We also briefly mention similar phenomena for other Lie incidence geometries. We begin with the latter.

6.1   Obvious examples

Example 6.1. Let $C_{n,1}$ be the natural Lie incidence geometry related to a symplectic polar space, $n \geq 2$. For two non-collinear points p, q we define the set $L_{p,q} = (p^\perp \cap q^\perp)^\perp$. Obviously $p, q \in L_{p,q}$, and it turns out that $L_{p,q} = L_{p',q'}$, for all $p', q' \in L_{p,q}$. If we add all sets $L_{p,q}$, for p, q ranging through the set of points, with p not collinear to q, to the geometry $C_{n,1}$ as additional lines, then we obtain $A_{2n-1,1}(k)$, the ordinary $(2n-1)$-dimensional projective space over k.

Example 6.2. Similarly as the previous example, we can construct the half-spin geometry $D_{n+1,n+1}(k)$ from the dual polar space $B_{n,n}(k)$, $n \geq 2$ (for $n = 2$, this is also the case $n = 2$ of Example 6.1).

Example 6.3. Let $G_{2,1}(k)$ be the Lie incidence geometry also known under the name of split Cayley hexagon. Let p and q be two opposite points and let $L_{p,q}$ be the set of points collinear with p and not opposite q. Then, adding these sets, for all choices of p and q, to the set of lines of $\mathsf{G}_{2,1}(k)$ produces $\mathsf{B}_{3,1}(k)$.

These examples all have the property that the point sets of the two geometries are the same, and only new lines need to be defined. In the next two examples we also have to extend the point set.

6.2   The line Grassmannian $\mathsf{A}_{5,2}(k)$ again

Let Δ be the Lie incidence geometry $\mathsf{C}_{3,2}(k)$. Then Δ is a hexagonal parapolar space with symplecta isomorphic to $\mathsf{C}_{2,1}(k)$, the symplectic quadrangle over k. Define the set $\mathcal{H}$ as the set of hyperbolic lines of Δ, where a *hyperbolic line* is a set $L_{p,q} = (p^\perp \cap q^\perp)^\perp$ in some symplecton Σ (see Example 6.1), for some non-collinear points p, q in Σ. If we add $\mathcal{H}$ to the line set of Δ, we obtain a geometry Δ' of diameter 2. The convex closure in Δ' of two lines that are opposite in Δ is a geometry isomorphic to $\mathsf{B}_{2,1}(k) = \mathsf{C}_{2,2}(k)$ and is called a *pseudo symplecton*. Let $\mathfrak{G}$ be such a geometry, and let x be an arbitrary point in $\mathfrak{G}$. Select an arbitrary point y of $\mathfrak{G}$ not Δ'-collinear with x. Let L be any line of Δ through y. Since x and y are opposite in Δ, there is a unique point z on L not opposite x and there is a unique point u Δ-collinear with both x and z. The set of pseudo symplecta defined by x and some point on $L \setminus \{z\}$ is called a *pencil of pseudo symplecta*. One can show that it only depends on u and one member of it (hence independent of L). We call u the (unique) *centre* of the pencil.

We now define new points as the pseudo symplecta, and new lines are the pencils of pseudo symplecta together with their centre.

The geometry of new and old points, and of old lines, hyperbolic lines and new lines is isomorphic to the Lie incidence geometry $\mathsf{A}_{5,2}(k)$.

6.3   Metasymplectic spaces and $\mathsf{E}_{6,1}(k)$

Now we construct $\mathsf{E}_{6,1}(k)$ out of the (split) metasymplectic space $\mathsf{F}_{4,4}(k)$. This construction prominently uses the $(4,1)$-equator geometries of $\mathsf{F}_{4,4}(k)$.

So we start with $\mathsf{F}_{4,4}(k)$. Our goal is to add two new kinds of lines and one new kind of points, just as in the previous subsection. One new kind of lines is given by the family of hyperbolic lines in the subgeometries isomorphic to $\mathsf{C}_{3,1}(k)$ (residues of elements of type 4). The new points are the *extended equator geometries*. We now explain what this is.

Consider two opposite points p, q and let $\mathfrak{E}_{p,q}$ be the corresponding $(4,1)$-equator geometry. We take the union $\widetilde{\mathfrak{E}}$ of all equator geometries $\mathfrak{E}_{x,y}$, for $\{x,y\}$ ranging over the pairs of opposite points in $\mathfrak{E}_{p,q}$. Endowed with all hyperbolic lines in it, $\widetilde{\mathfrak{E}}$ is a Lie incidence geometry isomorphic $\mathsf{B}_{4,1}(k)$, called an *extended equator geometry*. The family of all extended equator geometries is the set of new points.

The last type of lines is not so difficult to define. Consider an extended equator geometry $\widetilde{\mathfrak{E}}$ and a maximal singular subspace U in there. Recall that U is a set of points of $\mathsf{F}_{4,4}(k)$ such that the hyperbolic lines contained in it render it the point-line geometry $\mathsf{A}_{4,1}(k)$ of a projective space of dimension 4 over the field k. One shows that there is a unique point x collinear in $\mathsf{F}_{4,4}(k)$ with all points of U. Then the set of all extended equator geometries containing U, together with the point x, is a generic new line.

One now shows that the geometry of old and new points, and of old, hyperbolic and new lines is isomorphic to $\mathsf{E}_{6,1}(k)$, see De Schepper, Sastry & Van Maldeghem [3].

7   Coxeter Complexes and associated Graphs

Let (W, S) be a spherical Coxeter system and $\Delta(W, S) = (X, \Omega)$ the corresponding Coxeter complex. We define $\Gamma(W, S)$ as the graph with set of vertices X and two vertices are adjacent if they form a respective chamber with the same panel. It follows that $\Gamma(W, S)$ has as many connected components as its rank is. Each node of the corresponding Coxeter diagram defines a connected component by considering the vertices of $\Delta(W, S)$ of that particular type. If the node is numbered i, then we denote the corresponding connected graph by $\Gamma_i(W, S)$. If (W, S) corresponds to the spherical diagram X_n of rank n, then $\Gamma_i(W, S)$ is precisely the thin Lie incidence geometry $\mathsf{X}_{n,i}(1)$, if we identify each edge with a (thin) line. We shall also view $\mathsf{X}_{n,i}(1)$ as a graph.

In many cases the automorphism group of $\Gamma_i(W, S)$ is precisely W. Prominent counter examples are $\Gamma_i(W, S)$ for (W, S) of type D_n and $1 \leq i \leq n - 2$, since in these cases, the automorphism group clearly contains the Coxeter group of type B_n. But this is only a special case of the general phenomenon that, if the node labelled i is stable under a graph automorphism group of the Coxeter diagram, then the automorphism group of $\Gamma_i(W, S)$ also contains that graph automorphism group.

The graphs $\Gamma_i(W, S)$ are also standard apartments of the buildings of type (W, S). In this respect a good description of the apartment helps one understand the geometry of the building. For instance, all possible mutual positions of certain flags can be deduced from those in an apartment.

As an example, the graphs $\mathsf{A}_{n,i}(1)$ are precisely the Johnson graphs $J(n + 1, i)$: the vertices are the i-subsets of an $(n + 1)$-set, two such i-sets being adjacent if they intersect in an $(i - 1)$-set.

We now concentrate on some constructions of apartments for the exceptional types E_6, E_7, E_8 and F_4, occasionally giving rise to similar constructions of apartments of certain classical types. The nicest constructions are those of $\mathsf{X}_{n,\ell}(1)$, where $\mathsf{X} \in \{\mathsf{E}, \mathsf{F}\}$ and ℓ is an end node of the diagram. Hence we will restrict to these cases. At the end, we will have covered the following cells of the Magic Square:

			$\mathsf{F}_{4,4}$
			$\mathsf{E}_{6,1}$
		$\mathsf{D}_{6,6}$	$\mathsf{E}_{7,7}$
$\mathsf{F}_{4,1}$	$\mathsf{E}_{6,2}$	$\mathsf{E}_{7,1}$	$\mathsf{E}_{8,8}$

7.1   Apartments of type E_6

It is well known that the Coxeter group of type E_6 is isomorphic to the orthogonal group $\mathrm{O}_6^-(2)$, which in turn is isomorphic to the unitary group $\mathrm{U}_5(2)$. These groups are automorphism groups of the generalized quadrangle $Q(2, 4)$. It is not a coincidence that $Q(2, 4)$ and $\mathsf{E}_{6,1}(1)$ both have 27 points. In fact, the complement of the collinearity graph of $Q(2, 4)$ is exactly the thin Lie incidence geometry $\mathsf{E}_{6,1}(1)$ considered as a graph. Example 2.9 implies the following construction of $\mathsf{E}_{6,1}(1)$.

Example 7.1. Let $N = \{1, 2, 3, 4, 5, 6\}$ and $N' = \{1', 2', 3', 4', 5', 6'\}$. Set $V = N \cup N' \cup \binom{N}{2}$. Let adjacency be given by $a \sim b$, $a' \sim b'$, $a \sim a'$ and $a \sim \{b, c\} \sim a'$ and $\{a, b\} \sim \{a, c\}$, for all distinct $a, b, c \in N$.

The previous example can be written down in a more systematic and general way as follows. First note that the graphs induced on N and N' are complete graphs, more exactly they are isomorphic to $\mathsf{A}_{5,1}(1)$. The graph induced on $\binom{N}{2}$ is isomorphic to $\mathsf{A}_{5,2}(1)$. The labelling using $1, 2, \ldots$ and $1', 2', \ldots$ boils down to the choice of an isomorphism between the various underlying $\mathsf{A}_5(1)$. To determine adjacency between vertices of N and N' we use the chosen isomorphism and connect corresponding vertices. We say that *adjacency is induced by natural isomorphism* (with symbol $\cong$). A vertex a of N is adjacent to a vertex $\{b, c\}$ of $\binom{N}{2}$ if $\{b, c\}$, as a line of $\mathsf{A}_5(1)$ is incident to the hyperplane of $\mathsf{A}_5(1)$ obtained from a by applying the natural duality $x \leftrightarrow N \setminus \{x\}$. We say that *adjacency is induced by natural duality* (with symbol $\asymp$).

We picture this construction as follows.

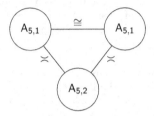

This way of writing the construction is of course not unique. One can try to use as many isomorphisms as possible, e.g., as follows.

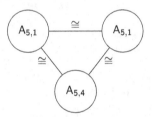

So this diagram means that we have three disjoint graphs, namely two copies of $\mathsf{A}_{5,1}(1)$, and a copy of $\mathsf{A}_{5,4}(1)$. We choose a standard thin building of type A_5, which we fix by defining the set of elements of type 1 as $\{a, b, c, d, e, f\}$. Then the vertices of each copy of $\mathsf{A}_{5,1}(1)$ are precisely a, b, c, d, e, f, and the vertices of the graph $\mathsf{A}_{5,4}(1)$ are all 4-subsets of $\{a, b, c, d, e, f\}$. Adjacency between an element $x \in \{a, b, c, d, e, f\}$ of a copy of $\mathsf{A}_{5,1}(1)$ and an element $X \subseteq \{a, b, c, d, e, f\}$, $|X| = 4$, of $\mathsf{A}_{5,4}(1)$ is given by $x \in X$; adjacency between an element $x \in \{a, b, c, d, e, f\}$ of one copy of $\mathsf{A}_{5,1}(1)$ and an element $y \in \{a, b, c, d, e, f\}$ of another copy of $\mathsf{A}_{5,1}(1)$ is given by $x = y$.

In such a case we could even omit the "$\cong$" symbols, and we shall do so. Note that, in this thin geometry, it is easy to see that $\mathsf{A}_{5,1}(1)$ is the $(2, 1)_1$-trace geometry of $\mathsf{A}_{5,2}(1)$ and $\mathsf{A}_{5,2}(1)$ is the $(2, 5)_1$-equator geometry of $\mathsf{E}_{6,1}(1)$.

It is a common feature of the constructions of Coxeter complexes in this section that the different parts are trace and equator geometries. But there are also exceptions, at least if we do not consider a generalisation of the notion of trace geometry with respect to opposite flags of different type. An example is the following.

Example 7.2. Let $\mathsf{D}_{5,1}(1)$ be the thin polar space of rank 5, i.e., the complete graph on vertex set $\{1, 2, 3, 4, 5, 1', 2', 3', 4', 5'\}$ minus the matching $\{(1, 1'), (2, 2'), (3, 3'), (4, 4'), (5, 5')\}$.

Let $D_{5,5}(1)$ be the thin half spin geometry with vertex set the 5-cliques of $D_{5,1}(1)$ which intersect $\{1,2,3,4,5\}$ in an odd number of vertices, and adjacency intersecting in three vertices. Let Γ be the graph with set of vertices a symbol ∞, the vertices of $D_{5,1}(1)$ and the vertices of $D_{5,5}(1)$, and define adjacency as follows. The vertex ∞ is adjacent to every vertex of $D_{5,5}(1)$; the adjacency inside the sets of vertices of $D_{5,1}(1)$ and $D_{5,5}(1)$ is the natural one; a vertex x of $D_{5,1}(1)$ is adjacent to a vertex $\{a,b,c,d,e\}$ of $D_{5,5}(1)$ if $x \in \{a,b,c,d,e\}$. Then Γ is isomorphic to $E_{6,1}(1)$.

Again, we can consider the adjacency between $D_{5,1}(1)$ and $D_{5,5}(1)$ as incidence given by the natural isomorphism of the labelling of the points of $D_{5,1}(1)$. We picture this construction as follows.

The construction in Subsecion 5.4 of $E_{6,1}(k)$ is a "thickening" of the above construction of $E_{6,1}(1)$. Note also that the construction of $D_{5,1}(1)$ as a complete graph minus a matching is equivalent to the construction pictured as follows.

which is the thin version of the case $n = 5$ of the construction given by Proposition 5.2. Generalisation to $D_{n,1}(1)$ is obvious. Likewise, the thin version of Proposition 5.3 is pictured as follows.

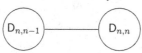

which provides a construction of $D_{n+1,n+1}(1)$.

Now we turn to $E_{6,2}(1)$. This is a thin long root geometry, and all thin long root geometries behave similarly. Hence the same pattern as we will see now will repeat itself for $E_{7,1}(1)$, $E_{8,8}(1)$ and $F_{4,1}(1)$. Long root geometries have equator geometries, and this will show in the construction of the corresponding apartments.

The first construction is performed using trace and equator geometries with respect to two opposite vertices.

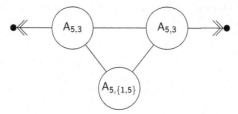

The next construction is performed using trace and equator geometries with respect to an opposite pair of half spin geometries of rank 4, i.e., opposite elements of type 1 and 6.

7.2   Apartments of type E_7

The thin Lie incidence geometry $E_{7,7}(1)$ is, still viewed as a graph, isomorphic to the Gosset graph. There are several constructions of this graph, but we will only mention those that fit in our approach. The first one is with respect to two opposite vertices. Then there are two trace geometries none of which is an equator geometry.

A similar construction exists for the thin Lie incidence geometry $D_{6,6}(1)$ (which is not coincidentally next to $E_{7,7}(1)$ in the Magic Square). It goes as follows.

The second construction of $E_{7,7}(1)$ is with respect to two opposite elements of type 2, hence two subgeometries isomorphic to $A_{6,1}(1)$. This yields the following diagram.

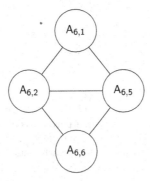

The next construction displays the $(1, 2)_7$-equator geometry. It is based on two opposite subgeometries isomorphic to $D_{6,1}(1)$.

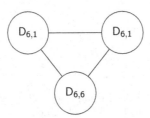

Now we consider $E_{7,1}(1)$, which is a long root geometry. The canonical construction is with respect to two opposite vertices, and we can read off the residue (as a trace geometry) and the equator geometry.

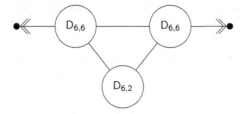

The second construction is with respect to two subgeometries isomorphic to $E_{6,1}(1)$. It reveals the $(7,2)_1$-equator geometry.

There is a striking similarity with the construction of $E_{6,2}(1)$ using $D_{5,5}(1)$ and $D_{5,2}(1)$, see above. We will encounter another case of the same shape below when discussing apartments of type F_4. Yet another example is given by a construction of $D_{6,6}(1)$, using two opposite thin projective 5-spaces, and it can be pictured as follows.

This construction can be used to derive the following, and third, construction of the thin long root geometry $E_{7,1}(1)$.

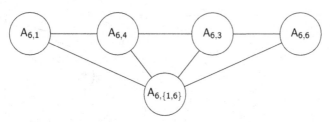

Note that this suggests that $A_{6,\{1,6\}}(k)$ be the $(2,\{1,7\})_7$-equator geometry of $E_7(k)$, for any field k.

Finally we would also like to mention a construction of $E_{7,2}$, which has 576 vertices and is therefore unpopular. We present the simplest one, using two opposite elements of type 7, i.e., two subgeometries isomorphic to a thin long root geometry of type E_6.

7.3   Apartments of type E_8

Here we mainly concentrate on $E_{8,8}(1)$. We provide three constructions, one for each end node of the diagram.

We begin with respect to two opposite vertices (elements of type 8). Then we get the usual long root picture.

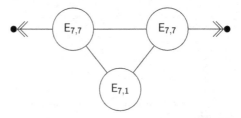

The second construction is with respect to two opposite elements of type 1, i.e., two opposite thin subgeometries isomorphic to $D_{7,1}(1)$. It is similar to the third construction of $E_{7,1}(1)$ and looks as follows.

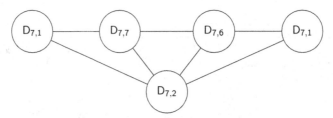

Finally, we construct $E_{8,8}(1)$ with respect to two opposite elements of type 2. This gives the following remarkable construction, where all of $A_{7,i}(1)$ are used, $1 \leq i \leq 7$, except that $A_{7,4}(1)$ is replaced with the thin long root geometry $A_{7,\{1,7\}}(1)$.

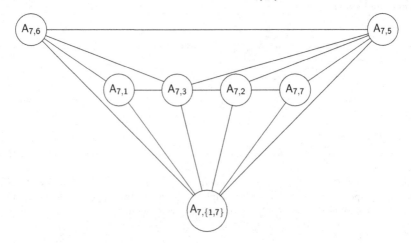

Note that this diagram is not entirely symmetric. There is indeed an isomorphism between $A_{7,3}(1)$ and $A_{7,5}(1)$, but no edges between $A_{7,2}(1)$ and $A_{7,6}(1)$ (unrelated detail: the latter keeps the graph in the picture above planar as otherwise it would contain a complete subgraph with the five vertices $A_{7,2}(1)$, $A_{7,3}(1)$, $A_{7,5}(1)$, $A_{7,6}(1)$ and $A_{7,\{1,7\}}(1)$).

We cannot resist to also give a less familiar example, namely, a construction of $E_{8,1}(1)$. The Lie incidence geometry $E_{8,1}(k)$, for a field k, turns up in several characterisation theorems of parapolar spaces, so the construction below gives some information about the structure of that geometry (see also Remark 9).

We construct $E_{8,1}(1)$ with respect to two opposite elements of type 8, i.e., two opposite subgeometries isomorphic to $E_{7,1}(1)$. It has 2160 vertices and degree 64.

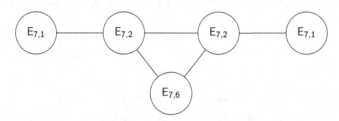

This suggests that $E_{7,6}(k)$ is a $(1,6)_8$-equator geometry of $E_8(k)$.

7.4   Apartments of type F_4

This is the first Coxeter diagram with a double bond that we encounter. The two end nodes of the F_4-diagram play the same role, except that the arrow on the double bond takes away the symmetric. The graphs $F_{4,1}(1)$ and $F_{4,4}(1)$ are nevertheless isomorphic. However, it is still more convenient to consider $F_{4,1}(1)$ because it corresponds to long root geometries.

There is a second special feature when dealing with Dynkin diagrams with double bonds. It concerns the long root geometries of buildings of type B_n, $n \geq 2$, and C_n, $n \geq 3$. The buildings of these types correspond to polar spaces. In general, one thinks of a polar space as a Lie incidence geometry of type $B_{n,1}$, and then the corresponding long root geometry has type $B_{n,2}$ and is a non-strong parapolar space of diameter 3. However, strictly speaking, long root geometries are only well defined for split buildings, and when the polar space is a symplectic one—hence belonging to the diagram C_n—the long root geometry is the polar space itself. Now, all other long root geometries have diameter 3, except this one. We will correct this and view the long root geometry of a symplectic polar space as a geometry without lines, where points have distance 2 and 3 (points collinear in the polar space have distance 2, the others distance 3). Also, the lines of the polar space will be conceived as the symplecta of the long root geometry. In this setting, the thin long root geometry $C_{3,1}(1)$ is an edgeless graph with six vertices endowed with an opposition relation which is a matching.

This opposition relation allows to well define isomorphisms between such structures, and also incidence preserving maps to other Lie incidence geometries of the same main type. For example, if we label the vertices of the graph $\{1,2,3,1',2',3'\}$ and declare a opposite a', for all $a \in \{1,2,3\}$, then the point set of $C_{3,3}(1)$ is, with self-explaining notation, equal to $\{123, 1'23, 12'3, 123', 12'3', 1'23', 1'2'3, 1'2'3'\}$. The natural isomorphism then connects 1 with 123, 123', 12'3 and 12'3', and it connects 123 with 1, 2 and 3. It is in this way that the below pictures must be read.

With these conventions, we have the following construction of $F_{4,1}(1)$. It is performed with respect to two opposite vertices, as a long root geometry.

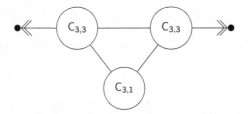

With respect to two opposite symplecta (elements of type 4), we have the following construction.

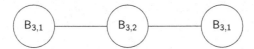

Here, $B_{3,2}(1)$ is the ordinary thin long root geometry of type B_3.

Remark 9. The diagrams that we have drawn in this section act as a kind of *floor plan* for the corresponding Lie incidence geometry (to stay in the terminology of real estate). The different components are exactly the types of the trace geometries one can find in these Lie incidence geometries. Such a trace geometry is an equator geometry if and only if it lies symmetrically in the diagram. In theory, there is a floor plan for each Lie incidence geometry and each choice of types of opposite elements. The most interesting floor maps are those for which these opposite elements correspond to irreducible subbuildings (which is automatic when the types correspond to end nodes of the Coxeter diagram).

8 Conclusion and open problems

In this paper, we have presented a lot of combinatorial and geometric constructions of spherical buildings, focusing on the Lie incidence geometries related to the Freudenthal-Tits Magic Square. However, some cells of that square always remained empty. The full Square looks as follows.

$A_{1,1}$	$A_{2,\{1,2\}}$	$C_{3,2}$	$F_{4,4}$
$A_{2,1}$	$A_2 \times A_2$	$A_{5,2}$	$E_{6,1}$
$C_{3,3}$	$A_{5,3}$	$D_{6,6}$	$E_{7,7}$
$F_{4,1}$	$E_{6,2}$	$E_{7,1}$	$E_{8,8}$

The cells that we left empty throughout the whole article contain $A_{1,1}$, $A_{2,\{1,2\}}$, $A_{2,1}$ and $C_{3,3}$. The first one is a rather trivial geometry consisting of points on a single line. Geometrically, it is convenient to think of it as a conic. The geometries of type $A_{2,\{1,2\}}$ are the flag complexes of a projective plane, i.e., non-thick generalized hexagons with thick lines. The representation they take in the Magic Square is related to triality of the hyperquadric of type D_4, although these only cover the case of characteristic 3. These and the others can also be obtained by intersecting a Segre variety $\mathcal{S}_{2,2}(k)$ with an appropriate hyperplane. The geometries of type $A_{2,1}$ are ordinary projective planes, and the corresponding representation in the Magic Square are the ordinary Veronesean varieties, which are intersections of Segre varieties with appropriate subspaces. The same thing holds for the Lie incidence geometries $C_{3,3}(k)$, which are symplectic dual polar spaces contained in $A_{5,3}(k)$.

We now mention the most important open problems related to the constructions here presented.

1. *Find an explicit set of quadrics in 77-, 132- and 247-dimensional projective space over the field k whose intersection is the Lie-incidence (long root) geometry* $E_{6,2}(k)$, $E_{7,1}(k)$ *and* $E_{8,8}$, *respectively.*

2. *Study and classify the inclusions of Lie incidence geometries.*

3. *Find projective constructions for the geometries in the third row of the Freudenthal-Tits Magic square.*

Concerning the second one, it is tempting to conjecture that, perhaps under mild conditions, inclusions will always arise from trace (in particular, equator) geometries, if not entirely contained in a singular (projective) subspace.

References

[1] N. Bourbaki, *Groupes et Algèbres de Lie*, Chapters 4,5 and 6, Actualités scientifiques et industrielles, Number 1337, Hermann, Paris, 1968.

[2] A.M. Cohen, *Point-line geometries related to buildings*, in: Handbook of Incidence Geometry, Buildings and Foundations, Editor F. Buekenhout, North-Holland, Amsterdam, 1995, pp. 647–737.

[3] A. De Schepper, N.S.N. Sastry and H. Van Maldeghem, *Split buildings of type* F_4 *in buildings of type* E_6, Abhandlungen aus dem Mathematischen Seminar der Universität Hamburg **88** (2018), 97–160.

[4] J.W.P. Hirschfeld and J.A. Thas, *General Galois Geometries*, Clarendon Press, Oxford, 1991.

[5] J. Schillewaert & H. Van Maldeghem, *Projective planes over quadratic 2-dimensional algebras*, Advances in Mathematics **262** (2014), 784–822.

[6] J. Tits, *Sur certaines classes d'espaces homogènes de groupes de Lie*, Palais des Académies, Bruxelles, 1955.

[7] J. Tits, *Buildings of Spherical Type and Finite BN-Pairs*, Springer Lecture Notes Series, Volume 386, Springer-Verlag, Berlin-Heidelberg, 1974.

Department of Mathematics
Ghent University
Krijgslaan 281, S8, B-9000 Gent
{hendrik.vanmaldeghem,magali.victoor}@ugent.be

Printed in the United States
by Bookmasters

Printed in the United States
By Bookmasters